Modern developments in
Flow
measurement

PPL Conference Publication 10

Atomic Energy Research Establishment, Harwell
National Engineering Laboratory, East Kilbride

Modern developments in
Flow
measurement

edited by
C.G. Clayton
Atomic Energy Research Establishment,
Harwell, Berks., England

Proceedings of the international
conference held at Harwell,
21st - 23rd September 1971

Peter Peregrinus Ltd.

Published by
Peter Peregrinus Ltd.
2 Savoy Hill, London WC2R 0BP, England

ISBN: 0 901223 30 1
PPL Conference Publication 10

Printed in England by
Creative Offset, Kingston upon Thames, Surrey

PREFACE

This book contains 30 p. rs presented at an international conference on modern developments in flow measurement held at Harwell during September 1971.

The object of the conference was to bring together scientists and engineers responsible for developing flow meters and flow measuring techniques with consultants, process engineers, hydraulic machine designers and others with a strong interest in measuring the flow of liquids and slurries in open channels of regular cross section and liquids and gases in pipes. The fact that over 150 delegates from 17 countries attended is some evidence that this object was achieved. The very comprehensive discussions which were held after nearly all the papers, and which are reported in full in the present volume, are further evidence of the strong and useful interactions which occurred between the different groups of professional interest.

As pointed out by Mr. R.H. Weir, Director of the National Engineering Laboratory, in his opening address to the conference, the papers presented emphasise the wide areas of science and technology now involved in the development and use of flow meters. Unlike many old-established procedures, the increasing complexity of our society and the advance in technology continue to stimulate interest in flow measurement: there is a need for devices and methods which will operate to an increasingly higher accuracy and precision in environments which were hardly credible a few years ago.

The papers have been gathered together in seven chapters. Chapter 1 is devoted to a review of the current state of the art. Chapter 7 is a commentary on the papers presented at the conference, and includes an opinion on present trends and future aims in flow measurement. The remaining twenty-six papers are subdivided into five chapters, and as far as possible papers with a common scientific basis have been grouped into the same chapter; but no merit is intended in the order in which these chapters appear.

Although the main emphasis was on the newer methods of flow measurement, contributions on well-established methods were not excluded. The only constraint was that these papers should indicate significant evidence of advance in the subject.

The number of papers devoted to each method of flow measurement is not the same, but the final balance was not intentional. It is more a reflection of the areas of interest in the papers presented for consideration: it is perhaps some indication of the distribution of effort in current developments.

The organisation of this conference was the responsibility of two committees (p. vi), and whatever success was achieved is to a large extent a result of the efforts of the members of these committees. We are much indebted to the work of the Conference Secretary, Mr. C.J.A. Preuveneers. Particular appreciation is also due to Prof. S.P. Hutton, Dr. E.A. Spencer and Dr. G.V. Evans for their very valuable work on the Paper Selection Committee. To my colleague Gordon Evans, I am especially grateful; not only for the heavy burden of correspondence which he carried on this committee, but also for his general support through all the stages of preparing for the conference.

My most sincere thanks go to my secretary Jill Mitchell, for her unflagging efforts before and during the conference and for the very considerable help she has given me in the final preparation of this manuscript.

C.G. Clayton
Harwell

April 1972

CONFERENCE COMMITTEES

Organising Committee

C.G. Clayton (Chairman)
G.V. Evans
R.A. Faires
J.F. Hill
S.P. Hutton
R.M. Longstaff
E.A. Spencer
C.J.A. Preuveneers (Secretary)

Paper Selection Committee

S.P. Hutton (Chairman)
E.A. Spencer
G.V. Evans (Secretary)

Contents

1   Review papers

Industrial demands for flow measurement
C.G. Clayton and G.V. Evans (United Kingdom Atomic Energy Authority,
A.E.R.E., Harwell, Didcot, Berks.)   *page*   1

Current practice in fluid flow measurement
E.A. Spencer (National Engineering Laboratory, East Kilbride, Glasgow)   13

Flow measurement by weirs and flumes
P. Ackers (Hydraulics Research Laboratory, Wallingford, Berks.)   21

2   Nuclear magnetic resonance, optical and microwave methods

Recent measurements of flow using nuclear magnetic resonance techniques
J.R. Singer and T. Grover (Department of Electrical Engineering &
Computer Sciences, University of
California, Berkeley, Calif. 94720, U.S.A.)   38

Development of the NEL laser velocimeter
K.A. Blake (National Engineering Laboratory, East Kilbride, Glasgow)   49

**Flow measurement in the presence of strong swirl using a laser Doppler
anemometer**
P.S. Bedi and M.T. Thew (Department of Mechanical Engineering,
University of Southampton,
Southampton SO9 5NH, Hants.)   61

**An evaluation of optical anemometers for volumetric flow measurement
of liquids and gases**
F. Durst, A. Melling and J.H. Whitelaw (Imperial College of Science &
Technology, Department of Mechanical
Engineering, Exhibition Road, London SW7   78

Flow metering using a microwave Dopplermeter
J. Harris (School of Chemical Engineering, University of Bradford,
Bradford BD7 1DP, Yorks.)   88

3   Thermal and ultrasonic flowmeters

A thermoelectric flowmeter of rapid response and small sensor size
A.G. Smith and W.A.K. Said (Department of Mechanical Engineering,
University Park, Nottingham NG7 2RD)   96

**The development of thermal flowmeters for the measurement of small
flows of corrosive and radioactive liquors in chemical plants**
A.L. Mills, C.R.A. Evans and R.A. Chapman (Experimental Reactor Establishment,
United Kingdom Atomic Energy Authority,
Dounreay, Thurso, Caithness)   105

**Ultrasonic method of flow measurement in large conduits and open
channels**
N. Suzuki, H. Nakabori and M. Yamamoto* (Central Research Institute of
Electric Power Industry, 1229 Iwato,
Komae-Shi, Tokyo, Japan)   115

Ultrasonics as a standard for volumetric flow measurement
S.G. Fisher and P.G. Spink (Westinghouse Electric Corporation, Underseas
Division, Ocean Research and Engineering
Laboratory, Box 1488, Annapolis,
Md. 21401, U.S.A.)   139

* Tokyo Keiki Company Limited, 16 Minamikamata 2-Chome, Ohta-Ku, Tokyo, Japan

4   Pressure-difference methods

The performance of long bore orifices at low Reynolds numbers
T. Cousins (Kent Instruments Ltd., Luton, Beds.)   160

Non-Newtonian flow measurement using conventional pressure difference
meters
J. Harris and A.N. Magnall (School of Chemical Engineering, University of
                            Bradford, Bradford BD7 1DP, Yorks.)   180

A differential pressure flowmeter with linear response
D. Turner (Gervase Instruments Ltd., Britannia Works, Cranleigh, Surrey)   191

Inferential flowmeters with photo-electronic transducer systems
J.A. Ryland (G.A. Platon Ltd., Wella Road, Basingstoke, Hants.)   200

Small diameter orifices in series
S.C. Okafor and R.K. Turton (Department of Mechanical Engineering,
                            University of Technology, Loughborough
                            LE 11 3TU, Leics.)   210

5   Tracer methods

Measurement of transient mass flow rate of a gas
N. Mustafa, I. Birchall and W.A. Woods (Department of Mechanical
                            Engineering, University of Liverpool,
                            Liverpool L69 3BX)   223

Development of a pulsed gas ionization flowmeter
T.J.S. Brain (National Engineering Laboratory, East Kilbride, Glasgow)   236

Measurement of gas flow by radiotracer methods
G.V. Evans, R. Spackman, M.A.J. Aston and C.G. Clayton (United
                            Kingdom Atomic Energy Authority,
                            A.E.R.E., Harwell, Didcot, Berks.)   245

Flowrate determination by neutron activation analysis
C.R. Boswell and T.B. Pierce (United Kingdom Atomic Energy Authority,
                            A.E.R.E., Harwell, Didcot, Berks.)   264

Testing and comparison of three methods of measuring flow through pipes
L. Roche, H. Andre*, S. Gamby† and B. Noirett (Électricité de France,
                            Chatou, France)   273

Experience in the use of the radioisotope constant rate injection method
in testing the performance of hydraulic machines
C.G. Clayton and G.V. Evans (United Kingdom Atomic Energy Authority,
                            A.E.R.E., Harwell, Didcot, Berks.)   276

Flow velocity and mass flow measurement using natural turbulence
signals
M.S. Beck, J. Coulthard, P.J. Hewitt and D. Sykes (Postgraduate School of
                            Studies in Physics, University of Bradford,
                            Bradford BD7 1DP, Yorks.)   292

**6   Turbine flowmeters, electromagnetic flowmeters and the velocity-
        area method**

Flow characteristics of turbine flowmeters
V.R. Withers, F.A. Inkley and D.A. Chesters (British Petroleum Company,
                            BP Research Centre, Chertsey Road,
                            Sunbury on Thames, Middx.)   305

Experimental, analytical and tip clearance loss studies in turbine-type
flowmeters
P.A.K. Tan and S.P. Hutton (Department of Mechanical Engineering,
                            University of Southampton, Southampton
                            SO9 5NH, Hants.)   321

* Électricité de France, Service Production Hydraulique, 3 rue de Messine, 75 Paris 8e, France

† Électricité de France, Division Technique Générale, 37 rue Diderot, Grenoble 38, France

Modern developments and new applications of magnetic flowmeters
R. Theenhaus (L. Krohne, 4100 Duisburg, Ludwig-Krohne-Strasse,
Postfach 493, W. Germany)   347

Electromagnetic flowmeters for liquid metals
G. Thatcher (United Kingdom Atomic Energy Authority, Risley,
Warrington, Lancs.)   359

Errors in the velocity-area method of measuring asymmetric flows in circular pipes
L.A. Salami (Department of Mechanical Engineering, University of
Southampton, Southampton SO9 5NH, Hants.)   381

7   An outlook on flow measurement

Flow metering and future aims
S.P. Hutton (Department of Mechanical Engineering, University of
Southampton, Southampton SO9 5NH, Hants.)   401

CONTRIBUTORS* TO DISCUSSIONS

AHAD, D.: G.A. Platon Ltd., Wella Road, Basingstoke, Hants.

AGAR, J.: Joram Agar, Alresford, Hants.

AU, S.B.: Department of Mechanical Engineering, University of Southampton, Southampton, Hants.

BLELLOCK, I.G.: Department of Trade & Industry, Nuclear Installations Inspectorate, c/o U.K.A.E.A., Risley, Warrington, Lancs.

BREWER, C.R.: C.E.G.B., Bankside House, Sumner Street, London S.E.1

CARRINGTON, J.E.: The Queen's University of Belfast, Department of Mechanical Engineering, Ashby Institute, Stranmillis Road, Belfast BT9 5AH

CAUSON, J.: Hydro Electric Commission, Hobart, Tasmania

CLAY, C.A.E.: C.E.G.B., Haslucks Green Road, Shirley, Solihull, War.

CROSSLEY, A.M.: Department of Public Health Engineering, Greater London Council, 10 Gt.George Street, London S.W.1

DRAIN, L.E.: A.E.R.E., Harwell, Didcot, Berks.

GEORGE, P.T.: B.N.D.C. Ltd., Cambridge Road, Whetstone, Leics.

GOLDRING, B.T.: Central Electricity Research Laboratory, Leatherhead, Surrey

HAYWARD, A.T.J.: National Engineering Laboratory, East Kilbride, Glasgow

HOELGAARD, J.C.: DISA Elektronik A/S, Herlev, Denmark

HUIJTEN, F.H.: Shell Chemicals UK Ltd., Carrington, Urmston, Manchester

JEPSON, P.: Gas Council Engineering Research Station, Killingworth, Newcastle

JONES, A.W.: Mullard Ltd., Southampton Works, Millbrook Industrial Estate, Southampton, Hants.

KINGHORN, F.C.: National Engineering Laboratory, East Kilbride, Glasgow

KOMIYA, K.: National Research Laboratory of Metrology, 1-10-4, Kaga, Itabashi, Tokyo, Japan

LICHTAROWICZ, A.: Department of Mechanical Engineering, University of Nottingham, Nottingham

LOOSEMORE, W.R.: A.E.R.E., Harwell, Didcot, Berks.
LOXLEY, G.E.B.: Rolls Royce & Associates Ltd., Derby
PAPPALARDO, M.: Instituto di Acustica, O.M. Corbino, Rome, Italy

ROUGHTON, J.E.: Central Electricity Research Laboratory, Leatherhead, Surrey

ST. CLAIR, F.W.: Foxboro-Yoxall Ltd., Redhill, Surrey

SANDERSON, P.R.: Water Research Association, Medmenham, Marlow, Bucks.

SCHUSTER, J.C.: U.S. Bureau of Reclamation, Engineering and Research Center, Denver, Colo. 80225, U.S.A.

SMITH, D.L.: I.C.I. Ltd., Agricultural Division, P.O. Box 6, Billingham, Teesside.

THOMAS, I.M.: British Nuclear Fuels Ltd., c/o U.K.A.E.A., Risley, Warrington, Lancs.

WALLES, K.F.A.: National Gas Turbine Research Establishment, Farnborough, Hants.

WHITE, K.E.: Water Pollution Research Laboratory, Stevenage, Herts.

WOLFMAN, D.: Israel Atomic Energy Commission, Soreq Nuclear Research Center, Yavne, Israel

YARWOOD, J.I.: Ministry of Power, Gas Division, Gas Standards Branch, 26 Chapter Street, London S.W.1

ZANKER, K.J.: Kent Instruments Ltd., Luton, Beds.

* Addresses of authors who spoke in the discussions are given in 'Contents', pp. vii-ix

INDUSTRIAL DEMANDS FOR FLOW MEASUREMENT

C.G. Clayton and G.V. Evans

Abstract: The growth rates of industries having a requirement for flow measurement are examined in relation to national economic growth rates. Current factors which affect the demand for flow measurement, such as the growing need for water conservation, pollution control, the increasing attention being given to the accounting of fluid movement, as well as the intro- duction of automatic process control systems, are examined. An attempt is made to con- sider the effects of the introduction of new techniques on the industrial demand for flow measurement.

1   INTRODUCTION

The need to measure the flow rates of fluids occurs in many branches of economic activity and the factors which govern the need are diverse and continually increasing.

The most important factors are attributable to an increase in world population, to a generally increasing standard of living and a more complex economic society, to a move from labour-inten- sive to machine-intensive activities and to the introduction of technological innovation into indus- try on a vastly increasing scale, especially during the past 30 years.

Although perturbations in the general economic climate will no doubt occur in the future, there is no reason to believe that the overall pattern of increasing intensity and complexity of flow rate measurement will not continue for the next few decades at least.

The growth of new industries, which are built almost entirely around the movement of fluids in pipes, can be expected to add considerably to the demand for flow measurement. The petro- chemical industry is probably the best example, and here, as in many other industries, it is worth noting that the need for flow measurement arises to satisfy the accountant as well as the engineer: the need to measure the 'cash flow' for accurate cost analysis may equal or exceed the need to measure mass flow for process control purposes. Although only yet in its infancy, the desalination industry of the future provides another example of a new industry which is expected to make a significant impact on the industrial demand for flow measurement.

Besides the newer industries, older industries are expanding and introducing automatic process control systems to an increasing degree: in doing so they are creating a demand for flow measure- ment and control which did not previously exist. The steel industry, especially iron-making and steel-making and the mineral-processing industry are important examples of this type of demand.

The increasing world demand for energy, and in particular the introduction of nuclear power and the discovery of new sources of oil and gas in geographical areas remote from centres of population, make their own particular demands on flow measurement. The transport of oil or gas, either continuously by overland pipeline or discontinuously by ship, poses its own particular inter- facing problems of estimating the amount transferred from one authority to another. The large amounts of fluid which are often transferred create an important demand for highly accurate flow measurement.

The increasing concern with environmental pollution, and the introduction of control legisla- tion in some countries, are beginning to affect the need to measure and control the effluent from industrial processes. The ability to bury pipes carrying oil and gas, compared with the economically enforced exposure of high voltage electricity supply lines, and the ability to use a high pressure gas distribution system as an alternative reservoir to a gas holder both appeal as offering a more aesthetic solution to the problem of energy transfer and increase the variety of demand for measurement of fluid flow rates.

The introduction of new industrial processes and new operational techniques are acting as a stimulus to the development of new types of flow meter to meet the demands for higher accuracy, wider environmental operation and lower cost. A large number of physical principles can be invoked as a means of measuring flow and, although many have been tried with only transient success in the past, it is a feature of the present 'state-of-the-art' that an increasing number of different methods of measurement are being introduced to satisfy the increasing industrial demand. Some indication of this change is given in Table 1.

TABLE 1: Introduction of important methods of flow measurement

Year	Method
1732	Pitot tube
1790	Current meter
1868	Variable area
1887	Venturi-meter
1902	Electromagnetic meter
1903	Orifice plate
1909	Thermal meter
1920	Gibson method
1920	Non-radioactive tracers
1922	Radioactive tracers
1931	Ultrasonic flow meter
1953	Dall tube
1959	Nuclear magnetic resonance
1964	Laser

This paper attempts to show that the need for flow measurement will continue to increase in the future. Measurements on a wider range of fluids can be expected in increasingly hostile environments. It is anticipated that the demand for higher accuracy will increase as the technology of flow measurement continues to improve.

2   WORLD ECONOMY

The expansion of the world economy during the past 25 years has been stimulated by three important factors: a rapidly increasing population, a high rate of technological innovation and a high level of world spending by the major industrial countries in the world. It seems likely that at least two of these factors will continue to operate for the next decade or two.

2.1 World population

The estimated growth of world population from 1650 to the year 2000 is shown in Fig. 1. From 1950, which is a convenient baseline for the present comments, the world population has risen by about 1.7% per year; that is from about 2,500 million in 1950 to about 3,500 million in 1970, and this has intensified the demand for material goods. However, as is well known, the distribution of material wealth is not uniform and, according to one source, about 10% of the human race lives off 60% of the world's income, whereas 60% subsist on 10% of the income. Although it seems generally agreed that a decrease in the rate of increase in world population is necessary in order to control poverty and hunger, little change appears likely during the next two or three decades.

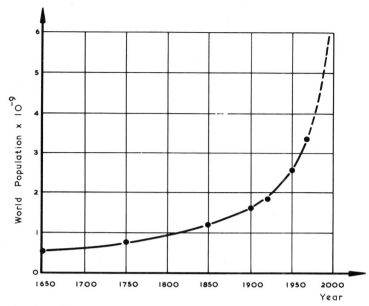

Fig. 1   Growth in world population

TABLE 2: Estimated distribution of world population in the year 2000 A.D.*

Asia	62%
Europe including USSR	15%
Latin America	9.5%
Africa	8%
North America	5%
Oceana	0.5%

*Estimated population in 2000 A.D. is about 6000 million

If the population continues to grow as expected, then there will be about 4,200 million inhabitants of the world by 1980 and about 6,000 million by the year 2000. The anticipated Continental distribution of this population is shown in Table 2. This continued expansion provides the basis for continued material production and industrial growth with a continuing demand on all industrial techniques, including flow measurement and control.

2.2 Technological innovation

The essential economic outcome of technology is that it lowers real costs so that the emphasis in production shifts from labour-intensive to capital-intensive industry. This results in the production of more refined and scientifically controlled materials and products, and more complex but more effective plant and machinery and production processes.

It is not suggested that the introduction of technological innovation into industry is highly efficient, although there are many examples where this is the case. But industry is now benefiting from the increased capital investment in scientific and technological education and in research and development which has taken place during the last 25 years. The first effects have been seen in the industrialised countries, but now the developing countries are also benefiting, as a direct consequence of closer political and economic communications in the world.

Whereas there are many examples of new techniques reducing, and in many cases eliminating the need for well-established methods, such is not the case in flow measurement. Advances in technology appear only to result in an increasing demand for flow measurement. Whilst, the

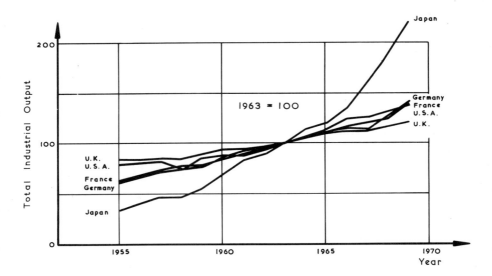

Fig. 2   Total industrial production in some industralised countries for the period 1955-1969

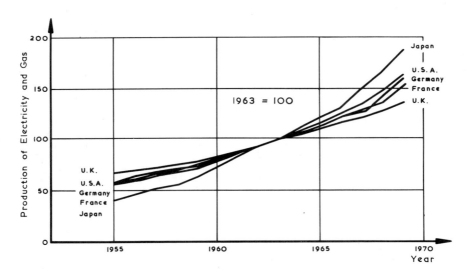

Fig. 3   Production of electricity and gas in several major countries from 1955-1969

introduction of new materials may compete with older materials in the fabrication of pipes, pumps and flow meters, in general, technological development has increased the complexity of the problem of flow measurement and has resulted, for example, in a wider range of fluids to be measured at temperatures and pressures and linear velocities and in environments which were hardly credible a few years ago.

2.3 Expenditure

The post-war reconstruction programmes, aid programmes to developing countries, the wars in Korea and Vietnam and the space programmes are examples of major activities which have involved large financial investments during the past two decades. Apart from contributing to industrial growth rates in the industrialised countries, a significant impact has also been made on industrial growth rates in the developing countries, and economic expansion, though stronger in many industrialised countries, is now world wide.

Very large economic growth rates have become apparent in some countries. Between 1950 and 1970 the annual economic growth rate exceeded 10% in Japan and 7% in the U.S.S.R.: in other Eastern European countries it varied between 5% and 10%. West Germany and France have advanced at a rate of about 6% and most other Western European countries at between 4% and 5%; but the economic growth rate in the U.K. was only just in excess of 3%. In the U.S.A., the rate was 3.8% and in Canada 4.5%. The increase in total industrial production in a number of industrialised countries for the period 1955 to 1969 is shown in Fig. 2.

There is some evidence that economic growth rates in the next decade may not be sustained at the same level as in the past twenty years, although any deceleration which does occur is not likely to be uniform. It is generally considered that an increase in aid to developing countries could prove to be a major factor in helping to sustain high growth rates as this would provide a major outlet for housing, transport, power sources, and industrial processing and manufacturing plant, all of which, in one way or another, continue to maintain a demand on fluid movement and hence on flow measurement and control.

3   GROWTH RATES OF SOME INDUSTRIES WHICH CREATE A NEED FOR FLOW MEASUREMENT AND CONTROL

The importance of fluid movement varies greatly from one industry to another and some industries are totally dependent on fluid movement and flow measurement for efficient operation. The power generation, gas and water supply and distribution, petroleum and petrochemicals industries are examples. By contrast, the heavy electrical machinery, electronics and building industries create virtually no demand. Some industries rely heavily on accurate flow measurement during some part of the process: examples include the beverage and cement industries, fuel supply in aircraft, stock control in paper making.

The increasing accent on water conservation is resulting in an increasing interest in flow measurement at all points in the supply system from source to consumer. It is perhaps worth noting that desalination plant, with its associated flow measurement and control requirements, can provide a more acceptable solution to a water shortage problem, even in countries with a relatively heavy annual rainfall, compared with the construction of new reservoirs with their attendant sociological problems.

The current interest in environmental pollution and control demands that consideration be given both to the mass flow rate of the pollutant in the effluent from an industrial process as well as to the flow rate of any fluid into which it might be dispersed. In countries where anti-pollution legislation is well advanced, as in the U.S.A., the requirements are now being translated into the specification of industrial equipment.

The growth patterns, and requirements for flow measurement in some of the industries which are highly dependent on flow measurement will now be examined.

3.1 The energy supply industry and world energy requirements

The world demand for energy is expected to accelerate to an annual rate of about 6% p.a. in 1980

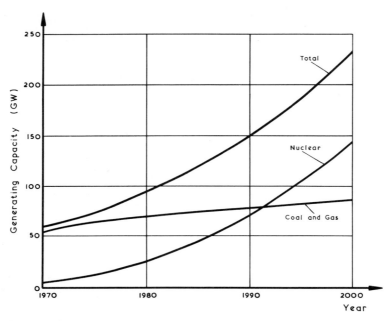

Fig. 4   Expected increase in generating capacity in the United Kingdom

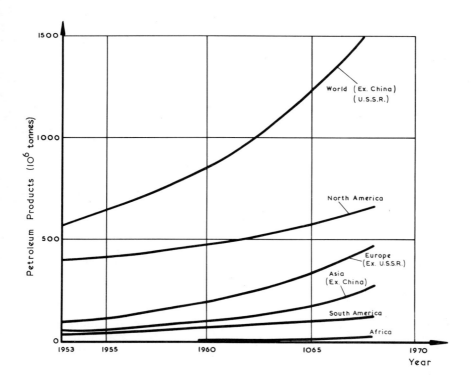

Fig.5   Volume production of petroleum products for the period 1953-1968

compared to a growth rate of about 4% p.a. since 1950: although in Western Europe the increase in growth rate is expected to be somewhat lower. The increase in production of electricity and gas, for example, in several major countries over the period 1955 to 1969 can be seen in Fig. 3.

If this prediction is true, the consequences in terms of fluid movement and flow measurement are considerable. Oil, which is expected to continue as the largest single source of energy will make increasing demands on flow measurement. The recent discoveries of natural gas in the North Sea and elsewhere will impose important new problems, both with regard to the supply and transfer of gas from one authority to another, but also with relation to a more sophisticated treatment of distribution problems. The dramatic increase in the use of natural gas in Great Britain can be seen in Table 3, which also shows the increase in total gas production over the past decade.

TABLE 3: Gas production in Great Britain

Year	Gas produced by gas boards	Natural Gas	Gas from oil refineries and coke ovens	Total gas production
		million therms		
1959	2226	12	557	2795
1960	2243	13	656	2912
1961	2199	13	688	2900
1962	2326	17	733	3076
1963	2392	21	847	3260
1964	2293	60	1033	3386
1965	2338	271	1143	3752
1966	2552	265	1209	4026
1967	2673	480	1186	4339
1968	2794	1062	1015	4871
1969	2589	1973	924	5486

*Indigenous and imported

The greatest increase in energy supply is expected to come from nuclear power, and Fig. 4 indicates the expected trend in generating capacity in the United Kingdom and in the main types of fuel likely to be employed. Many difficult problems of flow measurement are encountered in nuclear power stations and not all of these have yet been resolved entirely satisfactorily.

In general, in power stations, the fluids measured include oil, natural gas, air, steam, feed-water, condensate, cooling water and, in nuclear stations, liquid sodium, carbon dioxide and other gases, often containing a high level of radioactivity.

The highest accuracy is required for steam flow measurement ($\pm$ 1% is currently achieved) and for measurements of the flow rates of cooling water during acceptance and performance tests on pumps and turbines ($\pm$ 0.5% is possible). The ability to measure large flows accurately on site is resulting in a move away from accepting the performance of pumps and turbines on the basis of model tests. The accuracy of measuring feed-water flow rates is only about $\pm$ 2% since uncalibrated conventional differential pressure meters are often used.

3.2 The petrochemicals industry

This industry produces chemicals from sources of hydrocarbons, such as crude oil and natural gas and from petroleum refinery products. Petrochemicals have had a very rapid growth rate in recent years and, since 1958 for example, the total value of world production has increased at a rate of approximately 15% p.a., and output is expected to treble during the next 15 years.

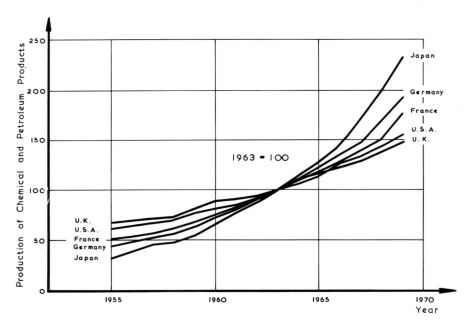

Fig. 6   Production of chemical and petroleum products in some industrial countries during the period 1955-1969

The volume production for the world and the Continental distribution is shown in Fig. 5. An accelerating growth rate in this industry is believed to be coupled to a corresponding increase in product movement and to be directly related to an interest in flow measurement and control.

Until about 1950, the petrochemical industry in the United States was significantly greater than in any other major industrial area, mainly because the U.S. petroleum and chemical industries were greatly stimulated during this period and were able to take advantage of available raw materials, a highly developed technology, available financial capital and a large and affluent domestic market. Together this led to the development of synthetic rubber, 'man-made' fibres and plastics.

During the decade from 1960, however, the United States was challenged by Western Europe and by Japan as a result of industrial reconstruction, economic integration and industrial growth in these countries. These factors, together with a supply of low-cost crude oil from the Middle East, has accelerated the growth of the petrochemical industry in these regions. The changing pattern of production in different areas of the world over the 14-year period from 1955 to 1969 is shown in Fig. 6.

For flow control within process plants, an accuracy of ± 3% is generally adequate with flow measurement being required over a range of 3 to 1, approximately. Orifice meters are used for most applications and are widely distributed throughout the plants. The greatest concentration of flow meters occurs in distillation units which are very similar to those used in petroleum refineries.

There is considerable interest in the accurate measurement of feed stocks and products and, as increased attention is given to product cost and to operating efficiency, the value of accurate flow measurement increases in importance. In the case of feed or product transfer on a large continuous plant, an error in flow measurement of a fraction of one per cent may represent a considerable amount of money. In a large modern ethylene plant, for example, with an output of about 500,000 tons per year, which would be worth approximately £20 million, an error of 0.5% corresponds to £100,000 p.a.

Invoice meters are also installed on service supplies, such as steam and water, as well as on plant feed stock and product lines.

In some processes a wide range of flow measurement, exceeding 20 to 1, is required and there is evidence that, as the size of process plants increase, a further slow but continuous increase in the operating range will be required.

Positive displacement meters are currently in use for the most accurate flow measurements in this industry, but turbine meters are finding increasing application, especially for gases.

Orifice meters are generally used for feed and product lines used to transfer fluids from one part of a plant to another. A measure of the emphasis on accuracy can be seen from the fact that sometimes several meters are operated in parallel to obtain accurate flow measurements over a wide range of flow rates.

Where highly accurate measurements are required and flow meters are used which are subject to upstream flow conditions, and to other factors which can affect performance, frequent, in-situ calibration is vitally important.

3.3 The water supply industry

An increasing population, a rising standard of living in most countries and a continuous increase in industrial demand has accelerated the need for water conservation and for flow measurement and control. This trend, which has become strongly apparent in the United Kingdom and in several other European countries during the past few years, is expected to dominate many industrial situations in the next decade or so.

In the United Kingdom, existing flow meters are mainly of the differential-pressure type and are usually Dall tubes and Venturi meters. These meters have a long life which may exceed 25 years. But eventually accuracy can deteriorate due to erosion, corrosion and to deposits of suspended sediments. To meet the requirements of higher accuracy in flow measurement, frequent calibration will be required in the future.

There is also an increasing interest in using electromagnetic flow meters in this industry, especially for pipes which exceed 1 m in diameter. Although significant errors have been associated with electromagnetic flow meters, most are attributable to the effect of installation conditions and especially to turbulence and swirl generated by bends in the pipe upstream of the installation.

The range of flow measurement which is required can be considerable and up to 50 to 1 has been reported. Environmental conditions can also be severe and meters can be subjected to temperatures corresponding to the full range from arctic to tropical conditions.

The increasing use of large pumps by the water supply industry has identified the need to consider the efficiency of the pumps being used. A decrease in efficiency by 1 per cent can equate to the cost of the pump so that regular calibration is required to ensure that optimum performance is maintained.

No significant variations in this pattern can be foreseen in the future.

3.4 Sewage disposal and pollution control

At the beginning of the 19th Century the population of the United Kingdom was 10.5 million and there was no sewage disposal system, although the Industrial Revolution had started 50 years previously. By the middle of the present century, when the population was 52 million, every town and nearly every village had its sewerage system. Capital investment in sewerage and sewage disposal systems is still very significant, as can be seen from Table 4 which gives the amount spent in England and Wales during the period 1958 to 1968.

TABLE 4: Capital investment in sewerage and sewage disposal systems, 1958-69

Year	£ thousand
1958-59	980
1959-60	1193
1960-61	1173
1961-62	1078
1962-63	1012
1963-64	1703
1964-65	1439
1965-66	1222
1966-67	1960
1967-68	1929
1968-69	2056

The increased awareness of environmental pollution has emphasized the need to measure the mass flow rates of gaseous, liquid and particulate pollutants and of the fluids into which they are dispersed. However, the full impact of pullution control on flow measurement has not yet been felt.

In general, a high accuracy in flow measurement is not required: ± 5% is acceptable and there appears to be no reason why this situation should change in the future.

3.5 Other industries

Several other industries have a major interest in flow measurement.

One of the most important of these is the Iron and Steel Manufacturing Industry where the greatest interest centres on the measurement of gas flows to blast furnaces and coke ovens, but no completely satisfactory measurement techniques have yet been devised. The industry is also a major consumer of water, and flow measurements are carried out to check water consumption and the thermal efficiency of various processes. A high accuracy is not required and this situation is seen as one which will not change significantly during the next decade or two. However, installed flow meters in both water and gas distribution systems are generally subject to severe corrosion and new instruments without contact with the working fluid would be particularly attractive.

The Paper and Pulp Industry is a major user of flow meters, and especially of high cost, high performance installations. Flow measurement problems are extensive and the fluids used range from water (> 50 m gal/day) to viscous stock, corrosive chemicals and steam. .

In the cement industry, variable area meters are used for small flows of water and electro-magnetic flow meters for measuring large flows and also for measuring the flow rates of chalk, clay and cement slurries. A high accuracy is generally required in measurements of slurry flows in long pipes. Fuel and fuel/air ratios are also monitored. There is an increasing tendency to introduce automatic control into cement works and this is expected to increase the need for flow measurement.

Special flow measurement problems occur in industries manufacturing cryogenic fluids, in the aircraft and aerospace industries, especially for fuel and lubricant monitoring systems, in bore holes used for both water and oil extraction and in two-phase flow systems in connection with heat-exchange problems. Not all these problems have yet received satisfactory solutions.

The pump and valve manufacturing industry also provides a yardstick of the industrial demand for flow measurement and the development of this industry in the United Kingdom during the period 1959 to 1969 is indicated in Fig. 7. The increase in supply for the same period of heating and ventilating equipment, gas compressors and exhausters, boiler plant and chemical plant are also included in Fig. 7. The general pattern is one of slow and continuous, if not dramatic increase and can be taken as a further indication of the future demand for flow measurement.

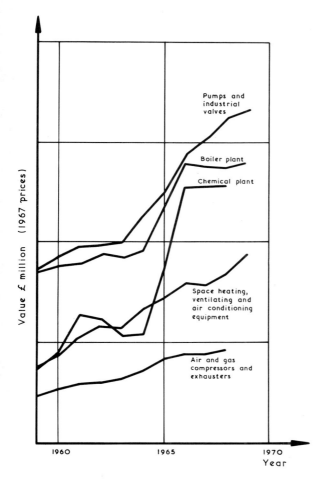

Fig. 7   Supply of equipment to the United Kingdom market

The production of pipes can also be used as an indication of the emphasis on fluid movement and hence of the demand for flow measurement. In the United Kingdom, for example, the production of concrete, pitch fibre, cast iron, steel, copper and stoneware pipes was 2.3 million tons per year over the period 1964-69.

4   NEW METHODS OF FLOW MEASUREMENT

The continuous interest in flow measurement can also be judged from the number of inventions which are offered for patent protection. From 1852, when the first flow meter was patented by Thomas Kennedy, up to the year 1890, a total of 989 patents had been registered on mechanical methods of flow measurement in the United Kingdom. During the past 5 years over 400 patents have been registered. However, if past experience is any criterion very few of these are likely to become successful commercially.

A large number of physical phenomena can be exploited to measure flow and current interest is high.

The principal difficulty, and the continued interest in flow measurement, partly arises from the fact that flow rate is a derived unit, expressed in volume flow or mass flow per unit time: there is no standard flow rate available as a reference. At the best, flow rate can be obtained from separate measurement of volume, mass and time, each of which is directly transferable to an international standard. Whereas it is not difficult to measure mass to an accuracy of 1 part in 10^6 and time to a few parts in 10^{10}, it is difficult to measure flow rates to 1 part in 10^3 in the conditions of a calibration laboratory and in industrial process plants an accuracy of 1 part in 10^2 is higher than is available from most flow meters.

A wide range of flow rate measurements are now required by industry in a wide range of applications which often demand a different combination of characteristics in the measuring system. In fact, the invention of new methods of flow measurement is not only a yardstick of total demand, but it is also an indication of unresolved problems, or problems with unsatisfactory solutions.

The current demands from industry for flow measurement are seen to include a need for higher accuracy, linearity in the operating characteristic, extended range, dynamic response, reliability, precision and low cost. These characteristics have to be compatible with a wide range of operational requirements, including operation in open channels or in closed conduits. The requirement may be to measure flow rate, or the volume or mass of fluid which has passed, in a system which may be steady or pulsating. The fluid itself may be a slurry or a light gas or a viscous liquid; it may be two-phase, non-Newtonian, explosive, radioactive or inflammable. The flow may be laminar or turbulent, sub-sonic or supersonic. The lateral dimension of the conduit may vary from less than 1 mm to several metres. The fluid temperature may be close to absolute zero, or it may be at the temperature of molten steel ($\sim 1350^\circ$C): the pressure may correspond to high vacuum or it may be at several hundred atmospheres. The situation may be accessible or it may not. The general environment may be earth-bound, inter-planetary or lunar, so far.

It seems not unlikely that continuing development in flow measurement will be required to meet the varied demands imposed by the increasing and complex society in which we live.

CURRENT PRACTICE IN FLUID FLOW MEASUREMENT

E.A. Spencer

Abstract: The transition from development to the acceptance of a flow meter is a gradual one which reaches different stages in different industries or organisations. Dominant in industrial flow measurement from earliest times have been meters based on the pressure difference principle. Though giving way in some areas to new methods, they are still used in well over 80 per cent of industrial installations, and are the only devices to have been standardised fully. Very little in flow measurement is static, and the paper aims to place emphasis on the dynamic adaptation of old and new which represents the situation today.

1   INTRODUCTION

An analysis of the present position of flow measurement, or an attempt to draw a line between the conventional and unconventional in flowmeters is inevitably arbitrary. The transition from development to acceptance is a gradual one and reaches different stages in different industries and organisations at any one time.

If conventional means static then there is little in flowmetering that can be said to be in this state. Perhaps one exception is the venturimeter which goes on being sold in small numbers to industry and probably has not changed in pattern in ten years. A Working Group of ISO/TC30, which is the committee of the International Organization for Standardization which is responsible for flow measurement in closed conduits, produced an International Recommendation on venturimeters (R781) in 1967. Three years later, after further meetings, they concluded that no new information was available to justify making any revisions to the original Recommendation and the Working Group disbanded itself. Here then might be said to be a static situation.

Another definition of conventional which leads from this example could be standardization. Pressure difference devices like the orifice plate and nozzle would then fit into the definition but many flowmeters which have been in production and general use for many years would be excluded. For instance, in Britain we have no standard on ordinary domestic water meters though vast numbers have been built in the last half century. Now the European Economic Community are drafting a directive on these meters for use in their countries. Much more standardization, if not of the metering devices themselves, but of the system in which they are installed is desirable and will come if the benefits of rationalization are to be gained. Meanwhile, very little information of this kind is available in this form.

In this paper, selection of the conventional has been on the basis of general acceptance. However, research and development is not confined to the unconventional and modifications and extensions are being made continuously to adapt present conventional meters and methods to tackle new problems.

2   CLASSIFICATIONS

Terminology in the classification of flowmeters, under such terms as positive, differential, inferential, total or partial can easily be interpreted differently. Simple groupings have been used later, therefore, in discussing characteristics, but one clear differentiation which can be used is on the basis of their application. Flow measurement could then be regarded as falling under two headings–quantity and rate-of-flow. It is well worth stressing this division, for not only does it really demand quite different qualities in the meter but its implications should be considered by the user before tackling any flow measurement problem. He can set out the reasons for his measurement and decide which of these two objectives he really wants to achieve. It is all too easy to try and force-fit a flowmeter into a situation because it is available or has been successfully used for some other project.

A quantity measurement calls for a measurement of the total amount of fluid passing, irrespective of the rate at which it is flowing. It is vital that a measurement should be unique which means that the same answer is given for the same quantity under any conditions, whether it is for the same volume or the same mass. The rate at which the fluid is flowing will often vary during the measurement but this is not relevant to the quantity measurement. Fundamentally, the mass flowing is almost always the required quantity, since it is the mass in the fluid which is actually of use, but for convenience or because it is less costly the user often only demands that he should get a particular volume. In practice, variations of temperature and density for example are certain to affect the actual flowmeter itself in some way which will not be entirely compensating, since true mass flowmeters are not commercially available.

In quantity measurements this uniqueness of the reading is clearly associated with its accuracy, even though in every application accuracy may not be of the highest importance. In the vast majority of cases the user of an instrument installed to measure the quantity flowing would be dissatisfied if the reading he obtained were not closely tied to a percentage limit from the true quantity. A linear flat characteristic which changes very little if at all over the flow range to be covered is a major objective in the design of a good flowmeter. Although it is recognised that present-day developments make it possible for any stable flowmeter characteristic to be linearised electronically, this has at least two major disadvantages—it is more expensive and each individual meter will need to be tailored to its lineariser for good accuracy. For example, such an electronic system would be entirely inappropriate for the relatively cheap domestic water meter referred to earlier, but this instrument is nevertheless expected to give the correct quantity for charging purposes, irrespective of whether it is used mostly at 5 per cent or 50 per cent of its highest rated flowrate.

Flowrate uniqueness on the other hand by its very name is time dependent: the value will relate to the amount flowing at a particular time. In these circumstances the most important characteristic of the flowmeter will be that it should give the same reading on all occasions when the rate of flow is the same. Whenever the meter indicates that the flow is $1.00 \, \text{m}^3/\text{s}$, then it is important that it should be the same $1.00 \, \text{m}^3/\text{s}$ as at other times. The range of use can be quite narrow and linearity is far less important.

Thus the question of whether quantity or flowrate is the controlling parameter in the particular installation being measured and whether on a mass or volume basis, must be answered. Most people in industry, when asked what are the most important factors which they require from a flow measurement or flowmeter, have said:

    (a) repeatability   (b) reliability   (c) accuracy

in that order of priority. No one denies the need for a knowledge of accuracy but it is not considered to be of either first or second importance. The term accuracy though is often used in many different senses: linearity, repeatability, response, calibration factor or, correctly, as the deviation from the absolute truth and there is a need for universal acceptance of the meaning of each of these terms. Efforts are now being made in the United Kingdom and elsewhere to standardise terms used in describing instrument performance.

Accepting that such confusion exists, accuracy as covered by the final definition given above is considered a desirable attribute which only becomes worthwhile when the answer is found to be wrong. If a 20 per cent error is found then no one would be likely not to take this into account if the error was against him. But it is difficult if not impossible to find an answer to the question of what benefit would result if the accuracy of all the measurements being made at the moment in the world were improved to, say, twice their present values. Where this can be assessed for specific situations it provides a guide to the relevance of the claims of accuracy made for the various flowmeters available.

Having given, as it were, an end-product division of flow measurement into quantity or flowrate, one can equally well use these terms as a starting point in the classification of flowmeters. Thus, a quantity meter can be defined as one in which the fluid in passing through the primary element is split up into virtually isolated quantities either of the same weight or the same volume, by filling and emptying successive containers. Positive displacement meters obviously fit into this category and can be used for both liquids and gases: reciprocating piston, nutating discs and so on have only been used for liquids but might well be used for gases at high pressures. For gases, the conventional wet gas meter is perhaps the best known example.

By contrast, a rate-of-flow meter can be defined as one in which the fluid passes through the primary element in a continuous stream. The movement of the fluid has an effect on the primary element according to some physical law, known or unknown and as a result the quantity of flow per unit time is derived. In practice the relationship between flowrate and the signal obtained is always an empirical one established for the type of device from experience or calibration. Examples include the orifice plate, turbine and electromagnetic flowmeters in which the pressure loss across the flowmeter progressively decreases to zero since in the last named there is no interference at all with the flow along the pipe.

It would be convenient if these groupings were consistent with each other and meters based on the quantity principle were used exclusively for quantity measurements and rate-of-flow meters for flowrate measurements. Such is not the case and a wide overlap between the two groups exists. It is however justified to conclude that the best accuracies are likely to be achieved if the principle is consistent with the application.

3   FLOWMETERS

Since there is no universally suitable flowmeter for all applications it is important to choose the one which meets most nearly the requirements of the user's particular installation.

Operational convenience and maintenance may often be important constraints, however, and it would be impracticable to use a wide variety of instruments in a group of installations on the same site. In this instance different quality levels can be achieved by using a single type of meter throughout but having vital installations calibrated to the requisite level of accuracy.

In the following sections a few widely differing types of instrument suitable for a range of applications are examined.

3.1 Variable area meters

Though not usually regarded as capable of high accuracy, variable area meters play a valuable part as indicators. The variable area is that between a tapered tube and the diameter of the floating element, and the tube must be vertical. The shape and weight of the float characterises its behaviour in a particular fluid and much theoretical work has been done on its design. For instance, it has been agreed that a thin disc float is nearly independent of viscosity effects over a wide range.

Grooves used to be cut in the sides of the float so that it would rotate and thus improve its self-centering ability, although the trend more recently has been to guide the float deliberately. There have indeed been many patents since its invention in 1868.

It should be noted that this is a flowrate meter based on a flowrate principle. In fact the flowrate is that at the instant of observation since the position of the float is directly governed by the cross-sectional area available for the flow of the fluid. This can be regarded as either a virtue or a vice since it means that if the flowrate is fluctuating then the area will be required to change and the float will move up and down. This 'bounce' can make it difficult to read the instrument even in good conditions to better than ± 2 per cent of full scale. It will be clear that the meter is unsuitable if there are severe fluctuations and will be damaged by shocks.

It is understandably not particularly suitable for large flowrates but limited to small flows. It has a major advantage in that it is virtually independent of the upstream flow pattern. While it cannot be jammed or stopped by foreign matter its calibration will change fairly rapidly if dirt particles score the float and it is normally recommended that an upstream filter should be used.

Sizes range from 1.5 mm diameter to over 300 mm with a maximum fluid velocity of about 3 m/s at inlet and the aim is for as nearly a linear characteristic as possible by the choice of taper of the tube. In fact near-linearity can generally be achieved from full flow to between 10 and 20 per cent of full flow.

Theoretical work has also been done on this instrument to predict its performance under different conditions. It is important to realise however that the calibration factor obtained for one fluid at particular conditions of temperature and pressure can only be converted for use in another situation if an additional tolerance is added to the calibration uncertainty. Whether it is worth further investigation to try and improve the confidence level on such conversions is a matter for the manufacturers and users to judge. It is suspected that effort will be concentrated on manufacturing techniques so that the variable area meter will remain essentially a flowrate indicator.

3.2 Positive displacement meters

By contrast, the most accurate flowmeter available at the present time is the positive displacement meter. It is a quantity meter based on a quantity principle. Claims of ±0.1 per cent over a very wide range of flowrates are reasonable for a calibrated meter while, uncalibrated, the accuracy probably need not fall to worse than ± 0.25 or 0.3 per cent. This means that manufacturing tolerances and inspection can ensure a very high standard of product.

Complete displacement however is not possible: there is bound to be some leakage which will increase as wear takes place. In the case of the wet gas meter the water seal is positive, but with liquids, contact between one metal and another, or between a metal and a non-metal inevitably leads to slip. This is affected by the properties of the fluid being measured, the rate of flow and other factors. Clearly dirt must be avoided and even particles of less than 0.010 inches will cause rapid wear. It is perhaps for these reasons that this meter has been used so extensively for accurate measurement with petroleum products with their good lubrication properties and extreme cleanliness.

The type of seal together with the external resistance of the counting mechanism will be the main source of the pressure drop across the meter. This pressure loss is relatively high but the meter usually has a very wide range of about 10:1 in flowrate so that appropriate sizing can be selected. Meters of this type have been used for more than one hundred years and, though the materials used in construction have changed radically, the principle of operation has been unchanged.

If he is looking for good accuracy of measurement why should the user not be satisfied then to buy a positive displacement meter and be done with it? Two factors—life and cost—can be put forward to account for choosing alternatives. As stated above the fluid must be clean for the moving parts to have small clearances. Dirt particles would jam the meter and thus stop the flow. Although this might be an advantage under certain circumstances, it will more often than not lead to other troubles. In any case it will mean that the meter will need to be stripped down for servicing. Whether this catastrophe occurs or whether there is a gradual change caused by wear, the meter will need regular servicing and re-calibration if its original quality is to be preserved. Thus, the user is not by any means done with it when he installs such a meter.

The second major consideration has been that of cost. A positive displacement meter is a precision instrument and its cost increases because of its bulkiness as the size of the meter increases. While this relatively high cost is justified for many applications, there will obviously be a balance point where it is no longer competitive. The increasing use of the turbine meter, which is basically a less accurate device, has come about because it occupied far less space and was more versatile than the positive displacement meter.

3.3 Turbine-type flowmeters

It took 20 years or more for the modern turbine meter with electronic counter to move from the research laboratory into commercial application. Ten to twelve years ago calibration tests on such meters were being carried out at the National Engineering Laboratory and it was not always possible to give the manufacturer or user the confidence he was seeking. There have been substantial improvements since then, for example, in bearing materials but it has to be recognised that the turbine-meter is still not regarded as capable of being left without frequent calibration in important bulk measuring situations.

It is interesting to examine how findings of agreement to high levels of accuracy of, say ± 0.05 per cent or better, are countered by experiences of hysteresis jumps in characteristics or sensitivity to changes with flow conditions or time. While the periodic checking of a turbine meter with a meter prover has built up a confidence which may justify a claim of say ± 0.2 per cent accuracy for the combination, there is surprisingly little information published on the reliability and fundamental accuracy of the meter prover itself and, with regard to the turbine meter in this situation, it is not its own accuracy but that of the meter prover which is the controlling element. Claims of ± 0.02 per cent accuracy are sometimes encountered; they can only be regarded as highly optimistic! Without individual selection and calibration, the predictable accuracy of a production run of turbine meters probably is not better than ± 2 per cent.

Although our present knowledge of the behaviour of turbine meters limits the viscosity range over which it can be used, it has nevertheless a wide flow range: the normal meter can be expected to work over nearly a 10:1 range. Within this range, a flat linear characteristic is possible at ± 1 per cent but certainly not within the 0.1 per cent level. Like positive-displacement meters turbine meters must be used with clean fluids since bearing troubles are the main cause of costly maintenance. This is the ultimate disadvantage of such devices which have moving parts. In fact it is still a relatively expensive meter which can be damaged by shocks or flow pulsations and requires long straight lengths of pipe upstream if its specified performance is to be achieved. The usual substitute which is recommended in place of a long upstream length of pipe is to instal a straightening device or tranquiliser as the French call it, but work carried out at the National Engineering Laboratory makes it apparent that this is a subject which has received too little attention. Evaluations to determine the interaction of different types of straightener with different upstream disturbances in relation to pipe length requirements are badly needed.

The turbine meter is thus still in a development stage and it is quite possible that novel variations will appear shortly. Its digital read-out, whether of the number of blades passing a point or the number of revolutions, gives it a very considerable advantage in the type of secondary equipment which can be used and its proportion of the total number of meters in operation is likely to continue to increase for the time being.

3.4 Electro-magnetic flowmeters

Before considering the ubiquitous pressure difference flowmeters it is important to refer at least to one other type which has become firmly placed in text books and catalogues.

The electro-magnetic flowmeter has proved to be a solution to many awkward problems which had to be evaded before its appearance or tackled quite inaccurately. For instance, as a volumetric flowmeter for handling slurries and corrosive liquids it has meant that a new level of measurement was possible for a wide range of difficult applications. The principle was suggested for measuring river flow many decades before an analytical analysis in 1930 showed its suitability for pipe flow.

Progress to a commercial instrument took nearly twenty years and even in the late 1950's there was over-confidence in the accuracy with which the characteristic of a particular meter could be estimated without calibration.

Long term repeatability has not been as high as initially seemed possible so that regular site checks remain desirable. Nevertheless it has placed a very valuable weapon in the instrument engineer's armoury.

3.5 Pressure difference devices

It would be very useful to have numbers of each of the different types of flowmeters which are in use even though clearly the ratios will be changing all the time. At the National Engineering Laboratory 12 years ago the author and a colleague made a general survey on flowmeters and contacted more than 20 different organisations in a wide spectrum of industries. From this sample it was estimated that over 90 per cent of the flowmeters in industry were pressure difference devices. It is probable that the number used in the process industries is still over 80 per cent.

Within the general category of pressure difference devices the orifice plate is by far the most frequently used in Great Britain. Three-quarters of a century of research and testing have been put into it and it is certain that this tremendous technical investment will not be thrown away though its dominance may decrease. In its various forms it can be used in almost every situation although it has to be recognised that quite often insufficient attention is paid to ensuring that the conditions just upstream of the meter are suitable. In practical terms most data are available for pipe sizes from 50-900 mm diameter.

It is said that the various national and international standards make full utilisation of the orifice plate too difficult. After all is it not just a plate with a hole in it pushed between two pipe flanges? The most frequent criticism is that the user does not want to pay by having to adopt the safeguards which the standard lays down to obtain the coefficients derived in controlled laboratory and field tests. He believes he would be satisfied with less but is not being given the necessary information. Perhaps the time has come to reassess the objectives of these standards and to see if the variety of different needs could be met, not by having a single standard, but with a series for different levels of tightness of specification. This will not be easy, for reliable information on relaxed conditions is very scarce.

The orifice plate has the advantage of cheapness and of reliability if no erosion takes place, though it must be recalled that an orifice plate by itself does not make a flowmeter. The pressure difference measurement outside the meter will often be a source of difficulty and the overall cost of the whole instrumentation may be just as substantial as with other types of meter. There is the advantage of no moving parts but the disadvantage of a significant pressure loss. The real problem of the orifice plate, however, is the uncertainties introduced by installation effects. It is easy to install in a pipeline but its repeatability and accuracy and its performance according to its prediction are tied intimately to the position it is placed with relation to the flow in the pipe.

The best accuracy which can be obtained with an orifice plate made and installed according to the standard specification is perhaps as good as ± 1.5 per cent. By calibration, an improvement to ± 0.75 per cent can be obtained. It has, however, a limited flow range of about 3:1 because it is based on a square-root relationship of pressure difference and flowrate. The square-edged type cannot be used below a throat Reynolds number of about 10,000-15,000 while at the higher Reynolds numbers there exists the puzzle of whether the coefficient continues to fall slowly in the range above 5×10^6 now being encountered in power station measurements.

Little has been published on the repeatability of the pressure difference readings from orifice plates installed in locations where there are only short upstream straight lengths of pipe. While experience at the National Engineering Laboratory suggests that in fully established flow conditions the steadiest readings are obtained from D and D/2 or vena contracta tappings and fluctuations increase with flange and corner tappings, data are needed to find noise levels for practical situations. How close, for example, can an orifice plate be located to a series of bends before controllability is lost?

The greatest point of sensitivity of the orifice plate in relation to time is the sharpness of the square edge. Studies made in recent years have shown that inspection to check the edge is difficult. Often the newly installed plate will not lie within the close tolerances which experience has shown to be necessary and, in addition, wear which leads to a rounding of the edge and a marked change of meter coefficient, will occur in service. The nozzle, which overcomes this disadvantage, although it is much more expensive to produce than an orifice plate, is being used more frequently in the United Kingdom. It has long been a standard flowmetering device in Germany with the VDI contour and in the USA with the ASME long and short radius patterns. Its predictable accuracy is almost identical to that of the standardised orifice plate.

3.6 Performance

There are many other types of flowmeter which have been commercially available for some years. It is doubtful however whether the numbers sold are large. The market is logically split between flowmeters required for accounting purposes and those needed as indicators and thus ultimately to be available for control purposes. The ratio is probably of the order of 1 to 10 of the latter and it is for this reason that it was stated earlier that industry wants reliability, and repeatability before accuracy.

The feedback of information on these factors, for example on reliability in use, both to the manufacturers and to instrument engineers in general is very restricted. To obtain data on the way in which a flowmeter has been influenced in service by its location and the fluid flowing through it would necessitate its withdrawal for detailed examination and calibration. Accuracy, though not the primary objective of the industrial user, is then essential in the equipment required in the evaluation of these other factors.

4   FLOW MEASUREMENT METHODS

The alternative to the withdrawal of the meter for calibration and checking is to obtain a reference measurement on site. Flow measurement methods of this kind are also needed for the occasional investigation of a circuit, for example, during commissioning or at acceptance tests.

As with flowmeters, the number of different principles which have been explored are numerous and can be divided similarly into quantity and flowrate groupings. The static volumetric or weigh-bridge tank into which the flow is diverted for a measured time has long been adopted as being fundamentally the most accurate method of quantity measurement, but more recently the very high repeatability of the meter prover has attacked this claim.

4.1 Meter provers

Undoubtedly the meter prover does make it possible for the flow in smaller diameters of pipe to be checked easily under conditions of use, but there is a practical limitation both in terms of size and cost. A 400 mm meter prover cannot be easily accommodated and must be incorporated as part of the permanent flow measuring equipment in the installation.

Investigations to evaluate the performance of meter provers are in progress, and in this field too the conventional meter prover is being modified and improved as new materials and novel methods of providing a curtain which will travel with the fluid are tried. This mutation from the travelling screen method, now archaic, is bound for major development in the future.

4.2 Bulk flow methods

Grouping the methods which are used for non-continuous measurement under the headings of bulk flow and integration methods provides a useful broad division. In the first named the basis is that a characteristic directly related to the total bulk flow is measured while in the latter local measurements are made in the cross-section which can then be integrated over the section to obtain the flow.

These methods have become well established in the last twenty years and have been included in many test codes and standards. Changes are incorporated into successive revisions but these in general have been adopted to improve the confidence level of the measurement carried out in accordance with the specification rather than to make any substantial improvement in accuracy. The salt velocity, dilution and Gibson methods are all bulk flow methods while use of the current-meter and pitot tube involve point velocity integration. Claims for accuracies of better than ± 1 per cent were being made 50 years ago for the dilution method using sodium chloride as the tracer. The standards of today claim no better than this but lay down sufficiently detailed rules to ensure that if followed there is at least a 95 per cent chance of success.

The amount of the tracers which were used for measurements had to be adequate for the analytical technique being used in their detection. This resulted quite often in very large quantities having to be transported to the site. Today the use of radioactive tracers and other chemicals in conjunction with more sophisticated instrumentation has meant that all the equipment necessary to measure very large flow rates could be carried by a two-man team.

4.3 Integration methods

Modern production methods have made possible relative cost reduction as well as improvements in the reliability of currentmeters but little apparent improvement has taken place in the claimed accuracy of measurement. When it is appreciated that currentmeter blockage effects were not known about twenty years ago but can result in a bias of 0.5 per cent or more, or that positioning a pitot tube according to the tangential rule must result in an over-estimate because of boundary layer effects, then high accuracies in earlier times were achieved only because compensating effects also not appreciated came into play.

The choice of location of the instrument has been studied intensively and stricter rules have been introduced so that it is possible that soon new levels of accuracy can be set. In fact, the object of the measurements listed in this chapter is primarily to achieve a given level of accuracy with high confidence. The expense of such special tests would not be justified if the calibration of the installed flowmeter or the measurement of the flowrate through the circuit remained indecisive. Indeed, if these methods are to achieve for the larger flowrates the confidence that the meter prover has given for the smaller flows, then built-in means of cross-checking to establish these limits of accuracy more precisely are essential.

5   CONCLUSION

The emphasis on the three qualities of reliability, repeatability and accuracy has been met in current flow measurement practice by a combination of improved materials and production, novelty as well as sophistication in instrumentation and calibration on the one hand together with better understanding on the other.

Claimed accuracy levels have not in general changed over the past twenty years. What has changed is the real statistical confidence level of the claims. The particular example of orifice plate coefficients, where the uncertainty tolerance in the most recent USA standard 'Fluid Meters' has been increased over that given in the 1959 Edition, illustrates the trend. For many other flow-meters and methods, research and evaluation have shown more clearly the boundaries, whether of installation effects or inherent performance, which limit repeatability and accuracy. Better test facilities have made this analysis possible and enhance the chances of further improvements.

ACKNOWLEDGEMENT

This paper is published by permission of the Director, National Engineering Laboratory, Department of Trade & Industry.

FLOW MEASUREMENT BY WEIRS AND FLUMES

Peter Ackers

Abstract: The performance of weirs and flumes is reviewed within the framework of a generalised flow equation. This contains coefficients which account for

(i)   fluid properties and crest roughness
(ii)   the profile of the structure
(iii)   the influence of approach velocity
(iv)   the shape of the cross-section
(v)   the influence of tailwater level

The types of gauging structure reviewed included broad-crested weirs, triangular profile weirs, V weirs, and trapezoidal flumes. Research carried out at the Hydraulics Research Station over the last decade has increased the understanding of the performance of such devices, and methods have been developed for designing them, for establishing calibrations theoretically including allowing for boundary layer effects, for assessing drowned flow performance and the effect of approach conditions including an upstream bend. The special problems in natural rivers, for example sediment transport, are excluded.

1   INTRODUCTION

Hydraulic structures such as weirs and flumes have long been used for measuring discharge in both natural rivers and artificial channels. The principles of use are much the same in both situations, although some of the difficulties met in irregular natural streams[1], for example the transport of sediment, may be absent from rigid-boundary prismatic man-made channels. This report reviews the present state of the science of flow measurement by hydraulic structures in artificial channels, such as lined irrigation channels, hydro-power conduits flowing with a free air-water surface, sewers running part full, the inlets to sewage purification plants, cooling water channels at generating stations and industrial plants. Water is assumed to be the fluid.

The structures concerned fall into two broad categories, weirs and flumes, although the border-line between them is blurred. The term "weir" is used to describe a raised construction in the floor of the channel, usually but not always occupying the full width in artificial channels. The term "flume" describes a contraction in width, often associated with a hump in the channel floor, which provides a prismatic contraction within which the flow accelerates and passes through the critical condition, defined by unit value of the Froude number, $v/\sqrt{gd_m}$ (terminology appears at the end of the paper).

Both of these categories of structure can have a range of cross-sectional shapes when looked at in the direction of flow. The most basic shape is the rectangle, but we might also use a triangular flow section, a trapezium or a U-shape to suit particular operational requirements. The shape when viewed in the direction of flow should not be confused with the profile when viewed from the side i.e. perpendicular to the flow section. A flume must have a fairly streamlined profile, but weirs range from thin-plate devices, through rectangular and triangular profiles to well-streamlined sections. The values of the coefficients of discharge and the detailed methods of applying equations to deduce flow rate from a measurement of upstream water level may differ from case to case, but generalizations are possible as will appear later. As this review concerns civil engineering applications in the main, large robust structures are required and hence little attention will be paid to thin-plate devices.

2   THEORETICAL BACKGROUND

The variables defining the two-dimensional motion of a fluid over a wide horizontal crested weir under the action of gravity are:

		Dimensions
M,	the mass flow rate per unit width	$ML^{-1}T^{-1}$
h_1,	the elevation of the upstream water surface relative to the crest	L
h_2,	the elevation of the downstream water surface relative to the crest	L
g,	acceleration due to gravity	LT^{-2}
ρ,	mass density of fluid	ML^{-3}
μ,	fluid viscosity	$ML^{-1}T^{-1}$

together with sufficient linear measures to define the boundary geometry fully. One such dimension suffices to characterise the geometry of a given weir installation, for example the crest height P, giving seven variables which by the theory of dimensions may be reduced to four dimensionless groups. A convenient functional relationship for a particular installation is thus:

$$\frac{M}{\rho h_1 \sqrt{gh_1}} = \text{function} \left(\frac{h_1\sqrt{gh_1}}{\mu/\rho} \ ; \ \frac{h_2}{h_1} \ ; \ \frac{h_1}{P} \right) \qquad \text{... (1)}$$

This is conventionally expressed as a volumetric rate of discharge per unit width:

$$Q/b = C_d h_1\sqrt{gh_1} \qquad \text{... (2)}$$

where C_d depends on the Reynolds number represented by the first term in the bracket of equation 1, on the drowning ratio represented by the second term, and on the flow geometry characterised by the third term.

Application of the concept of minimum energy to the flow on the horizontal crest of a stream-lined weir or in the throat of a certain limiting proportion of h_1, the volumetric discharge is related to the total energy level upstream of the weir or flume, and an even more fundamental equation thus results

$$Q = \frac{2}{3}\sqrt{\frac{2}{3}} \ b H_1\sqrt{gH_1} \qquad \text{... (3)}$$

where H_1 is the total head relative to crest or invert defined as

$$H_1 = h_1 + \propto v_1{}^2/2g \qquad \text{... (4)}$$

However, h_1 is the quantity measured directly from a water level recorder and this differs from H_1 because of the velocity of approach. Equation 3 above is thus not very convenient in application, even after incorporating a coefficient C_{b1} to allow for the frictional effects of a real fluid. Consequently it is preferable for the effect of the velocity of approach to be allowed for by means of a factor in an equation in terms of the gauged head h_1, usually denoted C_v. If a weir is not of streamlined horizontal broad-crested form, there must also be a coefficient to allow for the influence of the shape of the crest profile on the flow, C_p. Furthermore, if the weir or flume is not of two-dimensional form, then the shape of the cross-section seen in the direction of flow departs from a simple rectangle, and this may be accounted for by yet another coefficient, a shape coefficient C_s. Then if tail-water levels are sufficiently high to affect upstream levels at a given flow—or alternatively to diminish the discharge passed at a given upstream level—the flow depends upon h_2/h_1 and is said to be drowned. This flow diminution is best allowed for by incorporating another factor in the equation, f, which is less than one when the weir or flume is drowned. The equation for weirs and flumes therefore may be written generally as:

$$Q = C_{b1} C_p C_v C_s f b h_1\sqrt{gh_1} \qquad \text{... (5)}$$

where

C_{b1} account for fluid properties and crest roughness, which result in the development of a boundary layer of slower moving fluid at the fluid-solid interface

C_p accounts for crest or throat profile, being $2/3\sqrt{2/3}$ for streamlined broad-crested weirs and flumes

C_v allows for the approach velocity, permitting us to use h_1 in the equation rather than H_1 through the definition equation

$$C_v^{2/3} = \frac{H_1}{h_1} = 1 + \frac{\alpha v_1^2}{2gh_1} \qquad \qquad \text{... (6)}$$

C_s permits one to use the same general form of equation even if the structure provides a non-rectangular flow section.

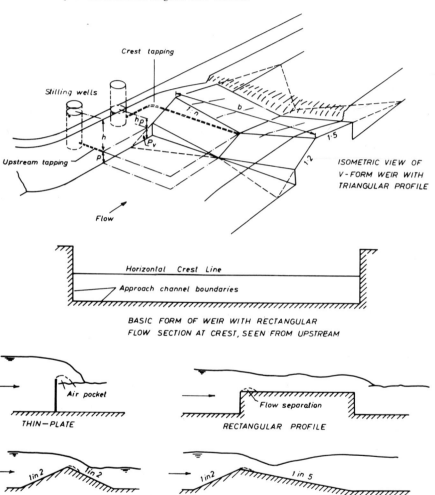

Crest tapping

Stilling wells

Upstream tapping

Flow

ISOMETRIC VIEW OF
V-FORM WEIR WITH
TRIANGULAR PROFILE

Horizontal Crest Line

Approach channel boundaries

BASIC FORM OF WEIR WITH RECTANGULAR
FLOW SECTION AT CREST, SEEN FROM UPSTREAM

Air pocket

THIN—PLATE

Flow separation

RECTANGULAR PROFILE

1in2 1in 2

TRIANGULAR WEIR PROFILES

1in2 1 in 5

STREAMLINED BROAD CRESTED WEIR

Fig. 1   Examples of weirs

f is the ratio of the actual discharge to the modular discharge with the same upstream head (i.e. when h_2/h_1 is below the limiting value for the weir or flume to be drowned by tailwater conditions).

Some recent developments and research results will be described by considering each of the above factors in turn, in relation to the types of weir and flume that are now finding wide application in engineering practice. Examples are shown in Figs 1 and 2 respectively, together with outline sections of the alternative forms.

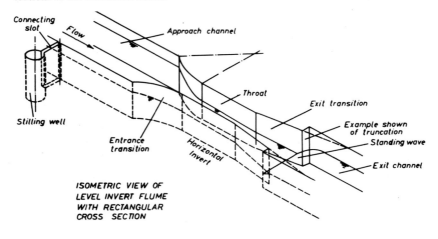

ISOMETRIC VIEW OF LEVEL INVERT FLUME WITH RECTANGULAR CROSS SECTION

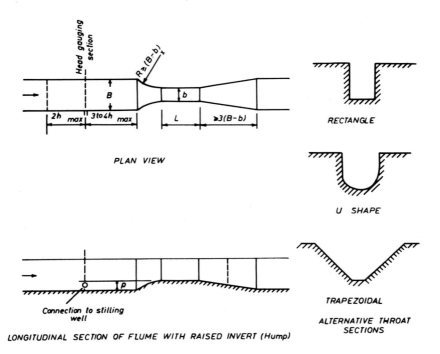

Fig. 2   Examples of flumes

3   THE BOUNDARY LAYER COEFFICIENT, C_{b1}

3.1 Streamlined broad-crested weirs and flumes

Referring to the definition sketch in Fig 3, showing flow over a weir or within the throat of a flume operating in the modular range (i.e. not drowned by high tailwater level), the boundary layer influence on discharge can be deduced theoretically[2]. Outside the boundary layer, the fluid velocity is given by

$$u = \sqrt{\{2g(H_p - d)\}} \qquad \ldots (7)$$

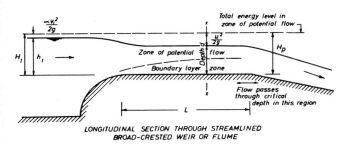

LONGITUDINAL SECTION THROUGH STREAMLINED BROAD-CRESTED WEIR OR FLUME

Fig. 3   Flow through a control section

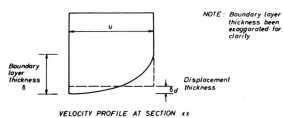

NOTE : Boundary layer thickness been exaggerated for clarity

VELOCITY PROFILE AT SECTION xx

where H_p denotes the total head at any point in the zone of potential flow (i.e. outside the boundary layer). From the definition of the boundary layer displacement thickness[3] the continuity equation can be written

$$Q = (A - P'\delta_d)u \qquad \ldots (8)$$

$$\text{Hence } Q = \sqrt{2g}\,(A - P'\delta_d)(H_p - d)^{1/2} \qquad \ldots (9)$$

Differentiate with respect to d to deduce the critical depth, d_c, at which Q is a maximum for a given value of H_p. Set $dQ/dd = 0$ and $dA/dd = W$, the water surface width. Thus

$$(H_p - d_c) = \frac{A_c - P'_c\,\delta_d}{2\left(W_c - \dfrac{dP'}{dd}\,\delta_d\right)} \qquad \ldots (10)$$

Inserting into equation 7 and solving equation 8,

$$Q = (A_c - P'_c\,\delta_d)\left\{\frac{g(A_c - P'_c\,\delta_d)}{\left(W_c - \dfrac{dP'}{dd}\,\delta_d\right)}\right\}^{1/2} \qquad \ldots (11)$$

and also

$$H_p = d_c + \frac{Q^2}{2g(A_c - P'_c\,\delta_d)^2} \qquad \ldots (12)$$

This pair of equations provides a parametric basis of calibration for any critical depth device. H_p, the total head in the potential zone is identical with the total head that exists just upstream of the contraction because there will be no energy loss between these successive points in a stream tube remote from the boundary.

Equations 11 and 12 may be written

$$Q = A_* \sqrt{\frac{g\,A_*}{W_*}} \qquad\qquad \ldots(13)$$

$$H_1 = d_c + \frac{1}{2}\left(\frac{A_*}{W_*}\right) \qquad\qquad \ldots(14)$$

where the suffix denotes the area and width at the critical section assuming that the boundaries are displaced inwards by δ_d. The relationship of total head to discharge can thus be deduced theoretically in this type of weir or flume, but with some algebraic manipulation the function can also be expressed as

$$Q = A_c \sqrt{\frac{g\,A_c}{W_c}} \qquad\qquad \ldots(15)$$

$$H_1 = d_c + \frac{1}{2}\left(\frac{A_c}{W_c}\right) + \frac{P'_c}{W_c}\delta_d \qquad\qquad \ldots(16)$$

These reduce to the general form of equation 5 with

$$C_p = \frac{2}{3}\sqrt{\frac{2}{3}} \qquad\qquad \ldots(17)$$

$$C_{b1} = \left(\frac{W_c\,d_c + \tfrac{1}{2}\,A_c}{W_c\,d_c + \tfrac{1}{2}\,A_c + P'_c\,\delta_d}\right)^{3/2} \qquad\qquad \ldots(18)$$

The advantage of this theoretical approach to the boundary layer coefficient of a critical depth device is that it may be predicted with some confidence from boundary layer theory, which relates δ_d/L to the relative roughness of the crest or throat and the length Reynolds number. L refers to the distance along the prismatic portion of the contraction to the position at which critical flow is reached, assumed equal to the full crest or throat length in the absence of research results that suggest a different measure of L. The value of δ_d depends on the Reynolds number at which transition from a laminar to a turbulent boundary layer occurs[4], as may be seen in Fig. 4.

At low Reynolds numbers (i.e. small laboratory and process control flumes) the boundary layer will be laminar throughout, and there is not much doubt in the estimate of δ_d, except that which arises from the application of results for non-accelerating flow to the accelerating flow in the throat of a gauging structure. At the other extreme, for $Re_L > 10^7$ (large civil engineering applications), the estimate again becomes insensitive to the point of transition. For $5 \times 10^5 < Re_L < 5 \times 10^6$, δ_d depends appreciably on the assumed value of $Re_{L\ trans}$. In constructions of concrete, evidence is consistent with $Re_{L\ trans} = 3 \times 10^5$, but in smooth installations especially those with gently streamlined entrance transitions and very good approach conditions, the transition will be delayed. There is evidence of this from the V-shaped broad-crested weirs studied by Smith and Liang,[5] whose data is re-analysed in Table 1.

3.2 Thin plate and triangular profile weirs

Let us assume two dimensional flow over a thin-plate (sharp-edged) or triangular profile (Crump[6]) weir, with a very deep approach channel. The geometry of modular flow is in this case defined by one length dimension only, namely the elevation of the upstream energy level

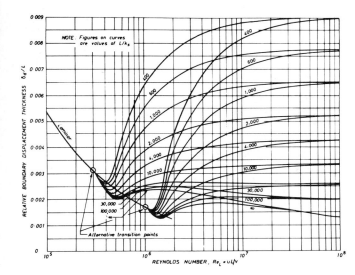

Fig. 4   Boundary layer
           displacement
           thickness

TABLE 1   Comparison of boundary layer theory with experimental coefficient triangular throat;
           data from Smith and Liang[5]
           (Throat assumed to be hydraulically smooth)

Throat length L ft	$\frac{H}{L}$	Throat Reynolds number Re_L 10^5	Calculated discharge, Q(cusecs) Re_L trans 3×10^5	Calculated discharge, Q(cusecs) Re_L trans 10^6	Calculated coefficient C_{b_1} Re_L trans 3×10^5	Calculated coefficient C_{b_1} Re_L trans 10^6	Experimental value of coefficient
1.33	0.60*	3.52	1.280	1.279	0.983	0.982	0.995*
1.33	0.30	2.49	0.219	0.219	0.958	0.958	0.965
1.83	0.40	4.64	1.029	1.024	0.980	0.976	0.982
1.83	0.20	3.28	0.174	0.174	0.947	0.946	0.969
2.33	0.30	5.78	0.910	0.907	0.975	0.972	0.976
2.33	0.15	4.09	0.155	0.153	0.945	0.936	0.965
11.00	0.024	17.1	0.0578	0.0705	0.689	0.812	0.795
11.00	0.029	18.7	0.0991	0.1145	0.739	0.832	0.828
11.00	0.039	21.6	0.225	0.246	0.803	0.861	0.867
11.00	0.049	24.2	0.417	0.442	0.842	0.883	0.893
11.00	0.059	26.5	0.684	0.714	0.869	0.900	0.910

* High value of H/L may affect coefficient

relative to crest height, together with the upstream and downstream angles of the crest profile. It
follows from similarity that for given crest angles the discharge coefficient for an ideal fluid would
not vary with h, i.e. C_p = constant. This principle provides a framework for analysing experimental
data, (Kindsvater and Carter[7], Shen[8], White[9]) because for full-width horizontal crest weirs
of these two types the modular discharge equation when both P/h_1 and b/h_1 are large reduces to

$$Q = \text{constant} \times C_{b_1} \, b \, H_1 \sqrt{g \, H_1} \qquad \text{... (19)}$$

The presence of a boundary layer on the upstream face of the weir structure may be thought of as
displacing the flow by a small amount, so there is a virtual crest at a slightly different elevation
from the actual crest. Thus

$$Q/b = C_p \sqrt{g} (H_1 - k_h)^{3/2} \qquad \text{... (20)}$$

where C_p is a constant

and $(1 - k_h/H_1)^{3/2} = C_{b_1}$ ... (21)

Strictly, k_h should depend on Re_h because this notional displacement arises from fluid properties (the weirs are assumed smooth), but Kindsvater and Carter[7] presumed that in practice k_h would be constant for a given weir. Research at the Hydraulics Research Station on triangular profile weirs has been analysed on the same basis[9]. Briefly this consists of trying various values for k_h and selecting that which, inserted into equation 20 with empirical values of Q/b and H_1, gives the most nearly constant value for the coefficient C_p.

A similar approach can be applied to weirs of V-shape in the direction of flow, e.g. V-notches as studied by Shen[8] and flat-V weirs studied by White[9]. Values of C_p and k_h are shown in Table 2, so that C_{b1} can be worked out as necessary from equation 21.

TABLE 2 Values of C_p, k_h and modular limit for various weir profiles and cross sections

Profile of weir	Cross section	C_p	k_h mm	H_2/H_1 at mod. limit
Thin-plate	2 dimensional	0.568(d)	− 1.0	< 0(a)
	90° V	0.545(e)	− 0.9	0(b)
	45° V	0.545(e)	− 1.4	0(b)
Triangular 1:2/1:5	2 dimensional	0.633	+ 0.3	> 0.75
	1:20 V	0.620(e)	+ 0.5	0.70
	1:10 V	0.615(e)	+ 0.8	0.70
Triangular 1:2/1:2	2 dimensional	0.683	+ 0.25	0.4 − 0.7(c)
	1:20 V	0.665(e)	+ 0.4	0.3 − 0.4(c)
	1:10 V	0.665(e)	+ 0.6	0.3 − 0.4(c)

(a)   Must be fully ventilated
(b)   Submergence not permissible
(c)   Depends on H_1/P_2. Values given for H_1/P_2 between 0.5 and 1.5
(d)   For small values of h_1/P_1
(e)   With C_s based on equation 25

4   THE PROFILE COEFFICIENT, C_p

It has already been shown how for streamlined broad-crested weirs and flumes C_p takes the basic value $(2/3)\sqrt{(2/3)} = 0.544$ given by critical depth theory. Table 2 shows that thin plate weirs and those of triangular profile have a more "efficient" profile in terms of providing a higher value of C_p. However, knowledge of the modular coefficient is not sufficient for the design of a weir for particular duties: the modular limit must also be observed, and this is also listed in Table 2. One of the merits of the Crump weir (1 in 2 upstream slope, 1 in 5 downstream slope) is its moderately high coefficient value coupled with a very advantageous modular-limit.

The lack of a sound theoretical basis prevents the separation of C_{b1} and C_p for rectangular profile weirs[10]. Recent research at Wallingford (carried out in association with Singer) has covered a very wide range of geometries (defined by H_1/P and H_1/L in this case). By carrying out the tests at fairly large scale, the hope is that C_{b1} will be close to unity and hence the empirically determined values may be applied to larger but geometrically similar conditions. The plot of C_p values in Fig. 5 demonstrates how far from constant is the coefficient of a rectangular profile weir as head, weir length or weir height is varied.

5   THE APPROACH VELOCITY COEFFICIENT, C_v

With a rectangular approach channel of width B and weir height P,

$$v_1 = \frac{Q}{B(h_1 + P)} \qquad \qquad \text{... (22)}$$

Combining this with equations 6 and 5

$$\left(\frac{2\,(C_v^{2/3}-1)}{\alpha C_v^2}\right) = C_{b1}\,C_p\,C_s\,f\left(\frac{h_1}{h_1+P}\quad\frac{b}{B}\right) \qquad \ldots (23)$$

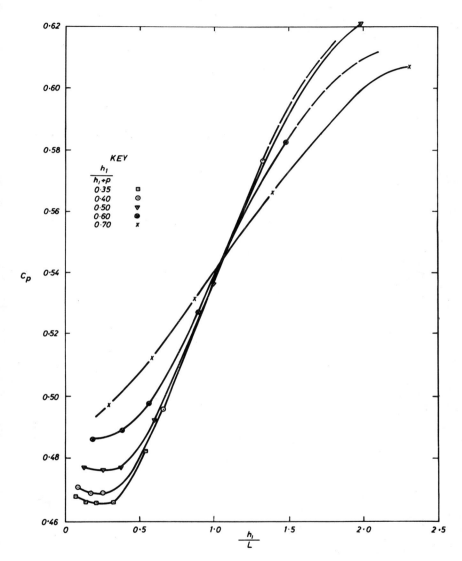

Fig. 5   Dependence of coefficient C_p on geometry of rectangular profile weir

This is a generalisation of an equation first derived by Jameson[11]; it is plotted in Fig 6 (for $\alpha = 1$) in terms permitting C_v to be read off when the right-hand expression is evaluated.

There has been some controversy over the sensitivity of weirs and flumes to non-uniform distribution of velocity in the approach channel, induced for example by a bend upstream. Little evidence on this topic was available before the Hydraulics Research Station's research[12] on the effect of a 180° bend upstream of a full-width weir of Crump profile. The weir was located at

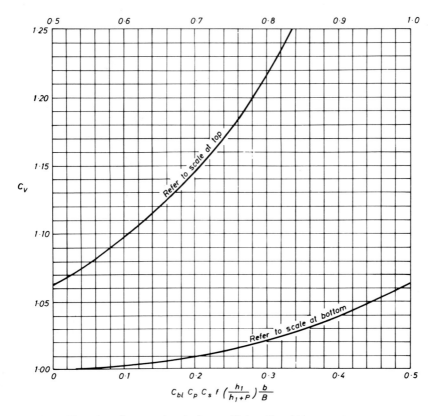

Fig. 6   Variation of approach velocity coefficient C_v with area ratio

various distances from the downstream tangent point of the bend. Velocity distributions were measured, together with water levels and discharges and the effect of the bend on the discharge was evaluated, both with the observed value α inserted in equation 6 to compute C_v and with α assumed equal to 1. The surprising conclusion is that full width weirs (at least those of triangular profile) in artificial channels are far less sensitive to a bend in the approach channel than had been thought. With a 180° bend (internal radius 0.6 B; external radius 1.6 B; b = B), the weir crest can be as close as 2 channel widths without influencing the coefficient (based on head measured at the outer wall) by more than ½ per cent, and there is no advantage in measuring the velocity distribution to evaluate α: the presence of the weir smooths out the velocity that would otherwise be very skewed beyond the bend, and it is sufficient to assume $\alpha = 1$ in practice.

6   THE SHAPE COEFFICIENT, C_s

The author demonstrated in a previous paper[13] how the performance of rectangular, trapezoidal and U-shaped flumes could be covered by an equation of the form of equation 5. The same principle applies to certain weir profiles equally well.

For rectangular sections,

$$C_s = 1.0 \qquad \qquad \ldots (24)$$

For trapezoidal sections,

$$C_s = \text{function} (nH_1/b) \qquad \qquad \ldots (25)$$

where n is the side slope of the throat and b its bottom width. For flat V weirs the equation

30   Ackers

deduced on the basis of integration in vertical slices is

$$C_s b = \frac{4}{5} n H_1 \qquad \qquad \dots (26)$$

whilst for critical depth devices of V form it may be shown that

$$C_s b = \frac{12}{25} \sqrt{\left(\frac{12}{5}\right)} n H_1 \qquad \qquad \dots (27)$$

thereby eliminating b and converting the three-halves relationship into a five-halves one. There remains scope for research into the transition between equations 26 and 27, but both have been used successfully for their respective cases.

For U-shaped flumes, C_s depends upon the flow depth in the throat, and the equation defining C_s differs according to whether this is greater or less than the radius of the invert[13].

The flat-V weir follows equation 26 when the head is less than P_v, the height of the V section. If H_1 exceeds P_v, then White[9] has derived an equation that corresponds to

$$C_s b = \frac{4}{5} n H_1 (1 - (P_v/H_1)^{5/2}) \qquad \qquad \dots (28)$$

The ability to express C_s as a function of $n H_1/b$ for trapezoidal flumes has particular value, because it enabled Harrison[15] to develop a very convenient graphical method of selecting the bed width and side slopes of a flume for a certain duty. The 'duty' is specified as two values of discharge (e.g. upper and lower points of range) associated with two values of upstream head. The latter may be determined from such factors as permissible afflux, accuracy of head measurement, the need to remain modular bearing in mind the tailwater levels induced by conditions downstream or a desire to control upstream levels within a certain range.

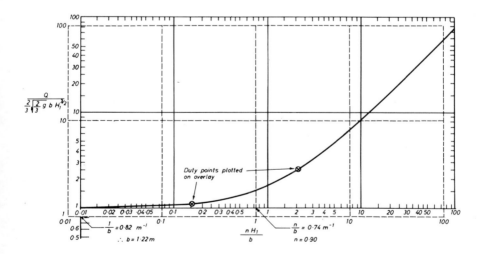

Fig. 7   Use of shape coefficient function for design of trapezoidal flumes

Fig. 7 shows C_s, plotted against $n H_1/b$. The method of selecting the geometry consists of preparing a transparent overlay with the same logarithmic grid drawn and annotated on it. The two duty points are plotted on the overlay, H_1 being in the x direction and $Q/(2/3) \sqrt{(2/3)g} \, H_1^{3/2}$ in the y direction. The overlay is slid over the C_s graph, keeping axes strictly parallel, until the two duty points lie on the curve. A solution is only possible if

$$\left(\frac{H_{max}}{H_{min}}\right)^{3/2} < \frac{Q_{max}}{Q_{min}} < \left(\frac{H_{max}}{H_{min}}\right)^{5/2}$$

The base chart scales are read at the unit x and y grid lines of the overlay, yielding n/b and b respectively. Hence, both the bed width and side slope values for a trapezoidal flume to meet these requirements are quickly deduced, eliminating the tedious trial and error procedures in use heretofore.

7   THE REDUCTION FACTOR FOR DROWNED FLOW, f

7.1 Modular limit

Most weirs and flumes are designed to operate independently of tailwater level, by selecting the structure and its crest or invert elevation so that downstream conditions do not cause it to drown. In other words, f = 1.0 in the modular range and the most important requirement is to know the maximum ratio of H_2/H_1 that can exist for f not to depart significantly from unity. A one per cent departure has been used in the work at the Hydraulics Research Station.

Recent research on two-dimensional triangular profile weirs[9] has shown that the modular limit depends on the angle of the downstream face of the weir and also on H_1/P_2 where P_2 is the crest height relative to the downstream bed level. With a 1:2 downstream slope, the limiting value of H_2/H_1 varies considerably with H_1/P_2, being below 0.3 where $H_1/P_2 < 0.4$. The picture is radically different however with a 1:5 downstream slope, when the limit is over 0.75 over the full range of H_1/P_2 values covered in the tests. This is a highly advantageous feature, permitting a modular triangular profile weir of gentle downstream slope to be set much lower, and hence causing less afflux, than one with a steep downstream slope. This feature was well appreciated by Crump when he recommended the 1:2 upstream, 1:5 downstream profile.

Harrison[2] studied the evidence on the modular limit of streamlined broad-crested weirs, and developed a theory based on continuity and momentum on the premise that drowning would occur when a downstream hydraulic jump was forced upstream until the foot of the jump was situated at the tail end of the crest. The back face of the weir was assumed either vertical or sloped, and the submergence ratio at the modular limit was again shown to depend upon H_1/P_2 (see Fig 8).

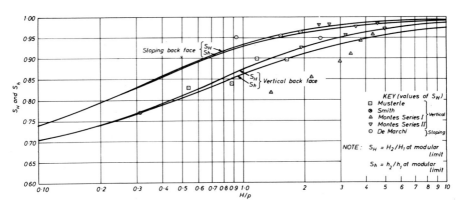

Fig. 8   Modular limit at broad-crested weirs

In respect of flumes, as with triangular profile weirs, the limiting submergence ratio for free discharge depends upon the gentleness of the expansion beyond the throat. Inglis[16] reported modular limits for rectangular throated flumes of 80, 85, 88 and 94 per cent respectively for expansions of 1 in 5, 1 in 10, 1 in 15 and 1 in 20, although Wells and Gotaas[17] claimed a figure of 85 per cent for a 1 in 3 expansion at a trapezoidal flume. Harrison[18] has extended his analysis to cover these devices.

7.2 Drowned flow at triangular profile weirs

Compared with other types of weir, perhaps the most accurate and comprehensive information on the reduction factor f is available for triangular profile weirs[6][9]. There are difficulties in measuring the tailwater level beyond a weir: just where should it be measured, how does the particular expansion geometry affect head recovery, is the water surface wavy or does it fluctuate etc? Consequently, as the flow over weirs is very sensitive to H_2/H_1 once they are drowned, accuracy is lost if the submergence is measured in the conventional way.

One of the great advantages of the Crump weir profile is that a new method of double gauging is made possible, the downstream gauging being of the piezometric pressure in the separation pocket formed in the lee of the crest. Fig 9 shows the reduction factor f in terms of H_2/H_1 and also in terms of h_p/H_1 where h_p is the pressure, relative to crest elevation, in the separation pocket. This provides a much preferable method of obtaining the reduction factor f, which avoids

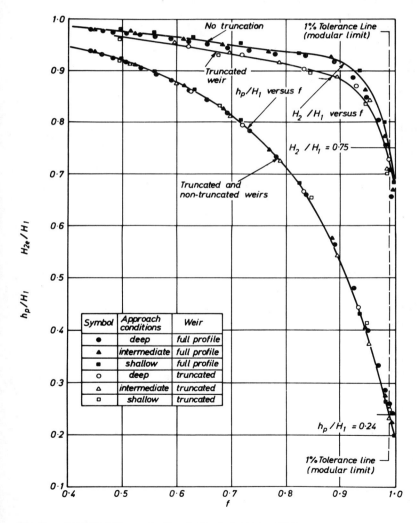

Fig. 9   Drowned flow reduction factor f at crump, weirs

the problem of head recovery and its effect on tailwater level. White[9] has produced similar information for flat V weirs of Crump profile as well, although the three dimensional nature of the flow in that case alters the relationship between f and h_p/H_1.

8   CONCLUSION

It will be realised that the information given in this review paper is not by itself sufficient for the complete design of weirs and flumes in lined open channels. The reports referred to should be read in full, together with relevant British Standards[19] and other specialist papers so that the limitations of the various devices are appreciated and may be complied with when designing a new installation.

The studies at the Hydraulics Research Station, have been of two types: (a) the development of techniques for design and calibration utilising experiments carried out by other researchers as confirmatory evidence; (b) precise experiments over very wide ranges of conditions on triangular profile weirs (both two-dimensional and flat-vee) in order to establish accurate coefficient values and to better understand many aspects of their performance. Some of this research has been incorporated in recent British Standards, and the information has been of value to the Water Resources Board and to British River Authorities in their hydrometric schemes. Of course all these measuring structures are equally applicable in man-made channels as in natural streams and rivers.

Although most research has been on a laboratory scale (albeit a large laboratory scale), it is important to obtain data on field performance. There are many reports on the successful use of the results of recent research, but it is beyond the scope of this paper to review these applications.

9   ACKNOWLEDGMENTS

This paper reports work carried out at the Hydraulics Research Station of the Department of the Environment and is published with the permission of the Director of Hydraulics Research. The author is grateful to his colleagues Messrs. Burgess, Harrison and White for their support in this programme of hydrometric research.

10   NOTATION

A	Cross-sectional area
B	Breadth of rectangular channel
b	Crest breadth (across channel), width of flume invert
C_{b1}	Coefficient allowing for boundary layer influence
C_d	Coefficient of discharge (dimensionless)
C_v	Coefficient of velocity
C_p	Coefficient allowing for effect of weir profile
C_s	Coefficient allowing for shape of flow cross-section
d	Depth of flow
d_c	Critical depth
d_m	Mean depth (A/W)
f	Drowned flow reduction factor
g	Acceleration due to gravity
h	Static head (general)
H	Total head
H_p	Total head in zone of potential flow
P_v	Vee height (F/V weirs)
k_h	Head correction factor
L	Length of throat or crest in direction of flow

M	Mass flow rate
n	Crest cross-slope, side slope of flume (1 vert : n Horiz)
P'	Wetted perimeter P Weir height
Q	Discharge (Vol/unit time)
Re	Reynolds number vl/ν
u	Point velocity
v	Velocity
W	Surface width

SUBSCRIPTS

1	upstream	
2	downstream	unless otherwise defined
e	effective i.e. corrected for fluid property effects	
h	based on head h as characteristic length dimension	
max	maximum	
L	based on throat length L as characteristic length dimension	
*	values adjusted for boundary layer displacement thickness	
α	Coriolis energy coefficient	
δ_d	Boundary layer thickness	
δ	Boundary layer displacement thickness	
μ	Dynamic viscosity	
ν	Kinematic viscosity	
ρ	Density	

11   REFERENCES

1   BURGESS, J.S.: 'Factors affecting the selection and design of gauging structures'. Symposium on river-flow measurements (Inst. of Water Engineers), pp.39-54, (1969)

2   HARRISON, A.J.M.: 'The streamlined broad-crested weir', Proc. Instn. Civil Eng., 38, pp.657-678, (1967)

3   SCHLICHTING, H.: 'Boundary layer theory' (McGraw-Hill 1962), p.27

4   HARRISON, A.J.M.: 'Boundary-layer displacement thickness on flat plates', J. Hydr. Div. ASCE, 93 (HY4), pp.79-91, (1967)

5   SMITH, C.D., and LIANG, Wen S.: 'Triangular board-crested weir', Proc. ASCE, 95, IR4, pp.493-502, (Dec. 1969)

6   CRUMP, E.S.: 'A new method of gauging stream flow with little afflux by means of a submerged weir of triangular profile', Proc. Inst. Civil Eng., Pt. I, Vol. 1 (March 1952), 2; pp.223-242

7   KINDSVATER, C.E., and CARTER, R.W.: 'Discharge characteristics of rectangular thin-plate weirs', Proc. ASCE, 83, HY6 (Dec.), pp.1-36, (1957)

8   SHEN, J.: 'A preliminary report on the discharge characteristics of triangular profile weirs', Water Res. Divn., US Geol. Survey, Washington (1959) (unpublished)

9   WHITE, W.R.: 'The performance of two dimensional and flat-V triangular profile weirs', Proc. Inst. Civil Eng., Suppl. (ii), Paper 7250S, (1971)

10   SINGER, J.: 'Square-edged broad-crested weir as a flow measurement device', Water & Water Engng., 68, (820), pp.229-235 (June 1964)

11   JAMESON, A.H.: 'The development of the Venturi flume', Water & Water Engng., 32, p.105, (1930)

12 'The triangular profile Crump weir: effects of a bend in the approach channel',
    Hydraulics Research Station, Wallingford, Report EX 518, (1970)

13 ACKERS, P.: 'Comprehensive formulae for critical-depth flumes', Water & Water
    Engng , 65, pp.296-306, (1961)

14 KALWIJK, J.P. Th: 'A note on the discharge of critical depth measuring devices
    with arbitrary shape', Proc Inst. Civ. Eng., 47, pp.227-238, (Oct. 1970)

15 'Critical-depth flumes for flow measurement in open channels', Hydraulic Research
    Paper 5, (HMSO, 1963)

16 INGLIS, C.C.: 'Notes on standing wave flumes and flume water falls', Govt. of
    Bombay, P.W. Delft, Tech. Paper 15, Bombay, (1928)

17 WELLS, E.A., and GOTAAS, H.B.: 'Design of venturi flumes in circular conduits',
    Proc. ASCE, 82, SA, Paper 938, (April 1956)

18 HARRISON, A.J.M.: 'Factors governing the choice of a hydraulic structure for
    flow measurements. Paper presented at the 7th Congress, International Commission
    on Irrigation and Drainage, Mexico, (R.8 Question 24) (April 1969)

19 BS3680: 'Measurement of liquid flow in open channels'
    Pt. 4A: Thin plate weirs and venturi flumes
    Pt. 4B: Long base weirs
    Pt. 4C: Flumes (not yet published)

DISCUSSION

J.C. Schuster: Would the Bend effect on the discharge of a weir depend largely on 'P' or on the velocity distribution?

Author's reply: Full details of the research on the effect of a bend are given in Ref. 12. The conclusions were that a weir should not be closer to the downstream tangent point of a bend than two channel breadths or six times the maximum head, assuming that the water level was measured at $2H_{max}$ from the crest. The measurement of head should be made on the outside of the bend, and no closer than $1\frac{1}{3}$ channel widths to the tangent point. There was no advantage in measuring the velocity distribution in order to correct for the Coriolis coefficient α : the coefficient changed by less than 0.5 per cent from its standard value within the range $0 < H/P_1 < 3.5$. There was no direct relationship between the error and the non-uniformity of velocity distribution nor with H/P_1. In fact the main effect of the bend arose through the influence of super-elevation, generated by the bend, on the water level at the head-measurement station. It must be emphasised that these results were obtained with a triangular profile (Crump) weir and a $180°$ bend of small radius.

K. Zanker: The parameters C_{b1}, C_p, C_v, C_s and f do not directly take into account surface tension. How is it taken into account?

Author's reply: Surface tension is important only at low heads and when appreciable curvature exists at the control section. Strictly speaking, the coefficient C_{b1} should allow for surface tension as well as for boundary layer effects. When the Kindsvater and Carter coefficient is used (see equations 20 and 21 of the paper) the empirical method of assessing K_h means that it allows approximately for both fluid property influences. This is further explained in Ref. 9. The questioner may also refer to G.D. Matthew "On the influence of curvature, surface tension and viscosity on flow over round crested weirs" in Proc. Inst. Civil Engrs. 25 511 (1963). The discussion of that paper also provided valuable contributions on fluid property effects (28 557 and 569 (1964)).

F.H. Huyten: What kind of accuracies may one expect? What are the main factors determining these?

Author's reply: The accuracy of measurement that can be achieved using weirs and flumes is obviously dependent upon the selected geometry and the flow. The overall tolerance is made up of several components, for example the tolerance on the geometry as manufactured, the assessment of each of the coefficients in equation 5, the tolerance on the head measurement, which in turn depends on zero-setting, instrumental and recording errors. In favourable cases such as the use of a V notch in a carefully designed laboratory installation, accuracy might approach one percent, but in situations typical of industrial and field conditions, tolerances would range from some 4 or 5% at the minimum flow gauged to about 2 or 3% at the higher end of the range.

A.T.J. Hayward: Is the coefficient of a weir or flume affected by an upstream flow disturbance (e.g. a bend near to the device) to the same extent as the coefficients of venturi devices in closed conduits?

Author's reply: Recent research has shown that a two-dimensional weir is less sensitive to an upstream bend than had been thought, using closed conduit devices as the analogy. The research on the subject at Hydraulics Research Station is described in reference 12 of the paper. The investigation covered the effect of a horizontal bend of very short radius, but there has been very little research on the effect of upstream disturbances since the work of Schoder and Turner [Precise Weir Measurements, Proc. Amer. Soc. Civ. Engineers, 53 1395 (1927)].

A.M. Crossley: There appears to be information available appertaining to rectangular and trapezoidal flumes, but little published on semicircular flumes. Is this information likely to be available in the near future?

Author's reply: A British Standard on flumes which is shortly to be published will cover U-shaped flumes i.e. those with a semi-circular invert. In fact, the methods of deducing the calibration of a U-shaped flume direct from theory are very similar to those used for other throat geometries: there is no difference in principle.

A.M. Crossley: We have a double-range trapezoidal flume which appears to be in error, at least when compared with a venturi-meter in a closed pipe.

Is it possible to check the calibration of the flume "ON-SITE" without shutting down the flow and which system is recommended. If the existing calibration curve were to be matched to an equation on a computer would it be possible to analyse the equation to deduce the various co-efficients, assuming the curve follows the "three halves" law?

Author's reply: If the trapezoidal flume appears to be in error, the first step should be to obtain an accurate independent measurement of head, paying proper attention to zero-setting, for comparison with the recorder head. Next, the downstream conditions should be checked to make certain the flume is not drowned. After checking the throat geometry, and ensuring that it is not partially blocked with silt or debris, the makers calibration should be checked by computations based on reference 15 of the present paper. Only if these steps do not bring to light the source (or sources) of error should one proceed to a calibration check.

The best method of checking the calibration in-situ depends very much on the local conditions. Chemical dilution may be suitable; or a careful velocity-area determination of discharge in the approach channel. Whatever method is chosen, it is important to check the zero-setting and measure the head independently of the recorder. The errors implicit in the calibration technique must also be assessed.

It is not advisable to use a three-halves law, because a trapezoidal flume does not follow such a law. In fact, unless the range of flow is narrow, the function can not be approximated by one straight line on a log-log plot. These points will be apparent from a careful study of Fig. 7 of the paper.

D.L. Smith: I get the impression, from reading general text-books on weirs and flumes, that much of the early work was done on a large scale—e.g. for river flow measurement. It is likely that increased effort to reduce polution of rivers by effluent from chemical plants, could result in increasing use being made of weirs or flumes, particularly with a linear characteristic, to enable the quantity of polutant to be measured by combination of continuous flow and analysis of the effluent.

Author's reply: Much of the recent research work was done on fairly small weirs of about 1ft to 3ft, with a head of the order of ½ft, although modern knowledge on boundary layer effects enable the characteristics of large weirs to be predicted accurately from tests at a small scale. Linear devices are available, e.g. profiled weirs, and the vortex drop meter, but modern electronic devices can effectively linearise the output characteristic of a given weir. The characteristics of the vortex drop are reported in a paper by Ackers and Crump; Proc. Inst. Cir. Engrs. 16 544-442 (1960).

2·1

RECENT MEASUREMENTS OF FLOW USING NUCLEAR MAGNETIC RESONANCE TECHNIQUES

J.R. Singer and T. Grover

Abstract: The use of nuclear magnetic resonance (NMR) techniques to measure flow rates is discussed. The velocity distribution function is introduced which gives the number of molecules per unit velocity integral, and it is shown that this concept has significant value in the flow measurement problem, especially in chemical, physiological and biological studies. The velocity distribution function for a human finger has been measured and compared to data obtained by counting the blood-carrying vessels in a dog. Data obtained from rats' tails is also presented and the associated velocity distribution function is compared to that for a human finger.

1   INTRODUCTION

Most of the work reported here has been directed towards the non-invasive measurement of blood flow in intact humans. Our earliest experiments[1,2,3] were the first to demonstrate the use of NMR to measure blood flow: in these the blood flows in mice tails, and in human fingers were measured. We have since measured blood flow in intact veins of humans.[4] More recently, we have measured the "velocity distribution function" in various types of complex flow channels including bundles of tubes and human fingers. In this paper we shall speak mainly about our recent work on the "velocity distribution function".

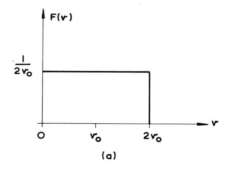

(a)

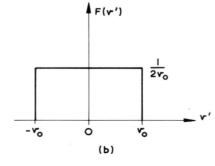

(b)

Fig. 1   a The velocity distribution function for laminar flow in a circular tube

   b The 'shifted' velocity distribution function with an origin at the average velocity

The "velocity distribution function," will be denoted "F(v)" hereafter. F(v) is obtained from our experiments and calculations as a function giving the number of molecules per velocity interval. By plotting F(v) versus velocity, we obtain a picture of the relative number of molecules flowing at different velocities. The simplest example will illustrate one such distribution. Consider the case of laminar flow in a circular tube. When we measure F(v) and plot F(v) versus velocity, Fig. 1 is obtained. We see that the flow is normalized to unity, that equal numbers of molecules flow in each velocity interval, and that the flow distribution is an even function, with half the flow greater than and half the flow less than the average velocity. Note that Fig. 1 does not tell us anything about the spatial distribution of flow.

This result is typical of our flow distributions. The velocity distribution function, F(v) has its most immediate applications to chemical, physiological, and biological studies, where velocity distributions are already in use. In order to understand the spin echo background of this work, reference should be made to papers by Hahn,[5] Carr and Purcell,[6] and Carr and Simpson[7].

The experimental procedure utilizes the equipment shown in Fig. 2. The protons in the flowing liquid make up the sample undergoing nuclear magnetic resonance spin echo.* The protons are subjected to radio frequency pulses of sufficient frequency, length and intensity to tilt the protons through $90°$ and after a delay t_1 through $180°$. The time between pulses (t_1) is systematically varied and the echo amplitude and phase $2t_1$ seconds after the $90°$ pulse are recorded using a memory oscilloscope. The sampled amplitude of the echo found in this manner is defined as $A(t_1)$ and the sampled phase of the echo is defined as $\phi(t_1)$. By taking a cosine transform of the echo amplitude the even part of the flow velocity distribution function is obtained. By combining the phase and amplitude information, the total velocity distribution is obtained.

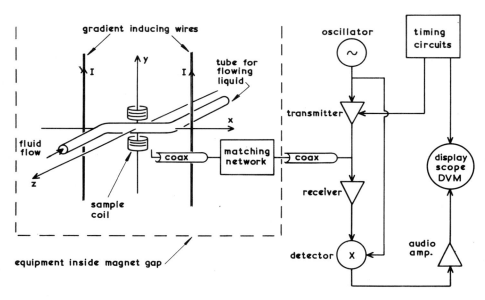

Fig. 2   Schematic experimental arrangement

An intuitive conception of the procedure is useful. Consider the groups of spins, (protons in the coil of Fig. 2), rotated through $90°$. The groups fan out. The flow of the fluid changes the relative phase relationships of the groups. Then after a time t_1, a $180°$ pulse is applied. Were there no flow and no diffusion in the fluid the $180°$ pulse would cause all the spin groups to come into phase and all of the spins would contribute to an echo pulse, (except for relaxation effects which are negligible for the times of our fluid measurements). The flow therefore shows up as a change in the echo amplitude $A(t_1)$ due to phase changes in the groups of spins. In the experiment, the time between pulses (t_1) is varied, and the function $A(t_1)$ is observed as a function of the time between pulses. Such an observation is plotted in Fig. 3. By means of the method, a number of interesting

*Other nuclei than protons may be utilized in different fluids.

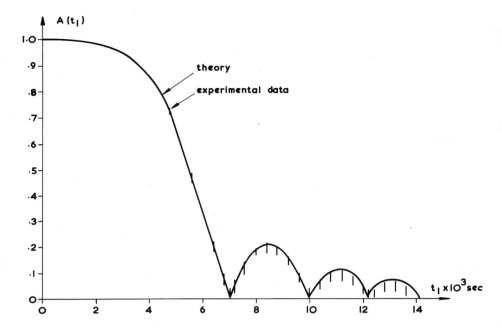

Fig. 3   The theoretical and experimental echo amplitude function for laminar flow in a tube.
90° – 180° experiment; average velocity: 5.2 cm/sec; field gradient: 0.2 gauss/cm

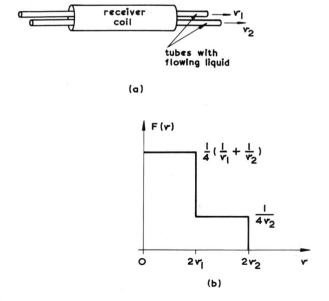

(a)

(b)

Fig. 4   a Mechanical arrangement for the two tube flow experiment

b Velocity distribution function for two tubes

flow distribution functions were examined. We reproduce here a velocity distribution function for two different circular tubes, both carrying fluid in the same direction, Fig. 4.

Of considerable interest to us was the measurement of the velocity distribution function in a human finger reproduced here as Fig. 5. The data is compared to data obtained by counting blood

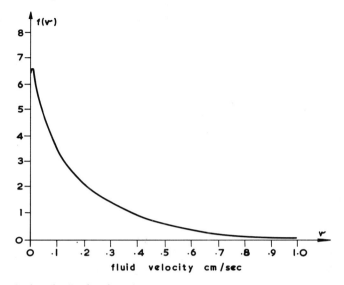

Fig. 5   The velocity distribution for a human forefinger

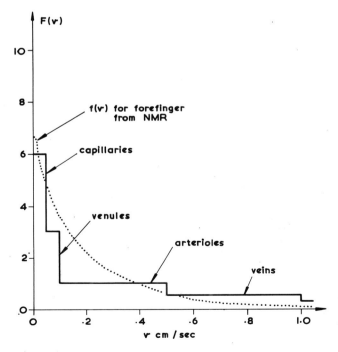

Fig. 6   Velocity distributions for human forefinger and dog intestinal mesentery. The dog data was obtained by counting vessels and dividing them into the classes shown. Laminar, Newtonian flow is assumed

carrying vessels in a dog shown in Fig. 6. Some of our data was obtained from rat tails. The velocity distribution function for these is shown (compared to the human finger function in Fig. 7).

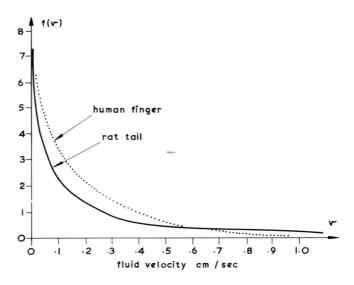

Fig. 7   Velocity distribution function $f(v)$ for human forefinger and a rat tail

2   THEORY

The conditions needed for a flow measurement are as follows.

1. The fluid has been in the DC field for a time much longer than the longitudinal relaxation time, T_1 [8,9]. The time between attempts to elicit an echo is also much longer than T_1 [8]. The time to echo, $2t_1$, is much less than T_1 or the microscopic transverse relaxation time, T_2.

2. The transmitter coil is arranged so that the radio frequency magnetic field, H_1, is uniform over the "sample", i.e. the volume of fluid contributing to the voltage induced in the receiver coil. [10]

3. The DC magnetic field over the sample is accurately represented by

$$H_z(x, y, z) = H_0 + Gx \quad ,$$ ... (1)

where $G = \delta H_z / \delta x$ is the field gradient in the x direction.

4. The H_1 field is sufficiently intense to meet the condition

$$H_1 \gg GL_s \quad ,$$ ... (2)

where L_s is the sample length in the x direction. In addition, the time between the pulses t_1 is much longer than the pulse lengths t_w. These are the "usual" conditins for spin echo experiments.[5]

5. The motion of any fluid molecule is constant and unidirectional over the time of the experiment $2t_1$. We assume that the position $\bar{x}(t)$ of any molecule at time t after the 90° pulse is given by

$$\bar{x}(t) = \bar{x}(0) + \bar{v}t \quad ,$$ ... (3)

where $\bar{x}(o)$ is the initial location and $\bar{v}$ the velocity vector of that molecule.

Write the DC field seen by an "isochromatic" group of molecules as a function of time after the 90° pulse as (assumptions 3 and 5)

$$H'_0(t) = H_0 + G[x(o) + v_x t] \qquad \qquad \ldots (4)$$

where v_x is the x component of the velocity vector of the group of molecules and $x(o)$ is its initial position. Another group of molecules would have a different $x(o)$ and v_x. The echo amplitude and phase at $2t_1$ are substantially independent of the distribution in $\bar{x}(o)$;[5] we now wish to find the effects of a velocity distribution.

We will examine the phase angles of a particular group during a 90°-180° echo experiment. The phases are measured relative to a frame rotating around the z axis at $\omega_0 = \gamma H_0$. Let x′, y′ be the transverse axes of this frame and let y′ be antiparallel to H_1. The phase of our molecular group's moment at $t = 2t_1$ seconds after the 90° pulse is given by[6, 11].

$$\varphi(2t_1) = \gamma G t_1^2 v_x \qquad \qquad \ldots (5)$$

Refer to Fig. 8 where the motion of the tip of our group's magnetic moment vector is shown in the rotating frame. After the first pulse the phase is

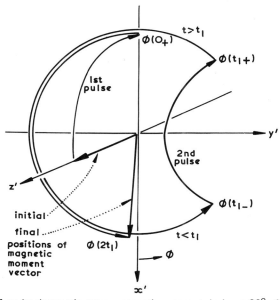

Fig. 8   Motion of an isochromatic group magnetic moment during a 90°– 180° experiment

$$\varphi(0_+) = \pi \qquad \qquad \ldots (6)$$

and after a time $t < t_1$

$$\varphi(t) = \int_0^t [\omega(t) - \omega_0]\, dt + \pi$$

$$= \int_0^t \gamma [H'_0(t) - H_0]\, dt + \pi$$

$$= \gamma G t \bar{x}(o) + \tfrac{1}{2}\gamma G t^2 v_x + \pi \qquad \qquad \ldots (7)$$

and after the second pulse

$$\varphi(t_{1+}) = -\gamma G t_1 \bar{x}(o) - \tfrac{1}{2}\gamma G t_1^2 v_x \qquad \qquad \ldots (8)$$

and finally, at t = 2t₁, the phase is

$$\varphi(2t_1) = \gamma G t_1^2 v_x \qquad \qquad \dots (9)$$

The group is no longer in phase with one which has a different velocity.

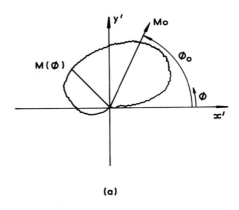

(a)

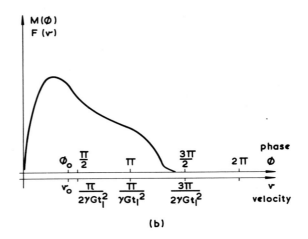

(b)

Fig. 9   a Distribution of group moments at $t = 2t_1$

b The angular distribution of group moments and the sample velocity distribution plotted together, illustrating the change of variable

To find the net moment of the sample and thereby the echo amplitude,[11,12] integrate over all possible phases. The situation at t = 2t₁ is depicted in Fig. 9 where the distribution of molecular group moments in the x′-y′ plane is shown. The distribution function M(φ) gives the fraction of moments with phase angle φ. The radial distance is proportional to M(φ). Assume that M(φ) is normalized

$$\int_{-\infty}^{\infty} M(\varphi)d\varphi = 1 \qquad \qquad \dots (10)$$

The net magnetic moment of the sample is M_0 and its phase is φ_0. To find M_0 perform the following integration:

$$M_0 = \int_{-\infty}^{\infty} \cos(\varphi - \varphi_0)\, M(\varphi - \varphi_0)\, d(\varphi - \varphi_0). \qquad \ldots (11)$$

To find φ_0 perform the second integration:

$$\varphi_0 = \int_{-\infty}^{\infty} \varphi\, M(\varphi)\, d\varphi \qquad \ldots (12)$$

The echo amplitude, as sensed by a coil whose axis is along the x or y axis, is proportional to M_0.[12] Note the shift in origin from 0 to φ_0.

We now wish to relate the distribution in phase to the distribution in velocity along the x axis. To simplify notation, drop the x subscript. The only velocity component of importance is that one parallel to the gradient in the H_0 field.[5] Molecular groups with velocity components perpendicular to the gradient will experience no changes in precession frequency due to these transverse velocity components (assumption 3). Define a normalized velocity distribution function $F(v)$. Due to assumptions 1 and 2, we now claim that these two distribution functions are related by nothing more than a change of variable, thus (see Fig. 9)

$$M(\varphi)d\varphi = F(v)\, dv. \qquad \ldots (13)$$

Because of this equality, the average velocity of the sample fluid is proportional to the phase angle of the net moment φ_0. Further, we can write the expressions for M_0 and φ_0 in terms of the velocity distribution function

$$M_0 = \int_{-\infty}^{\infty} \cos\,(\gamma G_1^2 v - \gamma G t_1^2 v_0)\, F(v - v_0)\, d(v - v_0) \qquad \ldots (14)$$

$$\varphi_0 = \int_{-\infty}^{\infty} \gamma G t_1^2\, vF(v)\, dv = \gamma G t_1^2 v_0 \qquad \ldots (15)$$

where v_0 is the average velocity of the fluid.

Consider, for the moment, the x' and y' components of the net magnetic moment written in

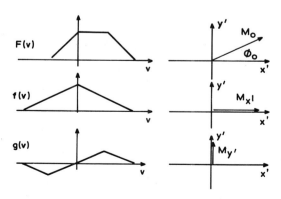

Fig. 10   Illustrative velocity distribution with even and odd parts and corresponding magnetic moments

terms of the velocity distribution function (see Fig. 10)

$$M'_x = M_0 \cos \varphi_0 = \int_{-\infty}^{\infty} \cos(\gamma G t_1^2 v) \, F(v) \, dv$$

$$= \int_{0}^{\infty} \cos(uv) \, f(v) \, dv \qquad \qquad \dots (16)$$

where $u = \gamma G t_1^2 v$ and $f(v)$ is the even part of the velocity distribution relative to an origin at $v = 0$:

$$2f(v) = F(v) + F(-v) \qquad \qquad \dots (17)$$

Similarly, for the y' component we have

$$M_{y'} = M_0 \sin \varphi_0 = \int_{-\infty}^{\infty} \sin(uv) \, g(v) \, dv \qquad \qquad \dots (18)$$

where $g(v)$ is the odd part of the velocity distribution function

$$2g(v) = F(v) - F(-v) \qquad \qquad \dots (19)$$

In order to find the even and odd parts of the velocity distribution, invert the two equations above:[14]

$$f(v) = \int_{0}^{\infty} M_0(u) \cos[\varphi_0(u)] \cos(uv) \, du \qquad \qquad \dots (20)$$

$$g(v) = \int_{0}^{\infty} M_0(u) \sin[\varphi_0(u)] \cos(uv) \, du \qquad \qquad \dots (21)$$

Finally, add the even and odd parts to find the complete distributions

$$F(v) = f(v) + g(v) \qquad \qquad \dots (22)$$

The echo amplitude and phase have the information needed to find the complete velocity distribution for the fluid under the given conditions.

The available experimental data consists of the echo amplitude $A(t_1)$ for various t_1's and a measurement of the phase for some t_1. Since $\varphi_0 = \gamma G t_1^2 v_0$ and v_0 is a constant for the time of the experiment, (assumption 5) we need only find φ_0 for one echo to generate $\cos[\varphi_0(u)]$ and $\sin[\varphi_0(u)]$. The echo amplitude can be found using a receiver and linear detector. The phase at $t = 2t_1$ can be found with reasonable accuracy by mixing a signal at the transmitter frequency and phase with the echo signal,[12] provided that the field is properly stabilized.[15] Note that while the integrals above have limits of 0 and ∞, in practice the echo amplitude becomes very small for large t_1 and the integral need only be performed over a finite range.

REFERENCES

1   J.R. Singer, Science **130**, 1652 (1959)

2   J.R. Singer, J. of Applied Physics **31**, 125 (1960)

3   J.R. Singer, J. of Applied Physics **31**, 406 (1960)

4   O.C. Morse & J.R. Singer, Science **170**, 440 (1971)

5   E.L. Hahn, Phys. Review **80**, 580 (1950)

6   H.Y. Carr & E.M. Purcell, Phys. Review, **94**, 630 (1954)

7   H.Y. Carr & J.H. Simpson, Phys. Review, **98**, 1201 (1958)

8   D.W. Arnold & L.E. Burkhardt, J. of Applied Phys. **38**, 568 (1956)

9   L.R. Hirschell & L.F. Libelo, J. of Applied Phys. **32**, 1401 (1961)

10 A.D. May & R.S. Timsit, J. of Sc. Instr., **44**, 636 (1967)

11 E.L. Hahn, J. of Geophys. Res. **65**, 776 (1960)

12 F. Bloch, Phys. Review **70**, 460 (1946)

13 C.J. Tranter, **Integral Transforms in Mathematical Physics**, J. Wiley & Sons, N.Y., pg 32, 1962

14 A. Allerhand, Rev. of Sci. Instr. **41**, 269 (1970)

3   ACKNOWLEDGEMENT

We are indebted to Professor I. Fatt for the loan of equipment, and to Professor E. Hahn for helpful discussions. We are grateful for the partial support of our work by Biomedical Sciences Support Grant FR-7006 from the General Research Support Branch of the U.S. National Institutes of Health, an Institutional Research Grant from the American Cancer Society, and for 615 Research Funds from the University of California.

DISCUSSION

S.P. Hutton: What sort of accuracy is obtainable when measuring laminar flow and comparing the integrated flow with the flow by collecting and weighing?

Authors' reply: We have carried out experiments to check the accuracy of our NMR flow measurement system and were able to achieve accuracies of two per cent. Some care is needed to achieve high accuracy, but fundamentally the NMR system is highly accurate.

F.C. Kinghorn: Is there any optimum relation between the separation of the two probes and the diameters of these coils? I am thinking particularly of the effects of mixing, and of the period for which the magnetisation is retained.

Authors' reply: In the design of the coils, one must consider the recovery time of the receiver from saturation levels to which it may be driven by a transmitter coil in near proximity. Generally using a diode network, or a transmit-receive switch, one can avoid problems with separation and size of the coils. It is necessary to keep the coils together close enough to consider the time of flow between coils. The time an average molecule takes to traverse the distance between coils should be less than T_1, the relaxation time of the fluid.

A.T.J. Hayward: Does the NMR technique for determining velocity distribution indicate total velocities, or velocity vectors in the longitudinal direction?

Authors' reply: The velocity distribution function provides total velocities. That is, we obtain a function giving the number of molecules per velocity interval.

A.T.J. Hayward: What are the advantages of the industrial NMR flowmeter?

Authors' reply: The major advantage of NMR flowmeters is that there is absolutely no interference with the flow. Another advantage is that materials of an inaccessible or highly corrosive nature may be easily measured.

D.L. Smith: Does the calibration of commercially-available NMR flowmeters depend on the fluid metered?

Authors' reply: No, most fluids are acceptable, e.g. water solutions, petrochemicals, other hydrocarbons. However certain fluids, e.g. liquid oxygen and fluids containing only even-even nuclei cannot be measured.

L.E. Drain: Is use made of the shape of the echo signals or just the amplitudes? Do you use pulsed field gradients?

Authors' reply: There are several modes of measuring flow using NMR. We often use the detection of the arrival of the pulse, hence the NMR pulse signal arrival time is important. For other modes, the amplitude is used, and sometimes the integrated area of the NMR signal is used. We have not used pulsed magnetic field gradients but they do have some applications, and we plan to use them in our future experiments.

2·2

DEVELOPMENT OF THE NEL LASER VELOCIMETER

K.A. Blake

Abstract: In this paper a novel form of laser Dopplermeter is described. The laser beam is split into two and the parts focused to intersect within the fluid being measured. Minute impurity particles cause light to be scattered from this intersection in all directions with a frequency change caused by the Doppler effect. The optical system ensures that only particles small enough to follow the flow closely are considered. A photodetector placed in either the forward or back-scattering direction converts the light energy to an electrical signal of wave-packet form. An electronic counter times the interval between cycles of the signal, this being directly related to the particle's velocity. A built-in computing facility enables the counter to process the results of many measurements and give a virtually instantaneous readout of mean velocity at a point. Tests are described in which air and water were used as test fluids, as well as a rotating disc assembly. Results indicate an accuracy of ±0.25 per cent over the range 0.3-10 m/s. Results are also compared with pitot-static tube velocity estimations, and are found to agree within ±0.2 per cent. Several practical examples of the instrument in use are given.

1   INTRODUCTION

Conventional methods of point-velocity measurement in fluids, such as the pitot-tube and hot wire anemometer, obstruct the fluid and alter the flow pattern. It is, therefore, desirable to develop an instrument which does not require the insertion of a probe into the fluid, while maintaining or improving upon the accuracy obtainable by pitot-tube traversing.

The advent of the laser, with its highly coherent output, has made possible the development of an almost ideal, obstructionless velocity meter based on the Doppler effect. It requires no calibration, and gives a reading linear with velocity in liquids and gases. Simultaneous measurement of turbulence and flow pulsations or unsteadiness are also possible, and the overall accuracy of the device compares favourably with conventional instruments.

2   THE LASER DOPPLER TECHNIQUE

Unless special precuations have been taken, all fluids contain small impurity particles such as specks of dust, algae, or air bubbles. The laser Doppler velocimeter makes use of contaminants whose dimensions are of the order of 1 micron or smaller and which, therefore, are small enough to follow very closely the flow pattern. These particles scatter light, which is slightly shifted in frequency by the Doppler effect, dependent on the particle velocity, and this change in frequency is detected and related to the flow velocity.

2.1 Principles of operation

The principles on which the instrument operates have been described in detail elsewhere[1] and so will be dealt with briefly here. Two approaches are possible. When light waves are scattered from moving objects they undergo a change of frequency, analogous to the familiar Doppler effect of sound. In the present case a particle within a fluid scatters light from the intersection of two beams. If $\underline{k}_{01}$ and $\underline{k}_{02}$ are unit vectors representing the beams, $\underline{k}_s$ is a ray of scattered light, and $\underline{v}$ is the particle velocity, then in Fig. 1 the frequency shift $\triangle\nu_1$ from the first beam is given by

$$\triangle\nu_1 = \frac{\underline{v}}{\lambda_0}(\underline{k}_s - \underline{k}_{01})$$

where λ_0 is the wavelength of the light.

Similarly
$$\triangle \nu_2 = \frac{\nu}{\lambda_0} (\underline{k}_s - \underline{k}_{02})$$

When these are combined, a square-law photodetector will see only the difference frequency

$$\triangle \nu = \triangle \nu_1 - \triangle \nu_2 = \frac{\nu}{\lambda_0} (\underline{k}_{02} - \underline{k}_{01}) \qquad \qquad \dots (1)$$

which is independent of the direction of viewing, $\underline{k}_s$.

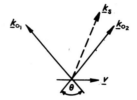

Fig. 1   Vectors representing incident and scattered beams

If θ is the angle between the beams, then

$$\triangle \nu = \frac{2\nu \sin \frac{\theta}{2}}{\lambda_0} \qquad \qquad \dots (2)$$

The same result may be obtained by imagining the existence of a set of light fringes where the beams intersect. A small particle will give a flash of scattered light as it passes through each bright fringe, the rate of flashing being proportional to its velocity. The equivalence of these two approaches has been demonstrated by Rudd[2].

2.2 Previously published systems

Early systems, such as those of Yeh and Cummins[3] and Goldstein and Kreid[4], had to be mounted on lathe beds as they required critical alignment. Rudd[2] has proposed a symmetrical system, which has less stringent requirements of alignment and is insensitive to vibration, but the masking wastes much of the laser power and traversing is still difficult since all the components have to be moved. Mayo[5] and Wilmshurst et al[6] have proposed systems in which parallel beams are brought to intersection and focused by a lens. These methods are similar to the NEL optical system where a beam-splitting prism is used, though they are less suitable for traversing.

The standard method of signal processing is to lead the photomultiplier output to a spectrum analyser whose output is recorded on a storage oscilloscope or chart recorder. The result is an approximately Gaussian-shaped curve centred on the Doppler frequency. The spread of width of the peak results from both optical and instrumental considerations. Finding the corresponding velocity necessitates estimating the centre of this peak, and consequently it is difficult to achieve accuracy better than 1 per cent or so at best.

Recently, several instruments have been under development using a commercial frequency-tracking unit, but again the overall accuracy is in doubt and is unlikely to be better than 1 per cent.

3   THE NEL LASER VELOCIMETER – PRINCIPLES OF OPERATION

The instrument developed at NEL has several advantages over the systems mentioned above: full use is made of the laser power, and traversing and optical alignment are particularly straightforward. Calibration is fixed by the choice of one lens and a prism, and thereafter the one multiplying factor converts output to velocity under all circumstances. The output is readily obtained in digital form.

3.1 The optical system

The basic optical circuit is shown in Fig. 2 the lenses, prism and photomultiplier being mounted on an optical bench. Light from a 5 mW helium-neon laser passes through a half-wave plate to a beam-splitting prism B which gives parallel beams of similar power. The lenses L_1 and L_2 are linked together so that they share a common focus. Hence, by moving them together, the pipe can be traversed without affecting the focusing. The direct beams are blocked at the aperture A and the scattered light is focused on a photomultiplier D by the lens L_3. The optical system can be readily modified to detect back-scattered radiation, see section 4.

The behaviour of a square-law detector such as the photomultiplier is described in Ref. 1. Because it cannot respond directly to signals at the frequency of light, the alternating signal in the detector output will consist of difference terms of the form indicated by equation (1). Inclusion of the direct beams to the photomultiplier, as in systems previously reported, does not enhance the output signal. From the scattering curve of a typical particle it can be shown that, at the Doppler frequency, the difference between the output signal levels when operating with direct beams blocked and the direct beams unblocked is insignificant. The signal-noise ratio, however, is greater when the direct beams are blocked. Another advantage of using only the scattered light is that in clean liquids and uncontaminated gases scattering particles are very infrequent, so that for all but a small fraction of the time, virtually the full beam power is passing uninterrupted through the fluid, and if this is allowed to focus on a small area of the detector there is an increased risk of photocathode fatigue. Furthermore, blocking of a small solid angle around the direct beams has the advantage of selecting particles of size comparable to the wavelength of light or smaller and ignoring those larger particles which may not be following the flow closely.

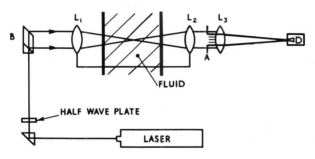

Fig. 2   Basic optical circuit

3.2 Signal analysis

The NEL flowmeter is intended primarily for high accuracy measurements of velocity in fluids with a low particle density where the chances of more than one particle being present in the measuring volume are slight. Good accuracy may be obtained by using a high frequency counter to time, say, 10 cycles from one wave packet with the triggering set so that only large signals are accepted. This also has the effect of tending to select only those particles which pass close to the centre of the intersection region where the light intensity is highest, and hence the spatial resolution is effectively increased.

The electronic processing system is shown in Fig. 3. From a preamplifier, the signals are fed to a band-pass filter, which removes the low frequency noise. The signals are then passed to an oscilloscope with a counter in parallel, and are of the wave packet form shown in Fig. 4. The shape of the envelope corresponds to the Gaussian intensity distribution across the beams. Because of occasional bursts of laser noise or the random signal amplitude, a count is sometimes begun which is not terminated till the arrival of the next wave packet. This results in a percentage of 'bad' counts which are much greater in magnitude than the true ones, and these have to be rejected before the true mean can be found.

The counter* used with this system has two advantages over conventional counters; firstly, its very high resolution at 0.1 ns, and secondly, its built-in computing facility, which enables it to perform calculations on the measurements as it makes them. It can, for example, take 10 000 readings and calculate their mean and standard deviation, all within a period of about 25

*Hewlett-Packard Computing Counter.

seconds. Hence it offers great advantages both in accuracy and in providing virtually instantaneous readings of velocity.

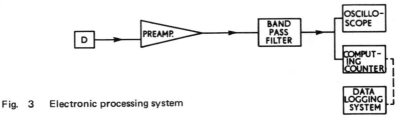

Fig. 3   Electronic processing system

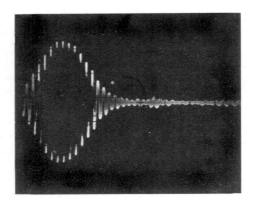

Fig. 4   Typical signal from filter

The most widely used programme for the tests described later was one in which counts more than 20 per cent below the mean value of the good ones were rejected. The selection of the cut-off value was made automatically and hence there was no danger of false readings being accepted as a result of operator error. Tests to ensure the validity of this technique are described in section 5.

4   INSTRUMENT DESCRIPTION

4.1 Arrangement of optical components — basic system

In the original system the laser and optical bench were mounted parallel on a metal plate. This was supported by a screwed rod at each corner to provide height adjustment. The optical components were mounted on a 1 m triangular optical bench, with the carriages of lenses L_1 and L_2 linked together by a metal rod. Subsequently an optical bench which has a moving nylon rack was incorporated, and L_1 and L_2 moved together by this means, (Fig. 5).

Correct alignment of the instrument is simple with the basic optical configuration. Where traversing in a pipe is required, it is essential that the light passes through the pipe axis, otherwise the changing refractive effect will cause the signal quality to deteriorate as the lenses are moved. Reflection from the opposite pipe/water interface helps in alignment. The correctness of the alignment may be checked by traversing the pipe; the position of the spots at the detector should not alter. Finally, the direct beams are blocked at L_2 so that only scattered light reaches the photomultiplier.

Fig. 5   Basic optical system in use with open jet wind tunnel

4.2 Other configurations

The basic optical configuration described in section 4 is suitable for use in many laboratory situations but requires further development. A more compact and robust system is desirable. High accuracy flow measurement requires traverses on several diameters, which necessitates a system capable of rotating about the pipe. In some situations access will be available from one side of the pipe only, so that back-scattered light has to be used; this also applies to the measurement of most moving surfaces.

A major step in achieving a more compact system is the replacement of the photomultiplier with a photodiode. Such devices have the obvious advantages of small size and robustness, and require only a battery as opposed to a bulky power supply. However, at high frequencies, good response time is gained at the expense of signal strength, and the poorer noise characteristics may become important.

The construction of a rotatable system is a fairly straightforward problem of engineering. The simplest technique is to rotate the laser along with the optics and this presents no difficulties for the lower-power lasers used here. A prototype system constructed in wood is shown in Fig. 6; in this the laser is slung beneath the carriage.

Fig. 6   Rotatable traversing system

A back-scattering system which is simple to align and is suitable for use with a photodiode is shown in Fig. 7. The scattered light is collected by L_1 (Fig. 2) and then re-focused on the detector either by a smaller lens placed between the direct beams or by a lens with two holes in it to allow the passage of the direct beams. An integrated system of this form is shown in Fig. 8.

Fig. 7   Basic back-scattering system

Fig. 8   Integrated back-scattering system

Fine adjustment of the position of the prism about three axes and also of the photodiode enables the beams and diode to be aligned correctly with compensation for any lack of parallelism of the two lenses or of perpendicularity of the incoming beam from the laser.

4.3 Position indication

Because of the bending of light on passing from one medium to another of different refractive index, the movement of the point of intersection does not correspond to the movement of the lens pair L_1 and L_2. It can be shown that the motion of the intersection point within a fluid of refractive index n_2 is related to the motion of L_1 by a factor $\sqrt{(1 - \sin^2 \theta/2) / (n_2{}^2 - \sin^2 \theta/2)}$ and is independent of the thickness and refractive index of the wall. Thus, its position may be found accurately once a single reference point has been established. The incoming beams passing through the pipe wall produce bright spots on the inner and outer surfaces. As the measuring point is brought towards the wall the pairs of spots will move together until those on the inner surface merge. At this stage the measuring point is located exactly at the inner wall and thereafter any position across the diameter can be found from the position of L_1. As a check the intersection point may be traversed to the opposite wall.

Automatic position indication may be achieved by attaching a rectilinear potentiometer to L_1 and calibrating a digital voltmeter against the wall positions as above. Thus remote control of traversing is relatively straightforward.

5   TEST RESULTS

5.1 Water tests

Early tests of the instrument were made in water in a straight length of 50 mm Perspex pipe supplied from a small head tank. The flow profile was generally laminar. The flowrate as found from integration of a traverse was compared with the gravimetric flowrate from a weigh-tank/timer system. Agreement was better than 1 per cent, and a typical result is shown in Fig. 9.

Later test traverses were performed in an 83 mm Perspex section in a 150 mm water line. The range of velocities was up to 10 m/s. The log-linear technique[7] was employed, so that the arithmetic mean of the eight or ten velocities from a traverse gave the mean pipe velocity and was, therefore, directly comparable with the mean velocity derived from the mass flowrate. Fig. 10 shows a typical result.

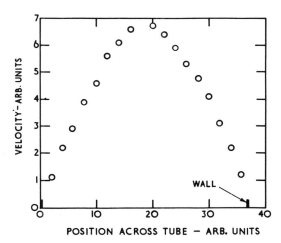

Fig. 9   Test result in laminar flow
      Mean velocity from gravimetric flowrate = 6.31 ± 0.01 mm/s
      Mean velocity from computer processing of laser data (best fit parabola)
      = 6.29 mm/s

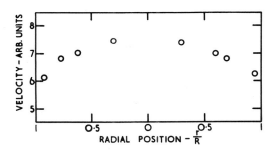

Fig. 10   Test results
      Mean velocity from gravimetric flowrate = 2.49 m/s
      Mean of log linear velocities = 2.49 m/s

5.2 Rotating disc tests

Tests were performed in which the rotation time of a Perspex disc was compared with its velocity, at a given radius, as measured by the laser. Inaccuracy in the measurement of radius limited the overall accuracy to around 0.25 per cent, but the consistency of repeated measurements was better than 0.1 per cent. The range of velocities was 0.1-10 m/s.

Use was made of the rotating disc to investigate system parameters such as filter settings. Fig. 11 shows the result of one such test, and indicates that the adjustment of the band-pass filter is not critical.

As described in section 3, the computing counter was programmed to reject automatically 'bad' counts, which were several orders of magnitude below the signal frequency, by imposing a limit based on the mean value of the previous test. After three or four cycles of this programme a steady situation would be reached. To test the validity of this technique the cut-off frequency was, instead, set by hand, and the variation of error with cut-off investigated.

Fig. 12 shows the result of one such test. For the velocities considered any cut-off limit in the range -5 per cent to -50 per cent of the mean value gave results within a band of ±0.1 per cent. In turbulent flow, however, the spread of velocities at a point may be considerable and the cut-off value clearly has to be set low.

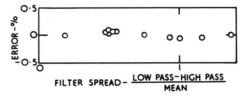

Fig. 11   Variation of discrepancy between laser velocity and rotation time with filter setting

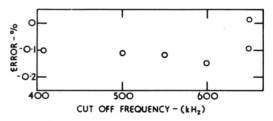

Fig. 12   Variation of discrepancy between laser velocity and rotation time with cut-off (signal frequency = 690 kHz)

6   EXAMPLES OF USE OF LASER VELOCIMETER

6.1 Calibration of free jet (air)

The speed of the wind tunnel shown in Fig. 5 is controlled by a dial graduated from 0 to 100.

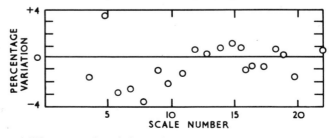

Fig. 13   Calibration of 150 mm open jet wind tunnel

A calibration at the lower end of the speed range is shown in Fig. 13 and indicates that the linearity of the calibration curve is within the specified limits. The velocity profile at the mouth

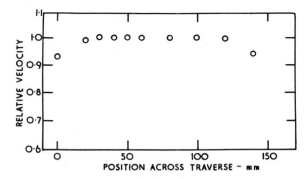

Fig. 14   Velocity traverse 150 mm from mouth of open jet wind tunnel

of the jet is nominally flat. Fig. 14 shows the results of a traverse carried out 150 mm from the mouth, and shows the profile beginning to decay at this distance.

6.2 Tests on water line

Tests to ascertain the reliability of a newly constructed water line revealed unexpectedly large variations in flowrate, as measured by the weigh-tank/diverter system, at the lower end of its range. Laser measurements made during and between diversions showed that each operation of the diverter affected the flowrate, which also had a tendency to decrease with time thereafter, see Fig. 15. A fault was subsequently discovered in the control valve of the line; a loose spindle allowed movement sufficient to give noticeable variation at low flowrates.

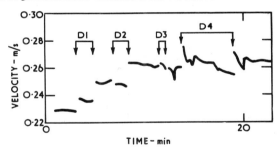

Fig. 15   Variation of axial velocity with time and operation of diverter (D↓ marks time of diversion)

6.3 Pitot tube investigations

The laser velocimeter provides a unique opportunity to investigate the performance of the pitot tube and its effect on flow. The tests described here are the initial stages of a comparative study on pitot tubes, which is part of a major study on flow measurement by integration techniques being carried out at NEL.

In the initial tests a 2.3 mm pitot static tube was inserted in the 83 mm Perspex section in the water line (Fig. 16). Comparison of point velocities was made both with the laser measuring simultaneously 25 mm upstream of the pitot, and measuring at the same point, with the pitot tube withdrawn while the laser was operating. Full corrections in accordance with draft BS 1042[8] were applied to the pitot results. Fig, 17 shows that up to 2.4 m/s, above which pitot static tubes in water are not recommended, agreement was within ±0.25 per cent, but above the limit errors increased to ±1 per cent. In an individual test this error was generally consistently positive or negative, but over the series of tests appeared random. Velocity profiles in the region of the nose of the pitot tube were also made and some are shown in Fig. 18.

Fig. 16   Test arrangement with 2.3 mm Pitot-static tube

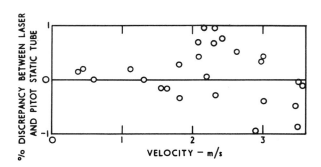

Fig. 17   Discrepancy between laser point velocities and corrected pitot static readings

7   CONCLUSIONS

Tests have been performed in a variety of situations and on moving surfaces as well as with several fluids. Where comparative tests were possible agreement was very good. A back-scattering system has been developed and operated successfully. The velocity range can be extended to 100 m/s by use of an alternative prism and lens. The instrument has been of practical use in several instances, and investigations of the behaviour of the pitot tube are being continued. Future developments will include measurements in two and three dimensions and turbulence investigations.

8   ACKNOWLEDGEMENT

This paper is published by permission of the Director, National Engineering Laboratory of the Department of Trade and Industry.

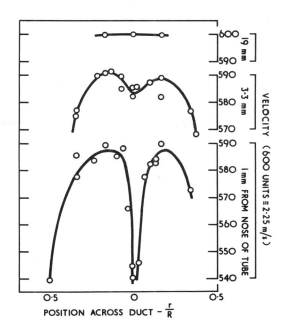

Fig. 18   Velocity profiles
(1 mm, 3.3 mm and 19 mm upstream of a 2.3 mm Pitot-static tube)

9   REFERENCES

1   BLAKE, K.A., and JESPERSON, K.I.: 'The laser velocimeter', NEL Report (in preparation)

2   RUDD, M.J.: 'A new theoretical model for the laser Dopplermeter', J. Scient. Instrum., 1969, **2**, (2), 55-58

3   YEH, Y., and CUMMINS, H.Z.: 'Localised fluid flow measurements with a He-Ne laser spectrometer', App. Phys. Letters, 1964, **4**, 176-8

4   GOLDSTEIN, R.J., and KREID, D.K.: 'Measurement of laminar flow development in a square duct using a laser-Doppler flowmeter', J. Appl. Mech., December 1967, 813-8

5   MAYO, W.: 'Simplified laser Doppler velocimeter optics', J. Scient. Instrum., 1970, **3**, 235-7

6   WILMSHURST, T.H., GREATED, C.A., and MUNNING, R.: 'A laser fluid flow velocity meter of wide dynamic range', J. Scient. Instrum., 1971, **4**, (2), 81-85

7   WINTERNITZ, F.A.L., and FISCHL, C.F.: 'A simplified integration technique for pipe-flow measurement', Wat. Pwr., 1957, **9**, (6), 225-234

8   BRITISH STANDARDS INSTITUTION: 'Methods for the measurement of fluid flow in pipes. Part 2: Pitot tubes', Draft British Standard 1042: Part 2: 1970

DISCUSSION

M.T. Thew: The laser system described in the paper is reported as being designed for systems with a low particle density, and the typical signal shown in Fig. 4 presumably is generated by one particle. However, in some cases the density of scattering particles will not be under the control of the engineer making the measurements. Especially for liquids, the particle density may be such that several particles are simultaneously contributing scattered light to the detector, by being in the high intensity region of the measuring volume. The amplitude and frequency of the summed signal will then tend to vary more and the accuracy of the frequency determination will presumably suffer?

To a considerable extent the problem may be overcome by increasing the number of counts or adjusting the trigger level of the counter, but it would be interesting to know if systematic variation in particle density had been carried out?

It may be noted that murky water may not mean a high concentration of scattering particles, since the particle size distribution is important.

Author's reply: It is certainly the case that a very high particle density causing individual wave packets to merge will affect the accuracy of individual measurements, though the accuracy of the mean value should not suffer greatly. However, it has been found in practice that with this optical configuration such a stage is difficult to reach. While concentration of scattering particles and 'murkiness' of water are not necessarily related, in this case it was evident from the oscilloscope that signal density increased as the water became dirtier. It is fortunate that typical water happens to contain a suitable number of particles to allow a high sampling rate, but few enough that there is seldom more than one present in the scattering volume at one time. In air, of course, the problem is always one of shortage of particles.

J.C. Hoelgaard: Would you please comment about the difference in signal to noise ratio when making use of backscattered light and making use of forward scattered light.

Author's reply: From viewing the filtered signals on an oscilloscope I would say, qualitatively, that signal to noise ratio was the same in the backscattering mode as in forward scattering.

A. Melling: Would you please clarify the statement that you obtained 10^4 measurements in 25 s. I should like to know whether this is dependent on the concentration of scattering particles in the flow.

Author's reply: Clearly this high sampling rate is possible only if enough scattering particles are present. Fortunately this appears always to be the case in liquids, where normal filtering does not affect such small particles. However, in gases the concentration of suitable material is much lower and if a high sampling rate is required slight seeding is necessary. The contrast is with the frequency tracking technique, where fairly heavy seeding is almost always needed.

K.F.A. Walles: You indicated that your instrument was designed for velocities up to 100 m/s. What is the highest velocity used − and could it be used in an 8 ft diameter duct?

Author's reply: The highest velocity measured thus far is about 30 m/s. However, the range is increased simply by narrowing the angle of intersection of the beams. In your case the two problems are solved together − a longer focal length lens increases the size of duct which can be accommodated and at the same time by decreasing the angle extends the upper velocity limit.

2·3

FLOW MEASUREMENT IN THE PRESENCE OF STRONG SWIRL USING A LASER DOPPLER ANEMOMETER

P.S. Bedi and M.T. Thew

Abstract: Measurement of total flow is frequently impracticable and recourse is made to integration of localised velocity values using procedures as given in B.S. 1042. A technique of laser Doppler anemometry is described which essentially is a localised velocity measuring technique but has been successfully applied to flow measurement with strong swirling flow; a difficult case to deal with using other available methods. Following a brief comment on the problems of metering swirling flow, a general introduction to the theory, description of optics and mode of operation of the laser Doppler Anemometry follows. The technique is further explained in the context of its application to flow measurement, especially where high swirl causes problems. Results obtained in flow measurement are discussed and possible extensions of the work to cover a variety of applications are mentioned. Conclusions are drawn in terms of versatility of the technique and its usefulness in flow measurement.

1   INTRODUCTION

1.1 The problem of swirl

Swirl, or more particularly unknown swirl, is still one of the more severe problems in flowmetering and even though Table 1 shows errors of 5%.or so, it was thought worth while to describe our preliminary results with laser anemometry for a flowfield that embodied axial recirculation in addition to strong swirl.

Neglecting methods where total flow measurement is performed, as with a venturi, the common methods of measuring local velocities (prior to integration) have been pitot tubes or current meters. The results of Salter, Warsap and Goodman[1] and of other workers have provided corrections for pitot-static tubes in swirling flow, but the amount of swirl must be known. The advent of the component runner current meter as discussed by Kolupaila[2] has removed some of the problems associated with swirl, but as for the pitot-static tube, interference effects of swirl with the support frame are difficult to predict.

With high turbulence intensity which may accompany swirl, the calibration factor for both types of instrument may change somewhat unpredictably, and this may be aggravated if structural vibrations are induced in the mounting frame.

Axial flow reversal may lead to very large errors and the two methods already mentioned would hardly be applicable to this case, even though it probably only arises near rotodynamic machines or in a recirculation bubble.

1.2 Background to the work

Our prime interest is more in the mechanics of swirling flow, but integration of the axial velocity component provides the only readily available check on the validity of the results.

Work with tubes of 50 mm or less in diameter has proceeded to maximum tangential water velocities of about 10 m/s. As considerably higher velocities are envisaged, laser Doppler methods for velocity measurement were investigated in the first place because of zero blockage effects and good directional precision. With length/diameter ratios up to 60 the experimental measurement of tangential and axial velocity profiles in the tube has generated much data, and in this paper only one aspect is examined.

The two basic requirements for laser Doppler are firstly, adequate transmission of the coherent laser light (here taken to include u/v and I/R) through the fluid, then secondly, an adequate concentration of scattering particles to give a usable signal to noise ratio. When laser Doppler is feasible a number of advantages ensue:

a) no blockage and no need to stop the flow to insert apparatus.
b) no instruments immersed in a corrosive and/or erosive environment.
c) the method is absolute and requires no calibration, though the optical configuration must be precisely known.
d) velocity is measured in a small zone that for many purposes is a point.
e) true velocity components are measured, with excellent directional precision.
f) very large ranges in velocity can be measured.
g) because measurements can be made in a very short time in a precise direction, information on turbulence as well as on mean velocities is obtainable.

It is not suggested that laser methods are a panacea for all flowmetering ills, but that they will provide a useful addition to existing techniques.

2   PRINCIPLE OF LASER DOPPLER VELOCIMETER AND VARIOUS OPTICAL SYSTEMS USED

Consider a particle A in a flowing fluid, having a velocity $\vec{V}$ and illuminated by incident mono-chromatic light of wave vector $\vec{K}_0$ (Fig. 1a).

Let the angle between the velocity vector $\vec{V}$ and the incident light be α. The scattered light from particle A is detected at an angle to the incident light, the wave vector of the scattered light being $\vec{K}_s$.

The following expression may be derived for the frequency of the scattered light, considering the Doppler effect, in terms of the incident light frequency, f_0:

$$f_s = f_0 \left(1 + \frac{V.\cos \alpha}{c}\right) \left(1 + \frac{V.\cos(\alpha + \theta)}{c}\right) \qquad \ldots (1)$$

where c is the velocity of light in that fluid. The total shift in frequency, f_d, to the order of V/c can be expressed as

$$f_d = f_0 - f_s = \frac{V.\mu}{\lambda_0} \left\{\cos \alpha - \cos(\alpha + \theta)\right\} \qquad \ldots (2)$$

where λ_0 = the wavelength of incident light in vacuum
μ = refractive index of the fluid.

As α in the above equation depends upon the velocity vector which itself is unknown in most cases, so the analysis is further extended to eliminate α from equation (2).

Considering the coordinate system X-Y as shown in Fig. 1(a), equation (2) can be reduced to the following

$$f_d = \frac{2\mu}{\lambda_0} . V_x . \sin(\theta/2) \qquad \ldots (3)$$

where V_x is the velocity component of V in the X-direction. Hence the measurement of f_d and θ only, with known values of μ and λ_0, determines the component of velocity in a particular direction.

Fig. 1(b) shows a simple optical arrangement proposed for realising the above configuration. One laser beam acts as a reference beam (source of reference frequency f_0) as well as the scattering beam (source of scattered light). The reference and the scattered beams are combined on a photo-detector which gives a Doppler signal (explained later).

Fig. 2(a,b,c) shows some typical variations in the optical arrangement used by Foreman[3] Yeh and Cummins[4] and Pike et al[5].

Since the frequency shift in the light frequency is proportional to the instantaneous velocity component of the particle, in turbulent flows information on both the mean velocity and fluctuating velocity can be obtained.

3   SIMPLIFIED OPTICAL ARRANGEMENT USED AND ITS ADVANTAGES OVER OTHER SYSTEMS

A simple optical system for Laser Doppler Velocimeter (L.D.V.) (Bedi[6]) was used in the present study. It was found to be very quick and easy to set up requiring no optical hetrodyning alignment as explained below.

Fig. 3 shows the optical arrangement for the simplified L.D.V. The laser beam is split up into two parallel beams, the scattering beam having much stronger intensity than the reference beam. When the ratio of intensities was about 20 the reference beam was just visible, which made the alignment convenient. But it had to be further attenuated after the optical alignment to make it equal to the intensity of scattered light from the other beam (approx. condition for a good Doppler signal). Both the beams are focused at a point A in the flow tube, the reference beam

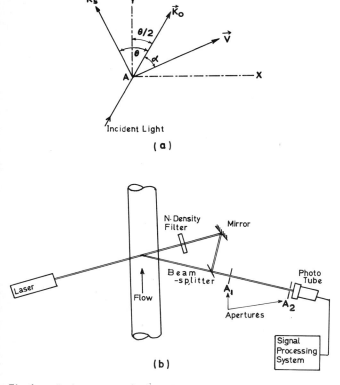

Fig. 1   Basic geometry for laser Doppler velocity instrument
a   Frequency shift for light scattered by a moving particle
b   Simple optical arrangement

falling on the photodetector through the receiving optics. An important characteristic of the system is that the reference beam goes straight through, from the laser to the photodetector while the scattering beam is parallel to and at a fixed distance from the reference beam, when both come out of the beam splitter. This geometry is preserved irrespective of any linear motion or vibration of the beam splitter. It can further be seen that the complete optical arrangement is based on the alignment of the reference beam (or the laser beam) on to the photodetector surface. Once the laser beam has been aligned on the photodetector, the whole optical alignment can be achieved by inserting the beam splitter, lens, flow tube, path length equaliser (where and if necessary), neutral density filters (not shown in Fig. 3) and pin-holes, one by one into the laser (or reference) beam. The passing of the laser beam along the lens axis or a tube diameter is very important and can readily be checked by ensuring that the beam is undeflected. In a large duct with plane windows it would not be necessary for the beam to pass along a diameter.

The rotation of the beam splitter about the reference beam permits a change in the direction of velocity measurement without disturbing the rest of the optical alignment. Since the separation of the two beams and the focal length of the lens remain constant, the angle and the point of intersection of the beams remain the same (neglecting the effects of refractions at the tube surfaces). Complete rotation of the direction of measurement can be obtained, thus permitting the measurement of any velocity vector in a plane.

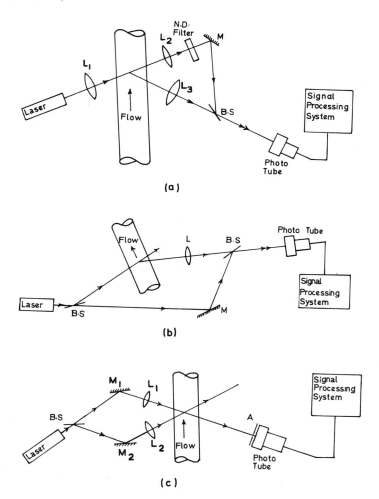

Fig. 2   Typical variations in optical arrangement for laser Doppler velocity instrument
   a   Single beam acts as reference beam and scattering beam; lenses used in transmitting
         and receiving optics[3]
   b   Laser beam split into two: 1 reference beam
                                    2 scattering beam
         Reference beam bypassing flow tube[4]
   c   Laser beam split into two: both passing through flow tube[5]

The movement of the lens only along the reference beam provides movement of the point of intersection without disturbing the optical arrangement. Traversing in this system is much easier and quicker than with those systems where the whole of the optics has to be moved with respect to the flow system.

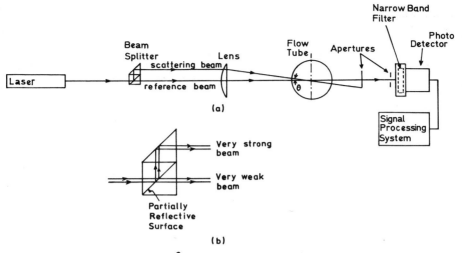

Fig. 3   a   Experimental arrangement
          b   Beam splitter

Apart from these there is no loss of the available power as in some other systems and no hetrodyning alignment is necessary.

The lasers used in two variants of this system were 5 mW and 15 mW.* The angle in the absence of the flow tube was $11.5°$ and $10°$ respectively.

4   LASER DOPPLER SIGNAL, ITS DETECTION AND PROCESSING

Since the laser Doppler signal corresponds to the instantaneous velocity component of the particle in the fluid, it contains information on mean and turbulent velocities of the flow. Due to the fluctuating velocities in the flow, there exist corresponding values of the frequencies in the signal. Fig. 4 shows a typical structure of the signal in terms of the flow parameters. The effects produced by other parameters in the system (as outlined in Section 5) are not shown.

The reference beam falling on the photodetector will have a frequency of about 10^{14}-10^{15} Hz. while the scattered light reaching the detector will have been Doppler shifted − for the configuration discussed here typically by 10^4-10^7 Hz. Since the detector receives both signals, the beat frequency in the total light intensity resulting from hetrodyning of the two signals will be the relatively slow Doppler frequency. There will be a much higher frequency resulting from the additive effect, but this will probably be too fast for the photomultiplier to respond. In our configuration the photomultiplier output is fed into a broadband fixed gain amplifier which has an upper cut-off at about 25 MHz so that only the Doppler signal is transmitted onwards.

Photomultipliers are capable of very high gain and have a very adequate bandwidth, but solid state devices are beginning to supplant them for some uses. A general discussion of photodetectors is given by Warner and Warden[7].

A tri-alkali 11-stage photomultiplier‡ was used for signal detection. Figs. 5 and 6 show the block diagrams for the hetrodyne detector and for signal processing systems used in experimental procedure. A photomultiplier and to a lesser extent a photodiode may be damaged by exposure

*Spectra Physics Model 120 and 124A.

‡E.M.I. − 11-stage − S.20-9558B.

to excessive light intensity.

For signal processing, the frequency spectrum of the photomultiplier output current can be analysed in several different ways depending upon the information needed and equipment available. Fig. 4b shows the stages of the photomultiplier signal processing in a simple manner. The elimination of amplitude variations at the initial stage makes the subsequent signal analysis better and easier. The signal is then left with a frequency spectrum which can be analysed in any of the following methods. A brief description is given for the methods more commonly used.

1.   Perhaps the simplest technique is to display the output voltage from the photomultiplier directly on an oscilloscope. The Doppler frequency can be determined directly from the scope screen if the velocity is steady or slowly varying. The accuracy obtainable is limited and only the time averaged velocity at a point can be determined in this way.

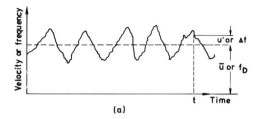

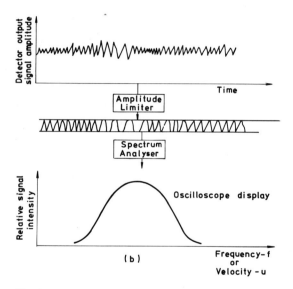

Fig. 4   a   Laser doppler signal
          b   Signal processing

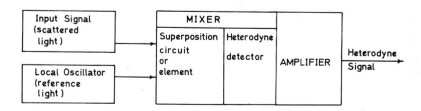

Fig. 5   Block diagram for heterodyne signal detector

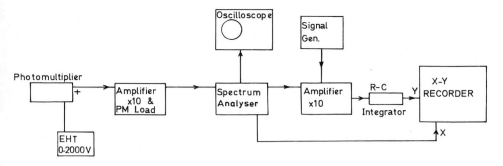

Fig. 6   Block diagram of signal processing system

2.   A counter can be used to give the mean Doppler frequency directly, though suitable gating can be difficult.

3.   Many oscilloscopes accept plug-in spectrum analyzers which allow visual display of the signal frequency spectrum directly on the screen. The time base can be made to represent frequencies while the vertical ordinate represents the intensity of signal at that frequency. The accuracy is limited to 2-5% due to inherent limitations of the scope, but can be improved if the data be permanently recorded on an X-Y plotter.

Two systems of plug-in Spectrum Analyzers*† was used in the present system for signals up to 25 MHz and the signal was recorded on an X-Y plotter‡. Fig. 7 shows a typical X-Y plot of a signal. In the system where internal markers were not available in the Spectrum Analyzer, an external signal generator was used to mark the reference frequencies on the plot.

4.   A wave analyzer can be used which performs basically the same function as a spectrum analyzer except that the turning mechanism is hand or motor driven and the voltage-frequency display is recorded on a strip chart or an X-Y recorder.

5.   Another technique uses a scanning plate Fabry-Parot interferometer to measure the Doppler frequency - James et al[8].

6.   Wide Band frequency Discriminator.

7.   Filter Bank.

8.   Doppler Frequency Tracker.

9.   Phase Locked Receiver.

Rolfe et al.[9] have discussed methods 6 to 9 in comparison with method 3.

5   SYSTEM PARAMETERS AFFECTING THE SIGNAL

While applying a laser Doppler velocimeter, some special points must be considered, otherwise the velocity signal may not be of as good quality as it should be. In some cases one may even lose the signal completely.

---

*Tektronix Oscilloscope No. 547.
  Tektronix plug-in Spectrum Analyzer No. 1L5 (50 Hz-1 MHz).
  Nelson-Ross plug-in Spectrum Analyzer No. 225 (1 KHz-25 MHz).

† Hewlett Packard Oscilloscope No. 140A.
  Nelson-Ross plug-in Spectrum Analyzer PSA-235 (1 KHz-25 MHz).

‡Hewlett Packard X-Y plotter No. 7035B.

5.1 Coherence and path length considerations:

While the coherence properties of the light source are met by the laser, the path lengths travelled by reference and the scattered beam must be small. Foreman[10] has analysed the dependence of heterodyne signal amplitude on optical path length difference for multimode lasers. Fig. 8 shows a typical dependence of hetrodyne signal amplitude on the path length difference.

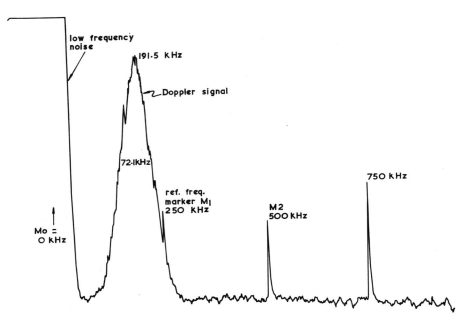

Fig. 7   A typical laser Doppler signal (frequency spectrum). (An xy plot)

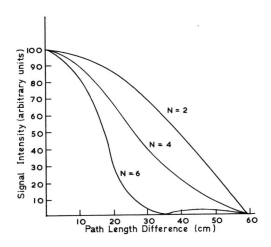

Fig. 8   Heterodyne signal intensity versus optical path length difference[10]
Laser = He-Ne
Cavity length L = 60 cm
Axial mode spacing = c/2L
                               = 250 MHz
c = velocity of light
N = number of axial modes excited

68   **Bedi and Thew**

5.2 Scattering volume:

The size of the scattering volume from which the scattered radiation is observed, can be estimated by the optics of both the laser transmitter and receiver systems. Apart from knowing the dimensions in the flow over which the velocity fluctuations and gradients are averaged out, it determines the particle number density required to ensure a good continuous signal and the ambiguity of the frequency spectrum. Born and Wolf[12] and Pike et al[4] have dealt with relevant cases to calculate the dimensions of the focal region for a lens.

5.3 Scattering particles:

Laser Doppler systems measure the velocity of a flowing fluid by measuring the velocity of particles suspended in the fluid. The validity of the assumption that the particle velocity is the same as that of the fluid which surrounds it, is of course, dependent upon the nature of the particles, the fluid and the flow. In past investigations Polystyrene spheres (Sp.Gr. 1.004) of diameter 0.55 μm at concentrations of 10-50 ppm. have been used for water velocity measurement; milk, P.V.C. (0.7 μm dia). and talc particles have also been used. Satisfactory results have been obtained (Foreman et al[11]) by naturally occurring particles in ordinary tap water. In the present study also satisfactory results were obtained by naturally occurring particles in the tap water.

5.4 Contribution of other sources in the resulting signal:

The frequency bandwidth which determines the turbulence characteristics of the flow is considerably affected by other sources, although the mean Doppler frequency corresponding to the time average flow velocity remains unaffected. It is not intended to discuss these factors and their effect here as it would be out of the scope of the paper, but a brief discussion could be found by Rolfe et al[9].

6 EXPERIMENTAL PROCEDURE – INITIAL ADJUSTMENTS – SOURCES OF ERROR AND ANALYSIS OF SIGNAL DATA

Fig. 9 shows the arrangement of the two laser beams (reference and scattering) for the measurement of the tangential and the velocity components. For measuring the flow through the tube, a radial traverse for the axial velocity measurement is necessary. It was found that the most convenient starting position for the radial traverse was the inside surface of the tube. It was noted that the accuracy with which the tube wall surface could be located with respect to the cross-section of the beams was approx. 0.25 mm. Refractions on the tube surfaces were taken into account while computing the true location of the beam intersection point within the tube. Since the optical arrangement used in the present system works more efficiently when the reference beam passes diametrically through the tube, it shifts the direction of velocity measurement slightly from the one defined theoretically. The direction of velocity measurement in this case is at an angle of $\theta/2$ from the true axial velocity. For $\theta = 10°$ the error introduced was the magnitude of the velocity measured to within ±0.38%.

Again the refractions at the tube surface may change the angle of intersection of the two beams. In the axial velocity measurement this real angle of intersection remains constant as the flow tube is traversed but in the case of tangential velocity measurement, the real angle of intersection varies with the position of the real point of intersection. In all cases, the real position and angle of intersection of the two beams was computed after taking into account the radii of curvature of the tube and the refractive indices of the media through which the beams travel.

Ordinary clear cast Perspex 50.8 mm. internal diameter tube was used. The tube, when measured near the ends was oval up to approximately 0.13 mm. to 0.25 mm. Although ovality was thought to be present over the tube length – (1.5 m. max single tube), it did not seem to affect the results. The maximum length of tube used was 3.1 m. made up of three different tube lengths. The ends of the tubes were carefully filed and matched to avoid sharp corners. Swirl in the flow was generated by having tangential entry at one end of the tube (Plate 1).

Since one of the beams passes through the lens, away from its centre, it may be affected by aberrations caused by the lens. A planoconvex which is an improvement over a double convex lens

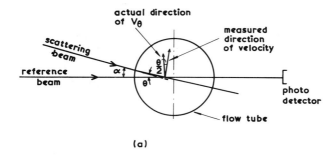

(a)

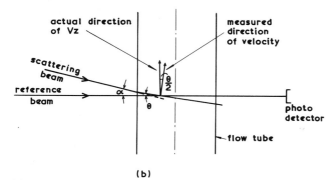

(b)

Fig. 9   Optical arrangements used for measuring
        a   Tangential velocity Vθ
        b   Axial velocity Vz

Plate 1   Optics and the inlet region of flow

for spherical aberrations, was used in the system.

Readings were taken at 17 points on one radius (25.4 mm.) for tangential velocity measurement and at 12 points on one radius for axial velocity measurement. There is no reason why the number of readings could not be increased.

Fig. 7 shows a typical graph. The mean velocity at that point is calculated from the frequency corresponding to the peak value of signal intensity, which in Fig. 7 is at 191.5 KHz.

The details of information on turbulence intensity is not within the scope of this paper. It is discussed in more details by Thew and Bedi[13] and Bedi[14].

7   DISCUSSION OF RESULTS

Table 1. Flow Measurement Data at different axial stations in the tube*

Axial Section	Axial Dist L/R	+ve flow rate	-ve flow rate	Net +ve flow rate	% -ve flow	Flowrate by Venturi-meter Q_V	% error wrt. Q_V
A	20	64.4	13.3	51.1	26.1	48.2	+ 5.59
B	60	53.3	6.1	47.2	12.1	48.2	- 2.08
C	80	50	3.6	46.4	7.85	48.2	- 3.78
D	100	50.2	0	`50.2	0	48.2	+ 4.25
E	118	47.0	0	47.0	0	48.2	- 2.64

* a)   Values of flow rates given in litre/min.
  b)   See Figs. 10, 11.

Figs. 10 to 12 show the radial distribution of the axial and tangential velocities at different axial stations along the tube. Reversal of flow near the axis at certain stations shows that an accurate (±1%) evaluation of the flow at these stations would be very difficult by most other available techniques. Table 1 shows the values of calculated flowrates at different axial stations in the tube. The flowrate was calculated using Simpson's Rule taking 21 points on a radius. The axial Reynolds No. based on the average axial velocity (0.396 m/s) and the tube diameter was 18000. A gravimetrically calibrated venturimeter was used to check the flowrate in the line. The venturimeter was made of stainless steel and dimensioned according to B.S.S. 1042 with an inlet diameter of 1.25" (31.750 mm.) and throat diameter of 0.840" (21.3360 mm.). It can be seen that the agreement between two measurements is fairly good.

As a check in one case the v.r:r graph was plotted and the area planimetered. The flowrate calculated was 0.5% nearer the value measured with the venturimeter, but the difference is not very significant.

It may be noted that there is practically no effect of tangential velocity on the mass flow measurement. Although theoretically, the axial velocity measurement is independent of magnitude of tangential velocity, one indirect effect may be anticipated by considering the influence of turbulence level on quality of signal near the point of flow reversal. It was noticed that the signal could not be distinguished from low frequency noise when the point of velocity measurement was very near to the point of flow reversal. The low frequency noise covering up the zero frequency marker is indicated on Fig. 7. This was anticipated as although the mean velocity in the axial direction at that point is decreasing to zero, the turbulence intensity is of the same order as in other regions of the flow. Secondly, when very near to the flow reversal point, the scattering volume covers a range of velocities encompassing zero (at the flow reversal point) and extending in both +ve and -ve velocity regions. The first condition as well as the second affects the quality of the signal. the first condition, i.e. the level of turbulence, can partly be attributed to the magnitude of tangential velocity. But at this stage it is very hard to say how the latter would affect the turbulence intensity in the axial direction. Otherwise a general effect of the level of turbulence in the axial direction would be to broaden the whole signal, which is not likely to have any appreciable effect on the mean velocity evaluation. Hence the only effect of level of turbulence would be to smear the signal very near the flow reversal point. In the present case normally 2 (with a maximum of 3) points out of total 12 on a radius could not be measured very near the flow reversal. Those three points had to be interpolated to give a smooth curve. Typically they are shown on Fig. 12.

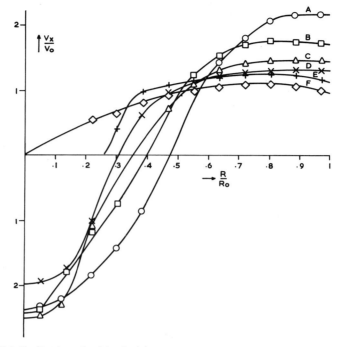

Fig. 10   Radial distribution of axial velocity •
Re = 18 000: Vo = 0.395 m/s (see Table 1)

Fig. 12 shows the axial and tangential velocity distributions in the same diameter tube but at an axial Reynolds No. of 85000. Also the ratio of tangential to axial velocity is different. In this case the flow was found to be asymmetric. Curves A and B show the radial traverses of axial velocity on either side of the tube axis. The calculated flowrates in the two cases were 240 and 232.5 litre/min. respectively whereas the venturimeter reading was 230 litre/min. Although the error introduced by asymmetry is hard to establish without taking traverses along two or more intersecting diameters, a first approximation can be seen from the difference in venturimeter measurement and mean of the flowrates from graphs A and B. This difference is 2.7% of venturimeter flow.

8   POSSIBLE EXTENSIONS OF THE WORK

8.1 Large ducts

In practical installations where integrating methods are in use for flow measurement, the path length for a laser beam would probably be in the range 1 - 10 m. Though lasers with a continuous output of well over 1 W were available, to cope with longer paths it is not merely a question of obtaining more power, but adequate scattered light and reference signal must reach the photo-detector. With large flows the concentration of scattering particles will normally be outside the engineer's control. Gas flows may not have enough particles, but in many water flows excessive particle concentration could gravely attenuate the signal, particularly when combined with natural dye, as in peaty water. The availability of lasers with output wavelengths varying from u/v to I/R could alleviate the problem. To our knowledge no work with large scale ducts has yet been reported. The results would be of considerable interest.

For flow metering upstream from rotodynamic machines there would be no danger of damage resulting from ingestion of debris from a failed meter mounting frame.

8.2 Velocity range

The research reported here had a velocity range limited to roughly 20:1, high velocities being curtailed by the test rig pump and low axial velocities both falling outside the range of the

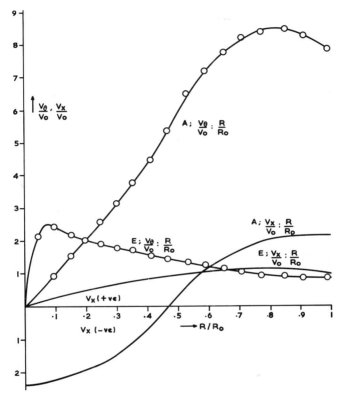

Fig. 11   Radial distribution of tangential (Vθ) and axial (Vx) velocities
Re = 18 000: Vo = 0.395 m/s (see Table 1)

spectrum analyzer and having the problem of the relatively larger turbulence intensity already commented on. The electronics mentioned here will process up to about 25 MHz, which with an optical system incorporating a scattering angle similar to the one used represents roughly 75 m/s. (Reference to section 2 shows the dominant effect of the scattering angle θ on the Doppler frequency.). Back scattering may be more convenient on occasion than the forward scattering used in this work, but the intensity of the scattered light is then much weaker. Laser systems have been successfully used at low velocities, where Greated and Manning[15] reported values of 40 mm/s in studies on water waves.

8.3 Speculation on practical installations

The measurements discussed in section 7 were limited to one plane, but most installations would require several planes particularly with non-square ducts. Salami[16] has discussed on theoretical grounds the relationship between various asymmetric velocity profiles and the optimum arrangement of measuring points. Unlike current meter installations, the number of points is not predetermined for a laser system, but the direction of traverses is limited by the position of windows. With an optical system similar to that reported here, each traverse requires a pair of windows nearly opposite. To minimise and simplify refraction corrections flat plates would be most convenient; those at the receiving end could be small, but at the laser end the window would have to be long enough to accommodate the maximum distance between the beams, which is a function

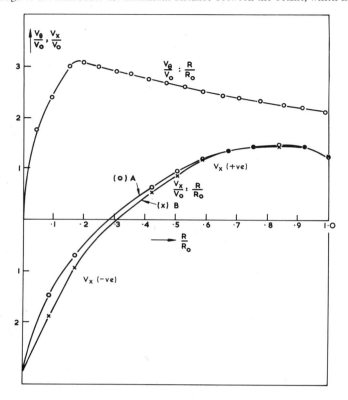

Fig. 12   Radial distribution of tangential (Vθ) and axial (Vx) velocities

Net flowrate: Q_A = 240
                        Q_B = 232.5   litre/mm
                        $Q_{Venturi}$ = 230
Re = 85 000: Vo = 1.925 m/s

of the minimum tolerable scattering angle. In contrast to the fixed current meter or pitot tube installation, a laser traverse would not be capable of giving data from each point simultaneously. Time, though not expense, would be saved by using one laser installed in an integrated optical system for each traverse. Since the swirling flow in the test rig was steady over long periods, it was reasonable to use a single long scan of 2-3 minutes on the spectrum analyzer for each point to gain accuracy. However, for a traverse in a large installation a development of the scheme developed by Wilmshurst et al.[17] could reduce the time to 10 secs or so, with mechanical lens movement as the limiting factor.

To avoid interference between scattering from several laser beams, slight axial spacing could be required.

Apart from deposits obscuring surfaces on the windows, a laser system should be reliable with no components immersed in the high pressure or corrosive fluid. With an unchanging calibration factor, results could be processed with an on-line computer.

8.4 Ambiguity in velocity direction

The method described earlier is incapable of discriminating between a positive and negative Doppler shift. Suitable visual presentation[15] of results or the physical phenomena will remove the problem. For automatic data processing Fridman et al[18] mention a technique of inserting single side-band modulation in the reference beam.

9 CONCLUSION

Results from small scale laser Doppler experiments with strongly swirling incompressible flow in a circular tube were obtained without undue problems with alignment; the medium was tap water with no additional tracer particles. The discrepancy between the integrated flowrate derived from a single traverse and that measured by a calibrated venturi was about ±5%. Since marked axial flow reversal was present in addition to the swirl, the method shows promise of power and versatility with an optical system that is easy to align.

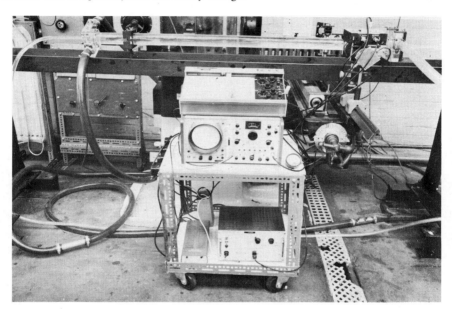

Plate 2   General layout of the apparatus

10 REFERENCES

1   SALTER, C., WARSAP, J.H., and GOODMAN, Miss D.G.: 'A discussion of pitot-static tubes and of their calibration factors with a description of various versions of a new design', 1965, A.R.C. R and M 3365

2   KOLUPAILA, S.: 'Significance of a component runner for the current meter method of flow measurement in closed circuits', Paper A-1 Proc. Symposium on flow measurement in closed conduits, N.E.L., 1960, H.M.S.O., 1962

3   FOREMAN, J.W., GEORGE, E.W., and LEWIS, R.D.: 'Measurement of localized flow velocities in gases with a laser Doppler flowmeter', Applied Phys. Letters (1965) 7, (4), pp. 77-78

4   YEH, Y., and CUMMINS, H.Z.: Localized fluid flow measurement with a He-Ne laser spectrometer', Appl. Phys. Letters, 1964, 4, (10), pp. 176-178

5   PIKE, E.R., JACKSON, D.A., BOURKE, P.J., and PAGE, D.I.: 'Measurement of Turbulent Velocities from Doppler shift in the scattered light', Jnl. Sci. Instr. (E), 1968, 1 (Ser. 2), pp. 727-730

6   BEDI, P.S.: 'A simplified optical arrangement for the laser Doppler velocimeter', Jnl. Sci. Instr. (E), 1971, **4** (Jan), pp. 27-28

7   WARNER, F.L., and WARDEN, M.P.: 'Superhetrodyne reception of optical frequencies', Paper presented at conference on Lasers and their applications, I.E.E.E. (UK & Eire section), 29 Sept. - 1·Oct. 1964

8   JAMES, R.N., BABCOCK, W.R., and SIEFERT, H.S.: 'A laser Doppler technique for measurement of particle velocity', A.I.A.A., 1968, **6**, p. 160

9   ROLFE, E., SILK, J.K., BOOTH, S., MEISTER, K., and YOUNG, R.M.: 'Laser Doppler Velocity Instrument', NASA, 1968, No. C.R.1199

10   FOREMAN, J.W. Jr.: 'Optical path length difference effects in photomixing with multimode gas lasers', Appl. Optics, 1967, **6**, (5), pp. 821-826

11   FOREMAN, J.W., LEWIS, R.D., THORNTON, J.R., and WATSON, H.J.: 'Laser Doppler Velocimeter for measurement of localized flow velocities in liquids, Proc. IEEE, 1966, **54**, (March), pp. 424-425

12   BORN and WOLF: 'Principles of Optics' (McMillan Co. Pergamon Press 1964)

13   THEW, M.T., and BEDI, P.S.: 'A technique of laser Doppler Anemometry applied to confined swirling flow', Symposium on "Flow Visualisation technique and applications" at Polytech of the South Bank (then Borough Polytech), London, June 1970

14   BEDI, P.S.: 'A study of flow patterns and the mechanism of oil water separation in a Hydrocyclone with a laser Doppler velocimeter', Mech. Engg. Dept. Report (in preparation), Southampton University, Southampton

15   GREATED, C.A., and MANNING, R.: 'Water waves measured with a laser flowmeter', La Houille Blanche, 1970, No. 6, pp. 567-570

16   SALAMI, L.A.: 'Errors in the velocity-area method of measuring asymmetric flows in circular pipes' (see p. 381)

17   WILMSHURST, T.H., GREATED, C.A., and MANNING, R.: 'A laser fluid flow velocimeter of wide dynamic range', J. Phys. E. (J. Sci. Inst.), 1971, **4**, pp. 81-85

18   FRIDMAN, J.D., HUFFAKER, R.M., and KINNARD, R.F.: 'Laser Doppler system measures three-dimensional vector velocity and turbulence', Laser Focus, Nov. 1966, pp. 34-38

11   ACKNOWLEDGEMENTS

The financial assistance of the Science Research Council is gratefully acknowledged and we also wish to thank the UKAEA for providing facilities used in part of the work.

For fruitful discussions and help, we are also indebted to Dr. C. Greated and Dr. T. Wilmshurst of the mathematics and electronics departments at the University of Southampton, also to Prof. P. Egelstaff of the University of Guelph, Ontario (formerly of the UKAEA) and to Mr. B. Moss of the UKAEA.

DISCUSSION

R.K. Turton: What comment have you on the variation of focussing accuracy and size of target with scattering angle θ ?

Authors' reply: Our comment will be easier to follow if we discuss first the size of the target, presumed to mean the volume from which a scattered radiation signal is received. The size of the target is defined by the transmitting and receiving optics and is affected by the refractions at the interfaces at the tube surfaces. Pike et al (Ref. 5 in the text) derived an expression for the waist diameter of a laser beam when it is reaching the lens almost parallel. It gives a waist diameter of $100 \mu m$ in the present case. The length of the scattering volume is defined by the coherence conditions on the receiving optics for good photomixing of the reference and the scattered beams. In the present system, the length is 0.5 mm. The scattering volume decreases as θ increases, up to $90°$, where the scattering volume is a minimum (other conditions remaining the same).

By focussing accuracy we presume that the contributor means the accuracy with which the scattering volume is located. This depends firstly, on the location of the starting point, and secondly on the accuracy of the subsequent transverse. The starting point in the present system was the tube axis and the tube wall, for tangential and axial velocity measurements respectively. The position of the tube axis was found by making the reflections from internal tube surfaces coincide and that of the tube wall located by adjusting the beams intersection point to be at the wall. Assuming the tube had no surface imperfections and ovality, the axis and the wall surface could be located to within about 0.3 mm. Although the accuracy of location could not be checked for each individual point, and estimation of the accuracy of traversing could be obtained, since the overall distance traversed coincided to within about 0.5 mm. of the measured tube diameter.

The accuracy could be improved by using truly cylindrical tubes; as noted in the paper, for our tubes the ovality near the ends was about 0.2 mm. Measurements in the middle could not conveniently be made. Other details of the calculations to locate the points, including refraction corrections at various values of θ may be found in ref. 14 (see text). The accuracy of traversing would be better and the effects of tube refractions less, for smaller values of the scattering angle θ.

R.K. Turton: What pinhole size is desirable with the system you use?

Authors' reply: The pinhole size was calculated on the basis of the coherence conditions for the optics receiving side. The diameter of the pinhole used in the present system was 2 mm. Although only one pinhole is necessary, a second pinhole in front of the first was needed in the present system to avoid stray reflections from the tube surfaces.

In addition to the references quoted in the paper, three others given below are helpful in discussing instrumental effects on signal broadening:

a   BOURKE, P.J., BUTTERWORTH, J., DRAIN, L.E., et al: 'A study of the spatial structure of turbulent flow by intensity-fluctuation spectroscopy', J. Phys. A: Gen. Phys., 1970, **3**, pp. 216-228

b   ADRIAN, R.J., and GOLDSTEIN, R.J.: 'Analysis of a laser Doppler anemometer', J. Phys. E: Sci. Inst., 1971, **4**, pp. 505-511

c   MAZUMDER, M.K., and WANKUM, D.L.: 'SNR and spectral broadening in turbulence structure measurement using a c.w. laser', Applied Optics, March 1970, **9**, (3), pp. 633-637

A.T.J. Hayward: The discussion on the laser flowmeters appears to have given the impression that this is an interesting research tool but that it has little potential in the field of industrial instrumentation. I suggest that this is not so. There is a need in industry for a fully automatic, rapid method of flow measurement by velocity traversing, to replace pitot traversing which is slow and cumbersome. The laser could, if suitably developed, fulfil this need.

Authors' reply: We strongly agree with Dr. Hayward that laser methods have much potential for industrial flow measurement systems. Our paper discusses one aspect of this potential for the difficult area of swirling flows. Because of time, the latter part of our paper touching upon industrial applications was but briefly mentioned in the presentation. The companion paper by Blake (see p. 49) discusses errors already lower than those present in many conventional measurement systems. The engineered, back-scattering integrated optics/receiver probe that he showed was impressive in its small size.

It has been suggested that integrated laser/optics systems may be made commercially available in the near future and this should give a boost to engineers in industry who are wondering whether to try laser methods. The direct electrical output and good range/sensibility characteristics are perhaps not yet appreciated. In our opinion laser methods are ready to leave the research laboratory and to enter use in those industrial applications where the advantages mentioned in the first section of our paper are significant.

2·4

AN EVALUATION OF OPTICAL ANEMOMETERS FOR VOLUMETRIC FLOW MEASUREMENT OF LIQUIDS AND GASES

F. Durst, A. Melling and J.H. Whitelaw

Abstract: The purpose of the paper is to provide an assessment of the applicability of optical anemometry to the measurement of volumetric flow rate of liquids and gases and of the precision attainable. The assessment is supported by measurements in air and water flow which allow the relative advantages of different optical and signal processing arrangements to be quantified in terms of precision, cost and convenience of application. Since optical anemometry depends on the presence of particles to scatter light, measurements are only possible where such particles exist. In gas flows, with low particle concentrations, frequency counting procedures are shown to be best and, in liquid flows, where particle concentrations may be high (e.g. in tap water) frequency tracking is preferable. In the case of frequency tracking an on-line signal is available and is suitable for monitoring changes in flow rate or, in some cases, flow rate itself. The effort required to apply optical anemometry is likely to be reasonable for those specialist applications where the equipment costs can be justified.

1 INTRODUCTION

The application of optical methods to the measurement of instantaneous velocity has been investigated intensively in recent years with the result that it is possible to assess the relative advantages of these methods for volumetric flow measurement. The purpose of this paper is to provide an assessment of this type and, in particular, to demonstrate the precision, cost and ease with which optical methods can be used to measure local mean velocity and, hence, volumetric flow rate. Since optical techniques measure local velocity, they are compared first with total-head probes which also measure local velocity and subsequently with volumetric flow devices such as orifice plates, rotameters and venturi meters.

The obvious advantage of optical methods is that they do not cause flow interference and, hence, a pressure drop. This can be of particular significance where pumping power is limited, e.g. in physiological systems, or where a flow disturbance gives rise to a change in the desired flow problem, e.g. in fluidic devices. An obvious disadvantage of optical methods is the requirement that provision be made for light beams to pass into and out of the fluid. In addition, they require light scattering particles in the flow; the most suitable substances, size and concentration of these must be indicated. Thus it is clear that optical methods are likely to be more suited to some applications than others and it is hoped that the present paper will provide the necessary guidance.

The succeeding section contains a brief review of the principles of optical anemometry and provides an indication of the equipment required to put these principles into practice. The principles are not described in detail since they may be found in references 1 to 4. Similarly, further details of equipment may be found in references 5 to 10. The third section describes experiments performed to assess the precision with which optical-beam methods can measure mean velocity in liquid and gas flow and describes some of the difficulties which can arise. The remaining two sections respectively discuss the results in the particular context of volumetric flow measurement and present recommendations.

2 MEASUREMENT TECHNIQUES

2.1 Principles

Conceptually, the simplest way to measure mean velocity using light beams is to locate two beams a known distance apart and to measure the time taken for a tracer or particle to pass from one to the other[11]. A more elegant, and generally useful, method is obtained by crossing two light beams of equal intensity; a fringe pattern is found in the region of the crossing and can be used in the same manner as the two distinct light beams. The number of light fringes and hence the distance between them can be controlled by the angle between the beams. It has been shown, for instance in reference 2, that the frequency of particles crossing light fringes is related to the instantaneous velocity according to the linear equation

$$\hat{\triangle \nu} = \frac{\nu_s}{C} \, \hat{U}_i \, n_i \qquad\qquad \ldots (1)$$

The same equation has been shown, for example in references 1 and 2, to apply to the signal obtained from light Doppler shifted in frequency by moving particles. In this case, light scattered from moving particles is observed at two different spatial locations; the frequency of the scattered light at each of these stations is different and this difference, equivalent to $\triangle \nu$, is related to instantaneous velocity according to equation 1.

It has been shown analytically in reference 4 that the fringe system is suited to flow systems where the particle concentration is such that there is only one particle in the measuring control volume at a given time. In contrast, a Doppler-scattering system provides optimum performance with a much higher particle concentration, when the difference between the frequency of scattered light and incident light is measured. These conclusions are borne out by the authors' experiments and indicate that the optical mode should be selected to suit the particular flow situation and its particle concentration.

In addition to the particle concentration, the particle size distribution is important. Since the optical techniques measure particle velocity, rather than fluid velocity, it is usually desirable that the particles follow the flow. For flow situations without rapid fluctuations comparatively large particles may be used. For example, low density solid particles with a mean diameter of 10 μm will follow frequencies up to 100 Hz with a precision of over 80%. In contrast, where turbulence intensity measurements are required, particles of diameter 1 μm or less must be used to follow the turbulence frequencies (10 kHz, say) to a precision of 1%.

The design of an optical anemometer based on the above principles is discussed in the following sub-section which includes comments specific to signal processing.

2.2 Equipment

Fig. 1 shows a typical measuring system consisting of a light source, an integrated optical unit, a light collecting system, a photosensitive device and a signal processing system. The following paragraphs discuss each of these in the context of mean-flow measurement.

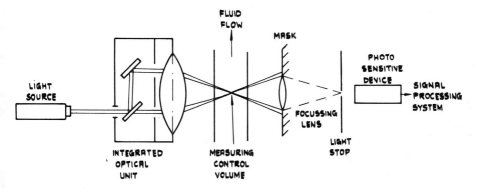

Fig. 1   Typical measuring system

A laser light source provides an intense, collimated monochromatic beam and is convenient for present purposes. Provided forward scattering is used, i.e. the light collecting system is located on the opposite side of the measuring control volume from the light source, a 1 to 5 mW He-Ne laser is satisfactory for use in water and air systems. In the case of flows with particles which absorb light a more intense source may be required.

The authors have directed much of their research effort[2,9,10,12,13] to the development of integrated optical units in order to combine flexibility of optical mode with robustness and ease of alignment. For most volumetric flow measurements only the robustness is required since an optical unit can be pre-aligned for a particular, fixed application and since the optical mode can be pre-arranged to suit the particular particle concentration. Optical units of this type, comprising a mirror, beam splitter, mask and lens can be constructed for small cost. Typical low-cost arrangement are described in references 2 and 12.

The light collecting system determines the volume of the flow observed by the photosensitive device and the appropriate arrangement depends on the optical mode. The mask-lens-light stop arrangement shown in Fig. 1 is appropriate to the fringe mode of operation and, in common with alternative light collecting systems, is easy and cheap to manufacture.

The photosensitive device may be either a photomultiplier or a PIN diode: the latter is to be preferred on grounds of low cost.

The system so far described must be prealigned and arranged on a stable frame. The entire arrangement, including a 1 mW He-Ne laser, the amplifier for the PIN diode output signal and a base plate could be built for a total cost of less than £400. This cost excludes the signal processing system and the windows in the duct; the former will add significantly to the total cost. It is, therefore, clear that optical methods are unlikely to be used for flow measurement unless they offer significant technical advantages over conventional and usually cheaper techniques.

The signal from the photosensitive device may be processed in a number of ways and the choice of a particular system will depend on the application. Three basic methods were used in the experiments reported in the next section; they are

    spectrum analysis
    frequency tracking
and   digital sampling:

Fig. 2 itemises the components required for all three systems.

Spectrum analysis requires that the analyser sweeps the frequency range of the signal to provide an output trace, say on a storage oscilloscope, proportional to the probability density distribution of the signal. The mean velocity corresponds to the mean frequency,

$$\overline{\triangle\nu} = \int_{-}^{+} \triangle\nu \, . \, p(\triangle\nu) \, d(\triangle\nu) \qquad \qquad \ldots (2)$$

with, from equation 1,

$$U = \frac{\overline{\triangle\nu}.\lambda_s}{2 \sin \varphi}$$

This method of analysis is, perhaps, best known but suffers from the disadvantage that it does not conveniently provide an on-line signal proportional to mean velocity.

Frequency tracking represents frequency to voltage conversion in situations which include those where the signal is of the same magnitude as the noise. It operates with the aid of a feedback loop which provides information to a voltage controlled oscillator. The oscillator, in turn, drives the signal through a band-pass filter and allows the frequency tracking demodulator to follow the signal from the photomultiplier or diode and to provide a voltage signal, directly proportional to instantaneous velocity. Instruments of this type can be based either on frequency-locked loops or on phase-locked loops and provide an on-line signal of the type required.

In common with spectrum analysis, frequency-tracking demodulators operate best in flow configurations where a scattering particle is present in the measuring control volume at all times. Spectrum analysis, with presently available sweep rates, is unsuited to particle-rare situations but tracking systems can be provided with drop-out control which allows them to measure mean velocity in particle-rare systems where the turbulence intensity is not excessive and the mean velocity not too high.

The problem of low particle concentrations can be overcome using digital counting systems provided the instantaneous signal from particles is of greater magnitude than the noise: a situation which can be achieved in many flow situations provided a fringe-mode optics is employed. The counting system requires the counting of zero crossings about a datum which is preset above the noise level; this yields the signal frequency and hence the instantaneous velocity.

3   EXPERIMENTAL RESULTS

The present measurements were effected in air and water to allow an assessment of the precision with which mean velocity could be measured in those media using the signal analysis systems discussed in the previous section and shown in Fig. 2*. The spectrum analyser used was the Hewlett Packard model 8552A/8553L; the prototype, frequency-locked tracker developed by A.E.R.E. for DISA Elektronik A/s was used in air and water; and the Hewlett Packard Computing Counter (Model 5360A) was used in air.

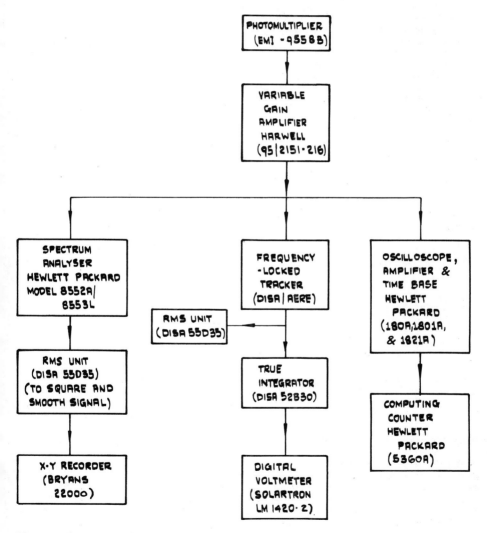

Fig. 2   Signal processing system

* For reasons of availability, a photomultiplier (EMI 9558B) was used rather than a photodiode.

3.1 In air

An assessment of the accuracy of optical anemometry may be obtained from earlier measurements[9] carried out in a fully-developed, laminar pipe flow. These measurements utilised the fringe-mode optics and the frequency-locked tracker. For this particular case, the exact result is known to be a parabolic profile and the measured values were within 0.2% of the exact values except for locations in the immediate vicinity of the wall.

The present measurements were obtained at the exit from a 0.93 m long, 13 mm dia. pipe, i.e. in fully-developed turbulent flow. Unlike water, there are insufficient particles available in natural air to allow the use of spectrum analysis and tracking. Consequently, the flow was seeded with atomised silicone oil at a rate which provided approximately one particle in the 0.5 mm long by 0.1 mm diameter measuring control volume at all times. The fringe-mode optics was found to be most satisfactory as is to be expected for particle concentrations of this order.

A typical signal spectrum is shown in Fig. 3. The width of this spectrum is due mainly to the 4% turbulence intensity present in fully-developed turbulent pipe flow; ambiguity and gradient broadening are equivalent to approximately 1% turbulence intensity. In contrast to the laminar pipe-flow results which allow accuracy to be assessed by comparison with an exact solution, turbulent pipe flow does not possess an exact solution and, therefore, only precision can be assessed. Since the optical anemometer is a linear instrument the precision may be judged from the repeatability of $\overline{\Delta \nu}$ values obtained from various independent measurements and analysed according to equation 2. For the present spectrum-analysis results, the rms deviation was 0.42%; most of this deviation is attributable to the uncertain shape of the spectra which were analysed.

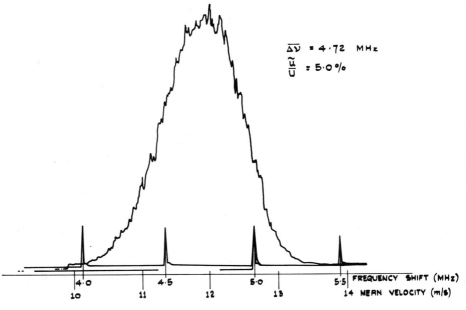

Fig. 3   Frequency spectrum measured in air

Measurements with the tracker proved to be repeatable to an rms deviation of 0.19% but, when the particle concentration was progressively reduced, the frequency-locked loop failed to lock on to the signal. Undoubtedly drop-out control will allow trackers to operate with low particle concentration but the lower limit is unknown. Unlike spectrum analysis, the tracker provides an on-line signal and will monitor rapid (> 10 kHz) fluctuations in velocity.

In flow situations with few particles, e.g. natural air, neither spectrum analysis nor tracking is likely to be of value. The best alternative appears to be the counting systems which, when each velocity measurement was the mean of 10^4 counts, proved to be reproducible to an rms deviation of 0.7%. Undoubtedly a substantial part of this deviation was caused by changes in mean velocity with time. In unseeded air a considerable time, of the order of 200s, is required to make 10^4 sample counts and it was impossible to maintain the flow velocity constant over such a time interval. This emphasises the difficulty of using counting systems and, indeed, optical anemometry in particle-rare flow situations: a single velocity measurement requires a long time to make.

3.2 In water

The water measurements were also carried out in fully-developed pipe flow; in this case, a 10 mm diameter glass tube was used and the measurements were made at a section 1.4 m from the tube entrance. The centre-line velocity was 2.0 m/s. No seeding was required and a fringe mode was again preferred. The resulting spectra were similar to that shown in Fig. 3 but the rms deviation improved to 0.24%. The tracker measurements were reproducible to an rms deviation of 0.13%.

4   DISCUSSION

The results presented in the previous section show that optical anemometers are best suited to mean-velocity measurement in fluid flows where scattering particles are present in the measuring control volume at all times; in practice this implies a particle concentration in excess of 1 mm^{-3}, even for comparatively large control volumes. In addition, the particles should be in the approximate size range from 0.2 to 5 μm. These criteria are satisfied with water. Thus an optical anemometer comprising the items indicated in Section 2 together with a frequency-tracking demodulator can provide an accurate, on-line signal proportional to mean velocity. Using a commercially available phase-locked loop, a suitable tracker unit has been constructed for approximately £400. The total cost of an optical anemometer capable of measurements in water and other transparent fluids is, therefore, around £800.

The cost of a suitable optical anemometer is likely to limit its use to flow situations where conventional alternatives cannot be used for technical reasons. In addition, the optical anemometer measures local velocity rather than volumetric flow rate and is likely to be of greatest value where its essentially instantaneous response to changes in velocity will permit rapid flow-rate changes to be monitored. The translation of the local velocity measurement to flow rate requires a knowledge of the velocity profile; for example either a fully-developed flow, say 40 diameters for a turbulent flow, or a flow constriction to form a uniform velocity profile could be used. The flow measurement could be effected with the same precision as the velocity measurement and, in the former case, this would be achieved without the usual pressure drop caused by an orifice plate, venturi or rotameter: in both cases, the precision of volumetric flow measurements is significantly higher than the conventional alternatives. However, in view of the dependence of the velocity profile in turbulent pipe flow on Reynolds number and wall roughness and the tolerance on the cross-sectional areas of commercial pipes, the accuracy of a volumetric flow rate based on a single velocity measurement on the pipe axis is around 4%, comparable with that of an orifice plate meter installed in conformity with B.S. 1042. This restriction would not apply to the monitoring of small, rapid flow fluctuations, since in this case the velocity profile could be considered invariant so that the fractional change in flow rate and in centre-line velocity would be equal.

These comments are also relevant to gas flows. In many circumstances, gas flows operate with a lower concentration of particles and at higher velocities. In such circumstances the replacement of the phase-locked tracker by a frequency-locked tracker with consequent total cost of approximately £1400 is likely to extend the range of operation, particularly if a form of drop out control is incorporated. For very low particle concentrations, such as that found in natural air, optical anemometers are unlikely to be used unless changes in mean velocity are unimportant within a time scale of approximately 3 minutes. In such cases, a counting system can be used but at considerable cost (approximately £4700). The precision obtainable using counting methods is undoubtedly very high provided the 3 minute time average is acceptable.

Particularly .at low velocities, it is possible to scan a duct flow and to record a velocity profile across the duct: the integral of such a profile provides the volumetric flow rate. A suitable arrangement was suggested in reference 14 and provides scan times of the order of 15 ms. The same principle can be applied to the flow configurations mentioned above but in cases where a counting system is required the scan time will, of course, be considerably longer.

Advantage can also be taken of the control which can be exerted over the size of the measuring control volume. The optical unit and light collecting system control these dimensions and can be adjusted to aid signal continuity and also to provide an average mean-velocity over lengths ranging from approximately 20 μm to 10 mm.

5   CONCLUSIONS

(a)   Optical anemometers are more expensive than conventional volumetric-flow measuring devices but can offer significantly greater precision and do not interfere with the flow or cause a pressure drop.

(b)   Optical anemometers, utilising integrated optical units, can be permanently aligned for use in industrial environments. They are also suited to use in fluid-and bio- mechanics laboratories and can provide accurate mean velocity measurement in, at least, the velocity range 1 mm/s to 100 m/s.

(c)   Continuity of signal is required for an effectively instantaneous (10 kHz frequency response) on-line signal and this requires the presence of a scattering particle in the measuring control volume at all times. The size of the measuring control volume is adjustable but the required particle concentration for instantaneous on-line operation is likely to be in excess of 1 mm^{-3}.

(d)   Where a continuous signal is present signal analysis is best achieved using the form of frequency to voltage conversion provided by tracking demodulators. The use of such devices, in conjunction with a fringe-mode optical unit, can permit mean velocity to be measured with a precision of the order of 0.1%. This can be transformed into a volumetric flow measurement with similar precision in different ways, e.g. measuring in a fully-developed flow region or introducing a contraction to provide a uniform velocity profile.

6   ACKNOWLEDGEMENT

The authors gratefully acknowledge financial support from the Science Research Council, Stiftung Volkswagenwerk and the Central Electricity Generating Board. The A.E.R.E., Harwell, kindly lent the prototype DISA frequency tracker and the Hewlett Packard Company the Computing Counter.

7   SYMBOLS

C	Velocity of light
n_i	Vector perpendicular to fringe pattern, $\lvert n_i \rvert = 2 \sin \varphi$
$p(\triangle v)$	Probability density distribution of $\triangle v$
U	Mean velocity component in direction of n_i
$\hat{U}_i$	Instantaneous velocity vector
v_s	Frequency of light source
$\hat{\triangle v}$	Instantaneous signal frequency
$\overline{\triangle v}$	Mean frequency of signal
λ_s	Wavelength of light source
φ	Half angle between the beams

8   REFERENCES

1   GOLDSTEIN, R.J. and KREID, D.K.: 'Fluid velocity measurements from Doppler shift of scattered laser radiation', University of Minnesota, Mech. Eng. Report HTL-TR-85

2   DURST, F. and WHITELAW, J.H.: 'Optimisation of optical anemometers'. To be published in Proc. Roy. Soc. 1971. Also available as Imperial College, Dept. of Mech. Eng. Report ET/TN/A/1

3 RUDD, M.J.: 'A new theoretical model for the laser Doppler meter', J. Sci. Instruments, 1969, **2**, p. 55

4 DRAIN, L.E.: 'Coherent and non-coherent methods in Doppler optical beat frequency measurement', Harwell report (to be published)

5 GOLDSTEIN, R.J. and HAGEN, W.F.: 'Turbulent flow measurements utilizing Doppler shift of scattered laser radiation', Phys. Fluids, 1967, **10**, p. 1349

6 PIKE, E.R., JACKSON, D.J., BOURKE, P.J. and PAGE, D.I.: 'Measurement of turbulent velocities from the Doppler shift in scattered laser light', J. Scientific Instr., 1968, **10**, 111

7 LEHMANN, B.: Projekt MHD — Stanstrahlrohr, AEG — Forschungs institut, 1968, Bericht 19

8 GREATED, C.: 'Effect of polymer additive on grid turbulence', Nature, 1969, **224**, pp. 1196-1197

9 DURST, F. and WHITELAW, J.H.: 'Measurements of mean velocity, fluctuating velocity and shear stress using a single channel anemometer' (to be published in DISA Information)

10 DURST, F., MELLING, A. and WHITELAW, J.H.: 'The application of optical anemometry to measurement in combustion systems', Imperial College, Mech. Eng. Report ET/TN/A/8, 1971

11 THOMPSON, D.H.: 'A tracer-particle fluid velocity meter incorporating a laser', J. Sci. Instruments, 1968, Ser. 2, **1**, p. 929

12 DURST, F. and WHITELAW, J.H.: 'Integrated optical units for laser anemometry', Imperial College, Mech. Eng. Dept. Report ET/TN/A/6, 1971

13 MELLING, A.: M.Sc. Thesis, University of London, 1970

14 BENDICK, P.J.: 'A laser Doppler velocimeter to measure "instantaneous" velocity profiles', ASME Paper No. 2-8-235, 1971

DISCUSSION

P.S. Bedi: Mr. Melling showed on a slide three different modes of operation of the optical arrangement shown in Fig. 1 and has emphasised that some modes of operation were found to be more suitable than the others under certain conditions.

(a)   Could he explain which mode of operation was better than the others and under what conditions? Could he also quantify the limiting conditions (as to particle concentration or size of particles, etc.) for the optimum operation in that mode?

(b)   Did he observe this only experimentally or has he some theoretical justification to support his observations?

Authors' reply:

(a)   The three modes of operation of the optical anemometer, described in ref. 12, are the fringe, Doppler, and reference beam modes. We have found the fringe mode illustrated in Fig. 1 to be most useful for our work so far, as it provides a good signal to noise ratio under the conditions of low scattering particle concentration experienced in flows of air or unseeded water. The Doppler mode, in which light scattered from a strong beam is collected in two directions, has no advantage over the fringe system unless simultaneous measurement of velocity components in two planes is required; however, alignment of the optics in the Doppler mode ensures it optimum performance in the fringe mode. In contrast, as pointed out in the paper (section 2.1), the reference beam mode (in which light scattered from a strong beam is combined with an undeviated reference beam) performs better with higher particle concentrations. This system requires careful alignment of the light-collecting section and selection of an aperture diameter in front of the photodetector such that scattered light is collected from only those particles which scatter waves in phase with the reference beam; under this condition the signal to noise ratio will increase with particle concentration. At low concentrations the signal to noise ratio from each wave packet is better for fringe or Doppler systems than for a reference beam system; it falls steadily, however, with increasing concentration since the waves from the particles are in random phase relationship and thus interfere destructively.

For optimum operation in the fringe mode, the particle concentration is such as to provide about one particle in the measuring control volume at all times. For our work we have used volumes of about 0.1 mm diameter and 0.5 mm length, requiring particle concentrations of about 10^{12} m^{-3}. The size of particles is likely to be limited by their esponse to the flow, e.g. to a maximum diameter of 1 μm when following turbulent fluctuations in gas flows, rather than from optical considerations, unless the fringe width is sufficiently small to be covered by a particle.

(b)   Theoretical justification for use of the fringe mode in flows with low particle concentration has been provided by Drain.[4] He considers contributions to the signal from "coherent" and "non-coherent" scattering and shows that the latter, obtained with fringe mode optics, permits large apertures to be used and, hence, strong signals to be obtained.

D.L. Smith: Would the optical laser technique described be adaptable to measuring the volume flow rate of gas in a steel pipe, 20 to 24 inches in diameter, pressurised to 1000 psi, at ambient temperature, assuming that translucent windows were sealed into the pipe walls? If this were feasible, the resulting system could perhaps be developed as an on-line meter prover for gas flows.

Authors' reply: Velocity measurements with an optical anemometer are independent of the temperature and pressure of the fluid, and so local velocities could be monitored under the conditions mentioned by Mr. Smith. If single lenses are used, then in view of the large diameter, long focal lengths would be required; this implies that lens diameters of 200 mm or so would be desirable to permit a large enough angle between the incident beams of a fringe system for good fringe definition, and to collect sufficient scattered light. Transparent rather than translucent windows would be needed, and a powerful laser (150 mW or more) would have to be used unless the gas flow were contaminated with particles. Unless the system were set up to traverse the pipe cross-section, the volume flow rate would be determined, as described in the paper, by a velocity measurement on the axis, an assumed velocity profile, and the density of the gas.

S.B. Au: The authors claim that their method permits the mean velocity to be approximately measured with a precision of 0.1% On what basis do the authors claim such precision?

Authors' reply: The estimate of precision of 0.1%, stated in conclusion (d), applies only when a frequency tracking demodulator is used; the actual rms repeatability obtained with 50 measurements was 0.19% in air and 0.13% in water as stated in section 3.

S.B. Au: The purpose of the paper is to provide an assessment of the applicability of optical anemometry to the measurement of volumetric flow rate. In doing so have the authors been able to compare their results with those generally obtained from accepted methods such as pitot tube traverses. How accurate is this method of measuring flow rate?

Authors' reply: We did not compare our measurements with pitot tube traverses or bulk flow measurements given by e.g. an orifice meter or a weigh tank, since we were directing our attention more to the precision of the measurements than to their accuracy. An estimate of 4% accuracy in flow rate from a single velocity measurement on the pipe axis was made in section 4 of the paper; but if the velocity profile were measured across one diameter an accuracy of 1% should be achievable.

K.A. Blake: Mr. Melling explained his relatively poor repeatability with the computing counter (0.7%) by

(i)   insufficient experience using the counter to achieve optimum triggering

(ii)   long test time − 3 mins. − allowing variations in flow.

I would say that if a counter technique is to be used, it must be accepted that there will be a percentage of 'bad' counts amongst the 'good' ones, as mentioned in my paper. A straight mean is therefore not sufficient for good results; the results must be processed to reject the bad counts.

On the matter of a cost estimate of £220, the small backscattering system I showed would cost about £50 including photodiode and the basic triangular optical bench set-up would provide versatility for under flow.

Authors' reply: Mr. Blake outlined in his paper a method for rejecting "bad" counts by excluding from the averaging process all measurements which deviated by more than an arbitrary fraction from the mean of a trial set of counts. In view of the limited computing capability of the Hewlett-Packard Computing Counter we could not program this rejection method and simultaneously evaluate mean and rms values of frequency. Instead we would hope to avoid bad counts by shaping the wave from the photodetector to provide very reliable triggering, and determine the Doppler frequency from a single period of the wave, or from timing a fixed number of counts.

We are very impressed by the remarkably low costs for optical equipment mentioned by Mr. Blake, but doubt whether £50 is a realistic commercial valuation of the cost of the integrated back-scattering optics.

F.C. Kinghorn: The authors quite rightly point out that in many cases what industry wants is repeatability, not accuracy, and they suggest measuring the local velocity at a single point (e.g. on the pipe centre line) and using this value as a measure of the flowrate. The success of this would of course depend on the velocity distribution within the pipe remaining unchanged.

I feel that this application of lasers would be using a sledgehammer to crack a nut, as a similar result could be obtained by, for example, measuring the change in static pressure over a long length of pipe. The great advantage of lasers in flow measurement is that they *can* make very accurate measurements, and this advantage should not be discarded.

A.T.J. Hayward: The discussions on the laser flowmeters appears to have given the impression that this is an interesting research tool but that it has little potential in the field of industrial instrumentation; I suggest that this is not so. There is a need in industry for a fully automatic rapid, method of flow measurement by velocity traversing, to replace pitot traversing which is slow and cumbersome. The laser could, if suitably developed, fulfil this need.

Authors' reply to F.C. Kinghorn and A.T.J. Hayward: We welcome the view of Mr. Kinghorn and Mr. Hayward that optical anemometry has a place in industrial instrumentation for rapid, accurate flow rate measurement by velocity traversing. While agreeing with most of what Mr. Kinghorn says, we would suggest that in certain cases a row of static pressure tappings would not match the performance of an optical anemometer, e.g. in responding rapidly to a sudden change in flow velocity caused by emergency closing of a valve.

2·5

FLOW METERING USING A MICROWAVE DOPPLERMETER

J. Harris

Abstract: Experiments have been performed on a flow meter designed around a microwave homodyne Dopplermeter. A description is given of the performance of this meter under industrial conditions. The work shows the feasibility of the application to several difficult areas of flow measurement. The effect of stratification is also treated.

1   INTRODUCTION

The early work on the applications of radar (radio detection and ranging) made use of magnetrons and klystrons as sources of electromagnetic radiation. These sources are relatively large, heavy and required high voltage electrical supplies. They were therefore not suited to industrial applications. However, with the advent of the Gunn diode, an inexpensive miniature source of electromagnetic radiation with useful power levels has become available, and still later devices (impatts) give increased power outputs. The Gunn diodes used to-date for flow measurement have been quite adequate with respect to power level and therefore they will continue to be used in the forseeable future.

A viable flowmeter[3] designed around a microwave homodyne Dopplermeter[1] has been described in a prior publication[4]. This meter is of use in measuring the flow of solid-fluid and liquid-fluid mixtures as well as for monitoring belt-conveyed material.

There are several different uses[2] for the instrument which for convenience are termed the Radascan series, a) Flow failure detection based upon either material density (i.e. the number of particles per unit volume) or speed. b) Flow metering for the speed of flow and c) Mass flow rate measurement[3] [4].

The instrument will not respond to liquids or gases alone, there must be discontinuities present which will reflect some of the radiation incident upon the stream. Water is also very opaque to the radiation frequency used of 13.4 GH_z which is in the J-band. However the electromagnetic flow meter performs very efficiently with aqueous materials so in this case a convenient meter is already available.

2   DESCRIPTION OF THE SYSTEM

The microwave homodyne system, which has been described in some detail previously,[4] transmits and receives through a "Y" junction. It is shown diagrammatically in Fig. 1. The purpose of this "Y" junction is to try to suppress noise in the signal arising from local fluctuations in velocity which may occur in turbulent flow. If there is no correlation between the fluctuations in the signal in both arms of the "Y" junction then these signals are suppressed whilst the main stream velocity, which is correlated in both arms, is not affected by the junction. Hence, there should be an enhancement in the signal quality. The well-known microwave "Hybrid-Tee" or "Magic-Tee" junction might also be adapted, but at the expense of some additional complication.

If the target has motion relative to the homodyne unit along its axis then a "beat" frequency is established proportional to the target velocity.

$$f \propto v \cos \theta \qquad \qquad \dots (1)$$

The frequency and amplitude of the beat signal (which is usually in the low audio frequency range) may be detected and the response to a change in either of these quantities is very rapid: this makes the instrument well suited to process control applications. The amplitude of the signal would typically be in the 50 mv p/p to 2.0v p/p range after two amplification stages.

3 APPLICATIONS

Two applications are treated here, namely:—

a)   Monitoring free flow of sand from a hopper
b)   Pneumatic conveying of alumina through a steel pipe.

In both these cases the diameter of particle/wavelength ratio was quite small [about 10^{-2} for sand and 2×10^{-3} for alumina] and both well inside the region of Rayleigh scattering. The ratio of

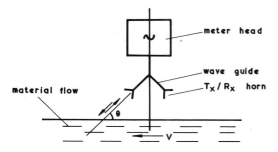

Fig. 1   Principle of operation of Radascan flow meter

scattering cross sectional area of individual particles might be expected to be of the order of 10^{-5} for sand and much less for alumina. It is remarkable therefore that in both these cases adequate signals were received, especially as the raw materials themselves are not very good reflectors in the bulk as compared with metals.

3.1 Sand flow from a hopper

The metering of powder flow from a hopper has hitherto required rather elaborate equipment, such as conveyor belt weighing. Current techniques have been reviewed by Reisner and Rothe[5]. Here a non-contact technique is described.

The basic arrangement was as shown in Fig. 2. A large steel hopper was filled with dry free flowing sand and the flow meter head mounted on a 0.14 in. bore (5.5 in.) perspex tube which had an outer coating of aluminium foil to enhance the signal strength.

The hopper was operated with four different orifices under full bore flow with the results shown in Table 1 and Fig. 3.

TABLE 1

Orifice Dia. (metres)	Estimated Flow Rate (Kg/hr.)	Voltage Output
0.0127	191	0.015
0.0254	1142	0.055
0.0381	3600	0.115
0.0508	6200	0.235

The voltage reading obtained in column 3 of Table 1 refers to the *amplitude* of the best signal not to its frequency. The flow rate was estimated from a catch and weigh system. It appears from this that the changes in mass flow rate occurred mainly as a result of increases in the fraction of the cross-section of the tube occupied by the material and very little from velocity changes, since this was not sensed. A very simple form of mass flow meter can therefore be applied to this situation.

3.2 Pneumatic conveying of alumina

The same pneumatic conveying plant was used for the present work as for the previously reported results on plastic flake.[4] However, there is a difference in the application of the meter in that previously a single transmitter/receiver horn (without the "Y" junction) was used as the radiation was propagated into the open (suction) end of the pipe and downstream to the powder flow zone.

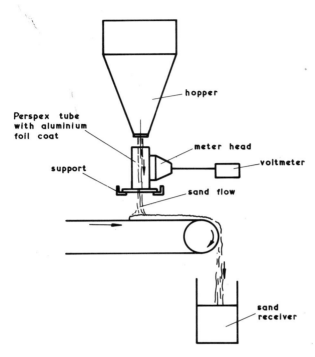

Fig. 2   Calibration of the Radascan for sand flow from a hopper

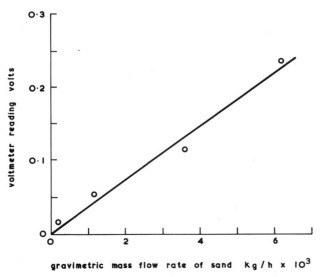

Fig. 3   Calibration curve for sand flow from a hopper (free fall)

Here the alternative side wall mounting in the powder flow zone was used together with the "Y" junction. Again, the amplitude of the beat signal was measured and not its frequency as in the sand flow from a hopper (section 3.1).

Due to instabilities in the flow mode of the powder, steady conditions were difficult to achieve and not many results could be obtained. These are given in Table 2 and Fig. 4.

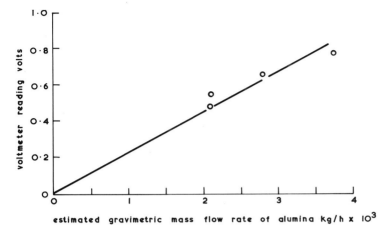

Fig. 4   Calibration curve for pneumatic conveying of alumina

TABLE 2

Estimated Flow Rate (Kg/hr.)	Voltage Output
2130	0.625
2130	0.690
2840	0.800
3810	0.915

Again it appears that changes in mass flow rate are mainly due to mass increases in the fraction of pipe cross-sectional area occupied by the material and not to velocity increases. The size distribution of the alumina is shown in Fig. 5.

4   OTHER EXPERIMENTAL RESULTS

Several other experiments have been conducted which have been directed towards supporting information on the behaviour of the flow meter and extending the field of knowledge of its properties. Some of these experiments are recorded briefly below.

4.1 Radiation pattern within a tube

No information was available on the distribution of radiation intensity within a duct to which the meter is attached. Therefore the same perspex tube was used as for the sand flow experiments in Fig. 2 and a small stirrer paddle attached to an electric stirrer motor was used as a probe target. This target was traversed along the pipe axis from a point outside the tube to approximately the mid-point of the meter head.

Results of the probe test are shown in Fig. 6.

In free space, an inverse square law should be obeyed resulting in the received power of the reflected signal being inversely proportional to the 4th power of the distance between the transmitter/receiver and the target. When examining Fig. 6, allowance has to be made for the angle of incidence, and then the pattern is roughly of the shape to be expected, with some cavity effects intruding in the region of the meter centre line.

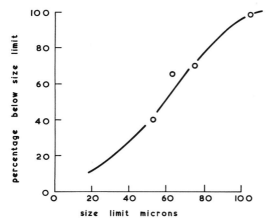

Fig. 5   Alumina size distribution

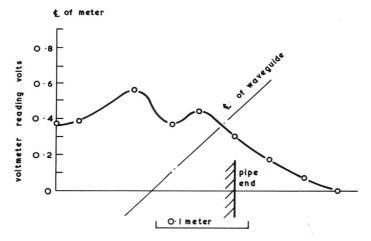

Fig. 6   The radiation pattern in a 0.14 m diameter duct

4.2 Stratification effects

It is possible in some circumstances that the flow in the field of view of the meter may be of an inhomogeneous nature and experiments have been conducted to determine whether any effect on the meter reading could exist due to stratification.

The meter head with one waveguide/aerial was mounted on a travelling microscope and placed to look vertically down on a shallow plywood tray. Under the tray a target paddle driven by a variac controlled electric motor was located to excite the meter. 3 mm diameter glass beads were loaded into the tray to give varying bed depth.

In the first place beads were loaded into the tray with the target and measuring head fixed in space. Secondly, beads were loaded into the tray and the measuring head moved vertically upwards to maintain a constant height between the top surface of the bed and the measuring head.

The results are shown in Fig. 7. It may be seen that movement of the bed relative to the head can have an effect and that varying the bed thickness does not give simple attenuation.

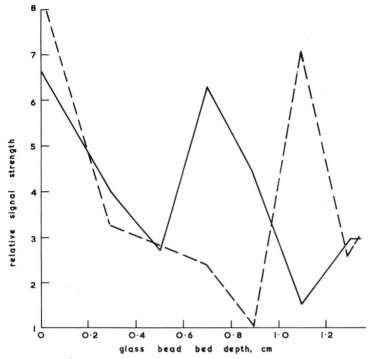

Fig. 7   The effect of bed depth on transmitted signal for propagation through 3 m diameter glass beads
———— bed surface to meter head constant
– – – – target to meter head constant

Keeping the target and tray containing a single layer of 8 mm glass beads fixed and withdrawing the head from the target gives results as shown in Fig. 8. Superimposed on the inverse square law effect are interference effects.

4.3 Linear bundles

Linear packages of 5, 8 and 12 mm glass beads were made up and allowed to slide through an inclined perspex tube past the measuring head. A linear bundle of up to four beads did not appear to make very much difference in the size of the output signal.

5   CONCLUSION

From the experimental work on pilot-scale plant it appears that variations in mass flow rate may be sensed by simply measuring the voltage output of the measuring head. Thus simplification of the instrument can be achieved over that described previously, in which a multiplication was involved to obtain mass flow rate.

On the basis of a highly idealised experiment involving small linear bundles of beads which were allowed to flow past the measuring head, no basis for the above result could be found.

Stratification of the flow could make calibration of the instrument difficult or impossible in the case of the Radascan just as in the case of many other types of flow meter. Further investigation is required on this point as on several of the other factors, but the applications indicated in this and prior reports can proceed with some confidence.

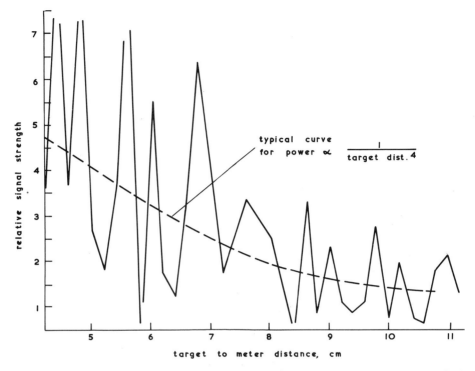

Fig. 8   The effect of varying target to meter distance for propagation through a constant bed depth of glass beads equal to 8 mm

6   ACKNOWLEDGEMENTS

The author wishes to acknowledge the facilities made available by Simon Handling Engineers Ltd., Stockport, U.K., and to Mr. C. Dickson for assistance in the pneumatic conveying experiments. Also to Mr. A.R. Cliff for assistance in carrying out some of the experiments.

7   REFERENCES

1   DRAYSEY, D.W.: Patent Spec. App. No. 14839/67

2   HARRIS, J.: 'Interrogating Flow Fields with Radar and Laser Sources', Paper presented to Symp. on 'The Measurement of Pulsating Flow'. Univ. Surrey, April 1970, Inst. Meas. and Control Meeting

3   HARRIS, J.: Patent Spec. App. No. 14757/70

4   HARRIS, J.: 'Flow Measurement Using Microwave Radar Techniques', Paper presented to Symp. on Flow Measurement of Difficult Fluids Univ. Bradford, April 1971, Joint I. Chem. E./Inst. Meas. and Control Meeting

5   REISNER, W., and ROTHE (M.V. Eisenhart): 'Bins and Bunkers for Handling Bulk Materials', Vol. 1 (1971), No. 1, Trans Tech. Publications, Cleveland, Ohio, U.S.A.

DISCUSSION

R. Theenhaus: Is the instrument already commercially available?

Author's reply: Units can be made available on a commercial basis on specification of the type of application and readout required. Costs can be supplied upon request and advice given regarding interfacing with given systems.

3·1

A THERMOELECTRIC FLOWMETER OF RAPID RESPONSE AND SMALL SENSOR SIZE

A.G. Smith and W.A.K. Said

Abstract: A thermoelectric flowmeter with a sensing element of size 2 x 2 x 1mm has been developed for monitoring petrol flows in the range 0.3 to 10 gallons/hr. Its 95% response time for a step change in flow is less than 0.2 s, and it is capable of an absolute error of less than 3% of its reading. The normal operation is in a fluid stream with velocity of order 50 mm/s. The sensing element is maintained at a constant temperature difference between its hot and cold ends, and the current necessary to do this is used as the fluid flowrate signal. Application of the flowmeter to the main jet of the carburetter of a car engine is described. The instrument must be calibrated in situ.

1 INTRODUCTION

The flowmeter described in this paper was developed to permit investigation of transient flows in the passages of carburetters. A performance specification of

- time of 95% response to step change of flow, 0.2 s
- absolute accuracy, 3% of reading
- fluid temperature range ±10°C from nominal was early decided as useful and feasible.

From study of papers, Grant and Kronauer[1], Rasmussen[2] and Bellhouse[3], a thermal flow-meter seemed a better prospect than a turbine-meter (difficulty of installation) or an orifice-meter (difficulty with pressure-drop, response-time and indication). From the available thermal field of hot-wire, hot-film and thermoelectric devices, the last of these seemed best for low boiling-point liquids, in view of its low operating temperature and high indicating signal.

The basic idea of a thermoelectric flowmeter is that a current is passed through a thermocouple, the Peltier effect heating one junction and cooling the other. If both junctions are in a fluid stream, the temperature difference between the junctions will be lower, the higher the stream velocity. The thermoelectric e.m.f., if it can be disentangled from the e.m.f. due to the sensor resistance, may then be used as an indication of stream velocity.

Obviously, operation may be in a constant current mode, a constant thermoelectric e.m.f. mode, or in any mode between.

Tests quickly showed (Fig. 1) that a p- and n-type bismuth telluride couple would produce a usable temperature difference. The mode of operation finally decided upon was: constant temperature difference across the thermocouple, direct current interrupted at 1 kHz, current amplitude controlled by the thermoelectric e.m.f. sensed in the interrupt period.

2 SENSOR AND CONTROL UNIT

2.1 Sensor size, sensitivity and response time

Size was governed firstly by ability to work the thermocouple material, and here typical dimensions 2 mm x 1 mm x 1 mm were found to be the limit. In the second place, however, the lower limit to size is provided by the balance between Peltier heats, Joule heats, internal heat flow within the sensor, and heat transfer to the fluid. Particularly, length of the probe in the flow direction is important; too low a length reduces temperature difference by internal heat conduction.

Fortunately, semiconductor materials tend to have low thermal conductivities. Whilst a low length will tend to reduce sensitivity, it will have the desirable effect of reducing sensor thermal capacity and thus reducing response time. The size finally adopted is shown in Fig. 2.

Mode of operation affects response time. The constant temperature-difference mode obviously conduces to quick response because thermal capacity effects are smaller than with a variable-temperature-difference mode.

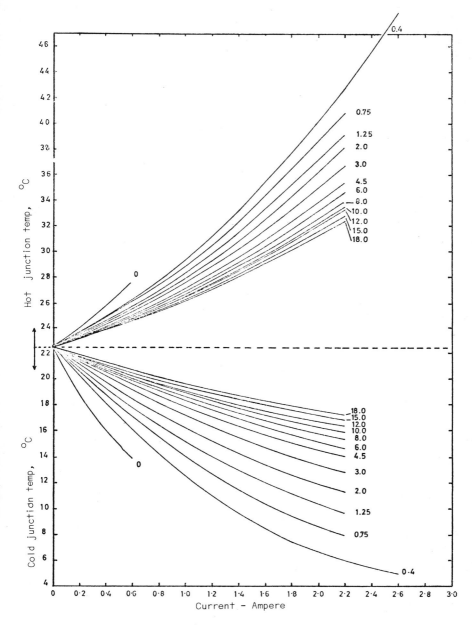

Fig. 1   First tests of sensor sensitivity
Sensor PP3 in Aeroshell 650 kerosene
Fluid temperature = 22.5° C
Cold leading edge
Sensor size = 4 x 3 x 1.5 mm
Numbers on graphs are flow rate (ml/s)
Flow velocity (mm/s) = 20 x flow rate

2.2 Control unit

The block diagram of Fig. 3 gives the system. The error between the memorised Seebeck e.m.f. (sampled in the interrupt period) and a reference voltage controls the D.C. amplitude. Temperature sensitivity of the Seebeck e.m.f. is cancelled by a compensator in the following way; the resistance of the sensor (which is temperature-dependent) is indicated by a residual current in the nominal "off" period, and this indication is used as a compensator of the temperature effect on Seebeck voltage. Output is linearised by a diode function-generator.

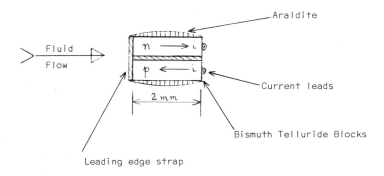

Fig. 2   Diagram of sensor, approximately to scale

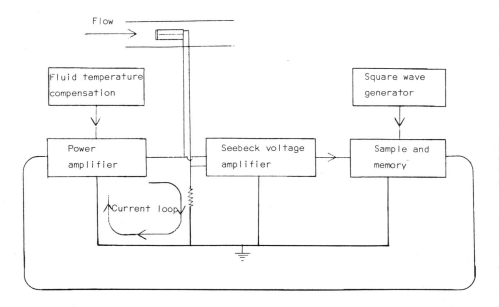

Fig. 3   Block diagram of the final system

3   DEVELOPMENT AND RESULTS OF TESTS

3.1 Manufacture of the sensor

The bismuth telluride p- and n-type blocks were 1.5 x 1.5 x 4 mm as purchased. Two such blocks were filed to required size, strapped at one end with a silver strip soldered to the block ends, and soldered to copper current leads at the other ends. The blocks were bonded together with araldite.

Sensors used for development only had wire thermocouples soldered to each end, tc give temperature information. Fig. 2 gives an impression of the device. Eleven sensors of various sizes were made. For obvious reasons much of the development was carried out with kerosine as the fluid.

3.2 Sensitivity

Fig. 4 shows some typical measurements of hot-cold end temperature difference with current (0 - 2.6 ampere) and flow velocity (0 - 36 mm/s) varied. As expected, temperature difference is directly proportional to current, but is very non-linear in its dependence on flowrate. The e.m.f.

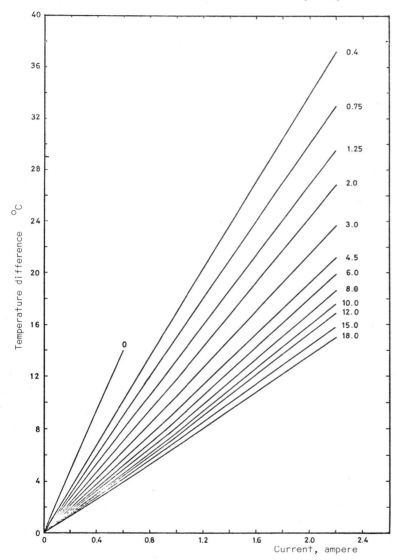

Fig. 4   Temperature difference versus current for various flow rates
         Sensor PP3 in Aeroshell 650 kerosene
         Fluid temperature = 22.5°C
         Cold leading edge
         Numbers on graphs are flow rate (ml/s)
         Flow velocity (mm/s) = 20 x flow rate

sensitivity of the sensor is 0.433 mv/°C temperature difference, so 10°C difference gives a convenient output. Current sensitivity in the constant temperature-difference mode of operation is shown in Fig. 5. The need for a linearising device is very clear.

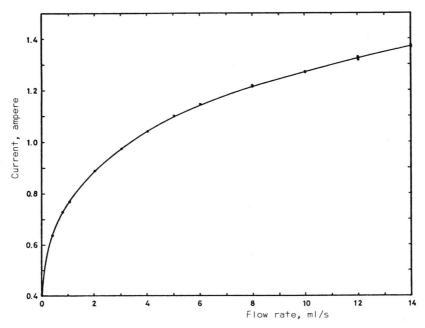

Fig. 5   Current response to flow rate for a typical sensor
Sensor PP3 in Aeroshell 650
Fluid temperature 22.5°C
Cold leading edge
Flow velocity (mm/s) = 20 x flow rate

3.3 Repeatability of sensor performance

Two calibrations of a particular sensor at an interval of 2 months, with 400 hours of testing in between, are shown in Fig. 6. There is little evidence of drift of performance.

3.4 The electronics

Some circuit details are shown in Fig. 7.

3.5 Performance of the complete system

Fig. 8 shows system output with the sensor in a calibrating rig. Output is satisfactorily linear, and dependence on fluid temperature is small, since varying the kerosine temperature from 18 to 30°C produced only about 5% change in reading. The electronic part of the system drifts a little, and a warm-up period of 30 min is advisable if the best accuracy is sought.

3.6 Measurements in a carburetter

A standard Ford carburetter was modified as shown in Fig. 9 to take a sensor before the main jet. The system behaved reliably under engine conditions. A typical fuel-flow record from these tests is shown in the same figure.

4   ACKNOWLEDGEMENT AND COMMENT

The authors are grateful to the Ford (Dagenham) Trust, who started and funded this work, for permission to publish. Other aspects of the work, including methods of manufacture, calibrations, and full details of the electronics may be found in Said[4].

5   REFERENCES

1   GRANT, H.P., and KRONAUER, R.E. 'Fundamentals of Hot-Wire Anemometry', Symposium on Measurement in Unsteady Flow, ASME, 1962, p.44

2   RASMUSSEN, C.G.: 'The Air Bubble Problem in Water Flow Hot-Film Anemometry', DISA Information No. 5, June 1967, p.21

3   BELLHOUSE, B.J., BELLHOUSE, F.H., and GUNNING, A.: 'A Straight Needle Probe for the Measurement of Blood Velocity', Journal of Scientific Instruments, **2**, 1969, p.936

4   SAID, W.A.K.: Ph.D. Thesis, University of Nottingham, 1970

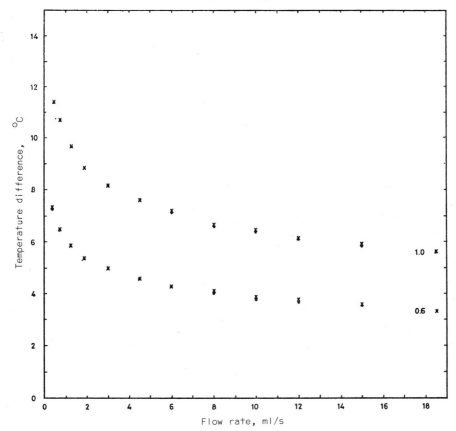

Fig.   6   Drift of sensor performance
Sensor PP2 in Aeroshell 650
Fluid temperature = 22.5°C
Hot leading edge
Numbers on graph are current (Amps)
· 4/10/68
x 7/12/68

DISCUSSION

N. Mustafa: How was the 95% response time for a step change in the flow concluded to be 0.2 sec or less? Was it derived experimentally?

Authors' reply: The 95% response time was measured in three ways.

(a)   The response time of the sensor to a step change in current was measured. A theoretical investigation showed that the time of response to fluid flow would not be significantly different. The sensor was found, under the conditions of step change of electric current, to have a response time (Seebeck voltage) of about 0.5 sec at a flow rate of 14 ml/sec in the test channel and about 1.6 sec at 0.4 ml/sec. Calculations from this data revealed that the 95% response time of the complete system at both flow rates would be rather less than 0.2 sec.

(b)   Some tests of the complete system were made by simply opening a tap between the fluid reservoir and the test channel as suddenly as possible. The response time of the complete system was found, for a nominal flow rate of 10.5 ml/sec to be less than 0.2 sec. Since there would be a finite acceleration time for the fluid in the channel this demonstrated that the response time of the overall system was less than 0.2 sec.

(c)   An apparatus was made which could give a sinusoidal flow rate in the test channel. The response of the overall system was measured and compared with the theoretical values for frequencies up to 6Hz and deductions from these tests revealed a 95% response time of less than 0.2 sec. Very full details are given in ref. 4 of the paper.

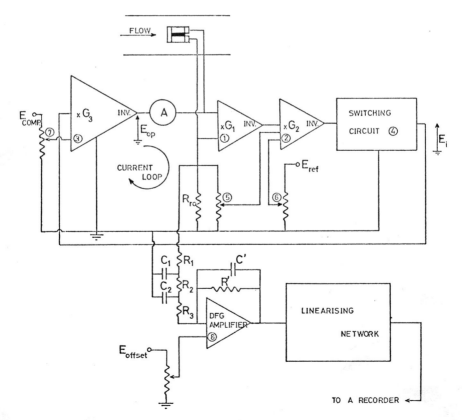

Fig. 7   Simplified circuit diagram of the final system

C.G. Clayton: Figure 8 in your paper shows a curve of voltage against flow rate in which the quoted error is ± 3%. Could you please comment on the origin of this error. In particular, could

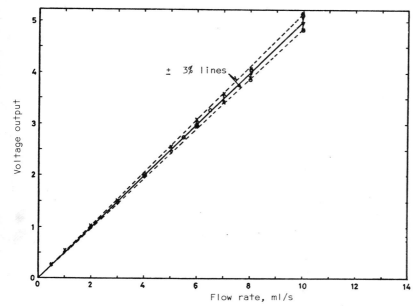

Fig. 8   Effect of the fluid temperature on the linearised output of the system
Sensor PP3 in Aeroshell 650
Fluid temperature
$\nabla = 18°C$, $\theta = 23°C$, x = 30°C
Cold leading edge
Flow velocity (mm/s) = 20 x flow rate (ml/s)

you say what part of the error was associated with the sensor and what part was associated with the electronic unit. To what extent does the variation in flow velocity caused by the blocking effect of the sensor in the stream reflect on the accuracy of the method?

Authors' reply:
(a)   The error was due to drift in the amplifiers. Actually the situation is quite complicated; once the system is calibrated with the sensor in a given tube or channel, the readjustment of a single potentiometer would compensate, over a period of weeks, for drift in the amplifiers. As far as could be detected, there was no significant change in performance of the sensor. The whole system, however, required skilled setting-up and use.

(b)   The system is substandard; it must be calibrated with the sensor in the position in which it is to be used, and with the fluid which is to be used. In a carburetter such calibration demands that the flow through the particular channel under investigation may be measured by non-standard external connections for calibration. This disadvantage is common to all devices which may be placed in awkward passages.

A.M. Crossley: Have the authors considered using this device in conjunction with the road speed of a motor car to produce a true "miles per gallon" meter? This is possibly of more interest to the average motorist than an accurate fuel flow meter.

Comment by other delegate: I understand that the meter to which my colleague refers measures petrol flow to the float chamber rather than actual petrol flow in the carburetter itself.

Authors' reply: The device would be too expensive for use in "miles per gallon" meter. Cheaper devices should suffice.

A.T.J. Hayward: May I supplement the answer to that question by saying that a miles-per-gallon device incorporating a fuel flowmeter, a take-off from the speedometer cable, and a computing circuit connected to a dial on the dashboard, is about to be marketed by a British firm at a price in the region of £8.

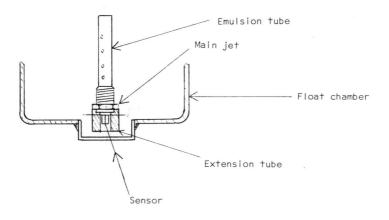

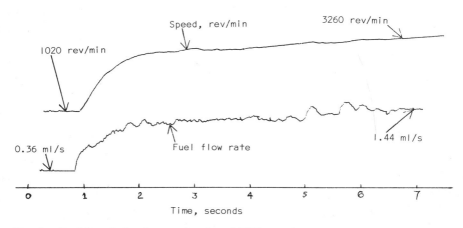

Fig. 9   Fuel flow during slam acceleration of 1600 cc engine

THE DEVELOPMENT OF THERMAL FLOWMETERS FOR THE MEASUREMENT OF SMALL FLOWS OF CORROSIVE AND RADIOACTIVE LIQUORS IN CHEMICAL PLANTS

A.L. Mills, C.R.A. Evans and R.A. Chapman

Abstract: The necessity to measure accurately low flows of nitric acid, organic solvents or radio-active liquids requires the use of very reliable flowmeters. The general requirement for such an instrument is that it should require no routine maintenance and should be corrosion free. There should be no restrictive orifice or moving parts, but an electrical output should be available for control or alarm. In addition the unit should be fail safe; particularly, liquor flow must not be obstructed under failure conditions. A series of thermal flowmeters which fulfil these requirements has been developed by the United Kingdom Atomic Energy Authority, at Dounreay. These have been installed in the Fast Reactor Reprocessing Plant where liquor flows, including 16 N nitric acid have been measured over a range 0.05-20 litres/hr with an accuracy of ± 2%. This paper describes the principle of operation, construction, application and operating characteristics of the Thermal Flowmeter.

1   INTRODUCTION

The main nuclear fuel reprocessing plant at Dounreay is geometrically eversafe; that is to say the dimensions and layout of the plant are such that it is not possible to obtain a concentration or configuration of fissile material that will give rise to a nuclear excursion. Such a plant is of necessity physically small, consequently the volumetric input of chemical reagents is small. This in turn makes the plant sensitive to changes in input volumes, hence a high degree of accuracy in volumetric flow metering is required.

In this particular plant some thirty different reagent input parts are required, each with a different flow rate in the range 0.5 to 40 litres/hr and handling a wide range of chemicals (0.05 - 22.0 N nitric acid; 0.1 - 0.5M sulphuric acid; 5% caustic soda; 5% sodium carbonate; 1.5M ferric nitrate in nitric acid; 1.5M ferrous sulphamate; 6M sodium nitrite; 3 vol % - 25 vol % tri-n-butyl phosphate in odourless kerosene). The metering device is required to be accurate, versatile, corrosion free, and cheap. It should not impede liquid flow in the event of a fault in the metering device and it should require no routine maintenance. From an operational point of view the metering device must provide an electrical signal that can be used to control the existing metering pump. The existing feed pumps are variable displacement pumps which provide a pulsating flow, changes in flow are made by adjusting the stroke of the delivery piston.

An examination of the then available methods of flow measurement and control; differential pressure devices, displacements of a float, inferential flowmeters, turbine meters, orifice plates and so on, all present problems when applied to these particular plant requirements. Other systems available as completely engineered flow control systems were rejected on the grounds of cost.

One was faced, therefore, with the development of a flowmeter that would fulfil the above requirements, and at a future time the unit should be capable of being scaled down to deal with flows of from 100 cc - 2 1/hr.

2   THE BASIC INSTRUMENT

If one considers a tube through which liquid can flow, and a heater is placed upstream of the flow, then the temperature rise across the heated section is a function of the flowrate if both the heater energy input and the specific heat of the liquid are constant (Fig. 1): T_1 and T_2 are temperature sensing devices placed in the liquid. With zero liquid flow the outputs of T_1 and T_2 will be the same and the differential output will be zero. As the flow rate increases T_1 will fall until such a flow rate is reached that the liquid is in thermal equilibrium with the heater. The temperature differential will then be a maximum. As the flow rate is increased still further the temperature differential becomes inversely proportional to the flow rate. Fig. 2 shows a plot of temperature differential as a function of flow rate for various heater input valves. Laboratory tests showed that the system could be developed into a viable flowmeter, and two general rules for zone 2, Fig. 2

were evident:

(a) The liquid flow is directly proportional to the heater power for a constant temperature differential.

(b) The liquid flow is inversely proportional to the temperature differential for a constant heater power.

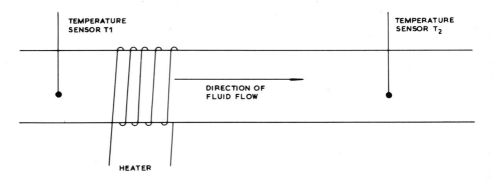

Fig. 1   Basic flowmeter

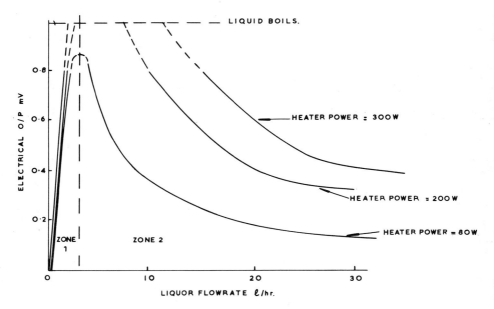

Fig. 2   Performance of thermal flowmeter

3   PLANT SCALE EQUIPMENT

The chemical plant and its associated reagent feed system is constructed in 18 : 8 : 1 or 18 : 13. : 1 stainless steel. 18 : 8 : 1 steel was used in the construction of the plant flowmeters since this was compatible with both the plant and the chemicals used. (Sulphuric acid corrosion can be inhibited by the addition of trace nitrate ion).

The heat exchanger assembly consisted of four sections of 18 : 8 : 1 bar two inches long with four holes drilled through the bar. A "rest" chamber is drilled out of one end of the section and the next section is placed so that its holes are misaligned with respect to the holes in the adjacent section by $45°$ (Fig. 3). The heater elements were wound on to the outside of the exchanger unit and clamped to it mechanically. Thermocouples were placed on each side of the heat exchanger in the liquid flow, a third thermocouple was inserted in the liquid flow to provide a cut out to the heater in the event of excessive temperature rise due to overheating. A sectionalised drawing is given in Fig. 4. Improvement to the original design included a modified heat exchanger section and the use of thermopiles rather than thermocouples, these consisted of pairs of Triple Ni Cr/Ni Al mineral insulated, stainless steel sheathed thermocouples.

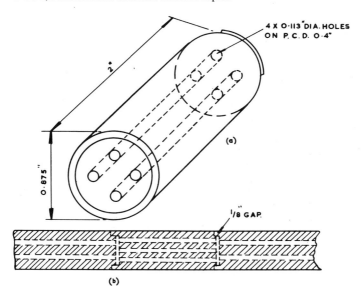

Fig. 3   Typical heat exchanger assembly
        a Standard section
        b General arrangement

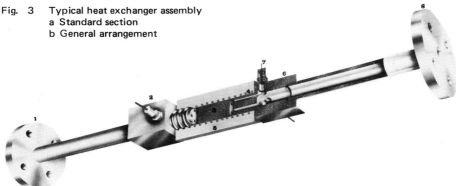

Fig. 4   Plant flowmeter

        1 Down-stream coupling flange
        2 Down-stream thermopile
        3 Heat exchanging section
        4 Electrical heater
        5 Lagging
        6 Up-stream thermopile
        7 Over-temperature trip thermocouple
        8 Up-stream coupling flange

In order to meet plant requirements three ranges of flowmeters were constructed. the electronics for each unit being identical. Visual output was displayed on a 1-6 mV reverse scale potentiometric three zone controller which also operated the small drive motor on the pump micrometer screw heat adjustment. Details of the electronics and associated control system is given in appendix I. Fig. 5 shows the typical characteristics of the plant scale instruments, Fig. 6 shows the flowmeters installed in the plant.

4   SMALL SCALE UNITS

The success of the larger equipment described above leads to a requirement for the development of small scale instruments with a maximum flow of 2 1/hr for use in an automated solvent extraction pilot plant.

As before, the instrument operates in zone 2 of the curves shown in Fig. 2. This gives a wide working range and does not require as accurate a method of construction as a unit operating in

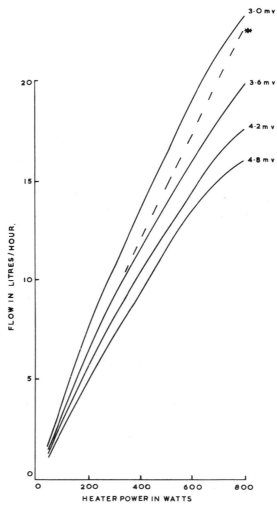

Fig. 5   Prototype thermal flowmeter characteristics

zone 1. Because of the low flow rates involved, it was not possible to machine heat exchanger sections as in the plant units, the flow tube of the instrument is 3 mm bore stainless steel tube. A polyamide silicone tape is wound on to the tube and this is then covered with a nichrome wire heater. Two stainless steel sheathed thermocouples were mounted in perspex annuli such that when the heater assembly was pushed into the annuli the ends of the thermocouples were centrally situated in the liquid flow. Inlet and outlet tubes were inserted into the annuli and thermal insulation was applied to the heater section. Fig. 7 shows the microflowmeters mounted on boxes which house the electrical connections. It will be noted that there is no cut out thermocouple on this instrument. Later versions of the small scale flowmeter use tape thermocouples placed externally on the tube rather than thermocouples mounted into the tube. This has

Fig. 6   Plant installation

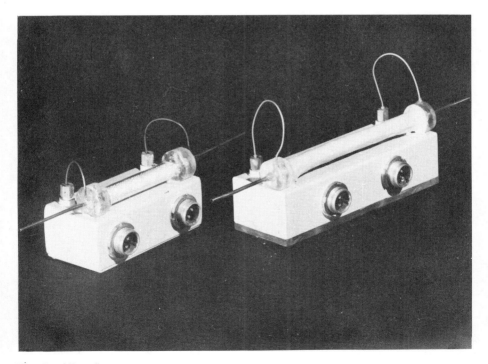

Fig. 7   Micro flowmeters

eliminated a possible source of liquid leakage at the perspex annuli. Fig. 8 shows typical operating curves for the three sizes of microflowmeters used at Dounreay. Appendix II gives details of the electronics and associated control system.

5   OPERATING EXPERIENCE

The plant scale flowmeters have worked successfully for five years or more with very little maintenance. They are situated in a building which has virtually no thermal insulation, the interior temperature of the building is usually the same as the external ambient temperature. No precautions have been taken to protect the units from draughts or acid fumes. The chemical plant in which they are installed processes nuclear fuel from the Dounreay Fast Reactor, Materials Testing Reactors and other miscellaneous reactors. This means that a variety of reagents may be fed in turn through one metering head. The heads are calibrated before a given plant run by adjusting

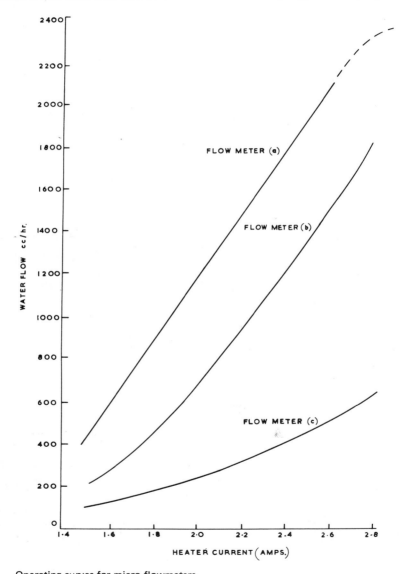

Fig. 8   Operating curves for micro flowmeters

the heater supply to give a required deviation on the reverse scale potentiometer which corresponds to a given flow rate. Any fine trimming is done during the plant run up period.

Use of these instruments has relieved the plant operator of monitoring his feeds by time consuming and sometimes inaccurate manual methods. The flowmeters are designed to "fail safe"; that is if the heater burns out or the electronics fail the liquid metering pump delivers at the setting immediately prior to the failure. Pump failure can also be rapidly identified and rectified.

The micro flowmeters are used in a somewhat different way to the larger units. If in a solvent extraction system there are "n" input variables and "n-1" can be fixed, then the system can be controlled by the one non-fixed input. For a given process it is possible to calculate the optimum input flows. On the automatic Pilot Plant the flowmeters are set to control these flows. The feed pumps are peristaltic pumps which, although being variable output pumps, operate best at a given rotor speed. It is usual to adjust the heater current of the micro flowmeters in such a manner that the pump operates at its optimum rotor speed.

Unlike the larger units the micro flowmeters are very sensitive to both draught and changes in ambient temperature and they are usually enclosed in a sealed box in order to eliminate these external influences.

6   CONCLUSION

Application of the simple concept of heating a liquid of constant specific heat in a tube for the measurement of fluid flow has enabled Dounreay to construct and use simple, robust, relatively inexpensive flow measuring devices that are used in both routine and development plant operation.

7   APPENDIX

7.1 Plant scale flowmeter, general technical description

A block diagram of the plant scale system is given in Fig. 9. The flowmeter heads are in three sizes of flow covering the range 0.5 - 40 1/hr, the electronics are common to all three flowmeter heads and are rack mounted as part of the general plant control console. The heater, control, sensing and over temperature circuits are energised by a common 110V 50 Hz stabilised supply.

The heater supply is controlled by a variac, a current of 0.5 amps is indicated on the controller unit panel and is manually varied.

The sensing unit ensures that when the temperature gradient across the heated section exceeds a certain prescribed limit the pump output will vary inversely as the error, thus providing a control cycle. The signal difference from the thermocouples is fed to a Servotronic 6-1 mV 3 zone controller indicator. This is a null balance servo instrument in which the input signal is continuously

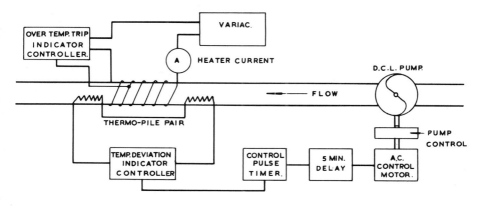

Fig. 9   Block diagram of plant-scale flowmeter control system

compared with a reference voltage. The difference signal is amplified and drives a servo which rebalances the system giving a control sensitivity of 0.25%. Limit switches in the instrument define the control zone. Should the pump output vary causing temperature excursions outside this zone, relays governing the direction of rotation of a synchromous motor will operate. The shaft of this motor drives a micrometer screw head which varies the stroke length of the DCL type M pump; pumping rate is, therefore, controlled by the direction of rotation of the synchronous motor.

Because of time lags in the system continuous control of the motor is undesirable. A cam switch is, therefore, used to limit control time to 10 sec/minute.

A Servotronic indicator controller monitors the output of the over-temperature control thermocouple. If a preset temperature is exceeded the heater and automatic control circuits are tripped. A timer in the unit prevents the trip operating due to temporary line blockages or electrical transients. The fault must persist for five minutes before the circuits can be broken. Once the heater and/or automatic control circuits have been tripped the pump output continues as at the last control position.

Manual control of the synchromous motor is also available.

7.2 Micro flowmeter, general technical description

A block diagram of the system is given in Fig. 10.

The control unit is rack mounted and contains all the electronics and control equipment. The heater is supplied with stable AC from a constant voltage transformer, variac and matching transformer. The heater current of 0-3 amps is adjusted and indicated on the front panel and a signal can be taken from a series resistance to a data logger.

The preferred temperature differential is 10-20° which corresponds to an input of 0.4 - 0.8 mV. This signal is amplified by a low drift, fixed gain chopper amplified with a gain of 10, followed by a variable gain amplifier. The resulting signal is converted by a DC amplifier and is displayed on a panel meter. It is also taken via a potential chain to a trip amplifier and the data logger. The trip amplifier operates relays which dictate the rotation of a motor driven variac, the output from which, after rectification, controls the speed of rotation of the peristaltic pump.

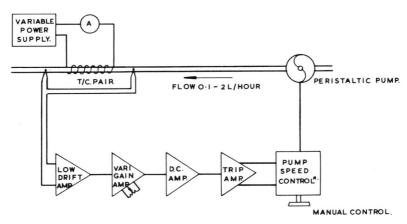

Fig. 10   Block diagram of plant-scale flowmeter control system
Automatic micro flow system

DISCUSSION

I.G. Blellock: Could this type of flowmeter be used at temperatures of 50°C to 60°C? Could it be adapted to measure flow in 4 in. to 6 in. diam. pipes at flowrates of 400/600 litres per hour?

Authors' reply: If this type of flowmeter was required to be used at a temperature of 50-60°C, attempts to achieve a comparable flow range, i.e. 6:1, would result in film boiling, particularly at the downstream end of the heated sections. This would increase the thermal resistance between the liquid and the heated wall of the tube, thereby reducing the thermal efficiency and, in the event of the flow ceasing, it would result in dangerously high temperatures.

Measurement of flowrates of 400/600 litres/hour in 4-6 in. diam. pipes using the method outlined in the paper is unpracticable since the large thermal capacity of the core of the liquid has to be heated to a measurable temperature gradient. This would call for an excessive power requirement with associated problems of maintaining a high thermal efficiency.

A.W. Jones: What range of gas flows did you measure in the Zone 1 mode?

Authors' reply: Zone 1 was described only to illustrate its unsuitability for measuring liquid flow. We have not used this type of instrument to measure gas flow.

J.C. Schuster: Could the efficiency of measuring non-corrosive fluids be increased by installing the heating element within the pipe?

Authors' reply: Increased efficiency would undoubtably result from including the heating element within the pipe. However, when applied to low flow measurement where bore diameters are small, increased practical difficulties and higher costs make this approach undesirable.

D.L. Smith: Presumably each flowmeter is calibrated individually and I expect that, in time, the calibration will be affected by deposition on the walls of the tube through which the liquid is flowing. Is the calibration checked at intervals, on-line, against the stock tanks?

Authors' reply: These flowmeters are calibrated on-line against the stock tanks only at the beginning of each plant run, which may last from four to six weeks. Examination of the tube walls has shown that deposition does not occur, since all reagent solutions are clarified and the tube is manufactured from compatable material.

J.C. Hoelgaard: Has the author investigated the characteristics of the thermal flowmeter when the mode of operation is such that the temperature difference between the two thermopile sensors is maintained constant — irrespective of flowrate — by automatic feedback control of the heater power, and then using the heater power or heater current as a measure of the flowrate?

Authors' reply: We have considered but not investigated this mode. The method is described in a paper by J.H. Laub entitled 'The Boundary layer flowmeter', Instr. Soc of America, Reprint No. 13.2.-1-65.

I.M. Thomas: In reply to the questioner requesting more information on the thermal gas flowmeter mentioned in the paper and as covered by British Patent 591,690 by Kronberger & Brown, this particular meter is designed for the measurement of small flowrates of gaseous UF_6 at subatmospheric pressures. The nature of the gas makes it desirable for the temperature difference to be measured by sensors placed outside the fluid and consequently this is achieved by the use of two matched continuous nickel resistance thermometers 47 s.w.g. thickness, wound around a long thin-walled gas-conveying copper tube and symmetrically disposed about a central heating coil again wound around the copper tube. The heating coil is supplied with constant power of about 6 watts from an external source at 50 volts D.C. while the resistance windings are connected to complete a Wheatstone Bridge external to the meter, the Bridge being powered by the same power source as the heater coil.

The whole flowmeter is enclosed in a thick aluminium tube with end flanges which serves both as a large thermal mass, to reduce the effect of ambient temperature gradients, and as a means of providing a vacuum tight annular chamber which is either connected to a vacuum line at about 30 microns or filled with foamed plastic. The insulating chamber eliminates convection effects and provides an adequate insulating medium to reduce the effects of ambient temperature variations.

In operation, the flowmeter may be used directly in the flow line to measure flow rate or for larger flowrates as a shunt meter i.e. in parallel with an appropriate number of tubes identical to the inner tube of the flowmeter, care being taken to ensure identical gas entry conditions to all parallel tubes. Correct choice of the number of parallel tubes allows operation in the linear range.

The meter output varies linearly with mass flow rate over the range 0 to 5 mg/sec with a sensitivity of 22 mV per mg/sec when used with UF_6 at an accuracy of 1 to 2% of range. At flow rates above 5 mg/sec, the calibration becomes increasingly non-linear until at 32 mg/sec the sensitivity is reduced to about 7 mV per mg/sec.

The meter may be calibrated over the linear part of the range with any convenient gas of known specific heat, and a factor applied equal to the ratio of specific heat of the calibration gas and the gas whose flow is to be measured. Calibration with the actual metered gas has been found necessary over the non-linear part of the range.

Response time is 7 minutes to reach 99% of equilibrium reading after imposing a step change in flow rate.

A more comprehensive paper on this flowmeter is to be published shortly.

ULTRASONIC METHOD OF FLOW MEASUREMENT IN LARGE CONDUITS AND OPEN CHANNELS

N. Suzuki, H. Nakabori and M. Yamamoto

Abstract: More than 400 installations of ultrasonic flowmeters have been made in Japan. Modified sing-around systems, consisting of one pair of probes and one electronic unit operate with high stability in all climatic conditions and with low sensitivity to temperature changes. One of the features of the ultrasonic flowmeter is that it is capable of measuring flow rate by using probes mounted on the outside of the conduit wall. Thus there is no pressure loss, it can be installed on existing pipes and installation costs are low compared with conventional meters. In this report, the authors describe the principle and practical applications of the ultrasonic flowmeter for conduits and open channels such as at water works and power plants.

1   INTRODUCTION

More than 400 installations of ultrasonic flowmeters have been made in Japan on large diameter water pipes, open channels and rivers.

The use of ultrasonic techniques for measurement of liquid and gas flow have been suggested by a number of workers. The most common technique is based on the apparent difference of the velocity of sound in still water and moving water. Techniques include the phase shift method, pulse propagation time method and the sing-around method. In addition, the beam deflection method and the Doppler method are known.

A very precise and stable measurement is required for practical applications of ultrasonic techniques in flow measurement. For water, the flow velocity in most cases is below several m/s, whereas the velocity of sound in water is approximately 1500 m/s and is affected by the flow velocity by a factor of approximately 10^{-3}. In order to achieve an accuracy in measurement of 1%, the flowmeter itself must maintain an accuracy to within one part in 10^{-5} to 10^{-6}. In actual installations, the above accuracy of equipment must be maintained day and night, summer and winter. In our approach, one pair of probes is mounted on the outside of the pipe without anything being inserted in the pipe. One electronic unit only is used in the system, and the direction of sound propagation is switched up-stream and down-stream at pre-set times. In this modified sing-around method, drift in the electronic unit is self-cancelling[1].

Differential sing-around frequency between up-stream and down-stream is often less than 1 Hz, which makes precise measurement difficult. In our approach, a frequency multiplication technique is incorporated, which makes precise measurement of sing-around frequency possible.

2   PRINCIPLE OF MEASUREMENT

2.1 Nomenclature

* C_1	Velocity of sound in the plastic shoe
* C_2	Velocity of sound in the pipe wall
C	Velocity of sound in water
* ϕ_1	Incident angle in the plastic shoe
* ϕ_2	Refractive angle in the pipe wall
ϕ	Refractive angle in the water
θ	$\theta = 90° - \phi$

* ℓ_1	Length of sound path in the plastic shoe
* ℓ_2	Length of sound path in the pipe wall
ℓ	Length of sound path in water
* w	Pipe wall thickness
d	Inner diameter of pipe
V	Flow velocity
τ	Fixed time in the sing-around period (Delay time)
τ_e	Fixed time in the electronic circuit
t_d	Sing-around period for down-stream transmission
t_u	Sing-around period for up-stream transmission
f_d	Sing-around frequency for down-stream transmission
f_u	Sing-around frequency for up-stream transmission
f_0	Sing-around frequency in still water
Q	Nominal flow rate obtained by the ultrasonic method
Q	True flow rate (after correction)

2.2 Flow velocity measurement

In Fig. 1, P_1 and P_2 are the transmitter and receiver of ultrasonic waves (probes) which are mounted on the outside of the pipe wall so that the ultrasonic sound path has a certain angle with respect to the direction of water flow. The oscillator disc in each probe is made of lead-zirconate ceramic and mounted on the pipe wall with a plastic shoe so as to face each other.

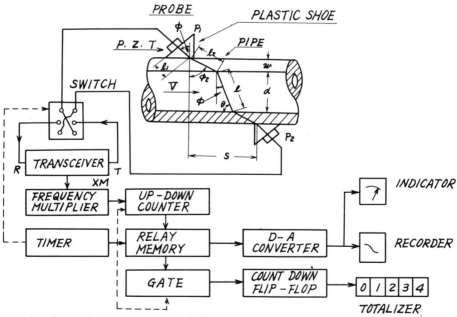

Fig. 1   Block diagram of the ultrasonic flowmeter

The sound propagates with an incident angle ϕ_1, as shown in Fig. 1. Owing to the difference in sound velocities in the plastic shoe and pipe wall, the sound is refracted at the angle ϕ_2. Similarly, the refraction angle in the water is ϕ. Reaching the opposite pipe wall, the sound travels to probe P_2 through a path similar to that described.

At P_2, the ultrasonic pulse is converted into a voltage pulse. This voltage is amplified in the transceiver and returned to P_1; thus the cycle is repeated.

When the sound velocities in the plastic shoe, pipe wall and water are C_1, C_2 and C, respectively, the following equation can be written, according to Snell's Law.

$$\frac{\sin \phi_1}{C_1} = \frac{\sin \phi_2}{C_2} = \frac{\sin \phi}{C} \qquad \dots (1)$$

The time required for the above cycle is called the "Sing-around period".

The fixed portion in the propagation time (τ) in the sing-around period is given by

$$\tau = 2 \frac{\ell_1}{C_1} + \frac{\ell_2}{C_2} + \tau_e \qquad \dots (2)$$

As indicated in 2.1, symbols marked with an asterisk are fixed values depending on materials used in the shoe and pipe wall.

When the sound is transmitted in the down-stream direction, the sing-around period will be:

$$t_d = \frac{\ell}{C + V \cos \theta} + \tau \qquad \dots (3)$$

The direction of propagation of sound is then switched towards the up-stream direction after a pre-set period of time, and the sing-around period is then:

$$t_u = \frac{\ell}{C - V \cos \theta} + \tau \qquad \dots (4)$$

In order to cancel out the effect of the change in sound velocity due to a change such as due to water temperature, reciprocals of these periods, which are the sing-around frequencies, are used for measurement. Here, $C^2 > (V \cos \theta)^2$, and

$$\triangle f = f_d - f_u = \frac{1}{t_d} - \frac{1}{t_u} = \frac{V \sin 2\theta}{d} \left(1 + \frac{\tau C \sin \theta}{d} \right)^{-2} \qquad \dots (5)$$

As previously mentioned, $\triangle f$ is small, and direct reading is not practical for accurate measurement of flow velocity. Therefore, sing-around frequencies f_d and f_u are multiplied M times to bring Mf around 200 Khz, where, from the difference, M$\triangle$f is obtained. Equation (5) is now:

$$V = \frac{d}{\sin 2\theta} \left(1 + \frac{\tau C \sin \theta}{d} \right)^2 \frac{1}{M} (M \triangle f) \qquad \dots (6)$$

Thus, the mean flow velocity along the sound path is obtained.

2.3 Flow rate measurement

2.3.1 Flow rate in full water pipe

Since the flow velocity obtained above represents only the mean velocity of the sound path, convertion to flow rate is required. As a first step, the mean flow velocity (V) obtained above has to be converted to a true mean flow velocity in consideration of the distribution of velocity in the pipe (V).

G.I. Birger of the USSR discussed this matter in 'Ismeritelnaya Technika', No. 10, 1962, and introduced his curve of correction coefficient as shown in Fig. 2. Here, the correction coefficient

K is defined as:

$$K = \frac{V}{\overline{V}} \qquad \qquad \qquad ... (7)$$

The correction coefficient incorporated in our approach is based on the above theory, which in our experience appears to be correct.

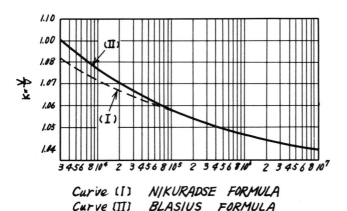

Curve (I) NIKURADSE FORMULA
Curve (II) BLASIUS FORMULA

Fig. 2   Curve of correction coefficient K

As the flow rate can be obtained by multiplying $\overline{V}$ by the cross sectional area, the following relations can be established:

$$Q = \left(\frac{1}{K}\right) \left(\frac{\pi}{4} \frac{d(d + \tau C \sin\theta)^2}{\sin 2\theta}\right) \left(\frac{1}{M}\right) (M\triangle f) \qquad ... (8)$$

but

$$\frac{f_d + f_u}{1} = f_0 \frac{1}{t_0} = \frac{1}{\frac{\ell}{C} + \tau} = \left(\frac{C \sin\theta}{d}\right) \Big/ \left(1 + \frac{\tau C \sin\theta}{d}\right) \qquad ... (9)$$

and equation (8) can be written as:

$$\overline{Q} = \frac{1}{K} \frac{\pi}{8} d \tan\theta \left(\frac{C}{f_0}\right)^2 \frac{1}{M} (M\triangle f) \qquad ... (10)$$

You will probably note C, the velocity of sound, in the equation. But, the accuracy of measurement is not affected in practice by the change of sound velocity in the case of large diameter pipes for the following reason:

As

$$f_0 = \frac{C \sin\theta}{d + \tau C \sin\theta} \qquad ... (11)$$

$$c/f_0 = \frac{d + \tau C \sin\theta}{C \sin\theta} \times C = \frac{d}{\sin\theta} + \tau C \qquad ... (12)$$

Here $d/\sin\theta$ is of the order of 1 m and τC is approximately 10^{-2}m. As τC is changed by several per cent the measurement is changed by a factor of approximately only 10^{-4}.

In the case of large diameter pipes, the flow is laminar only when the linear velocity is below several mm/s. Therefore, it can be assumed that in practice the flow is always turbulent.

In an actual flow meter, the electronic circuit is controlled by the timer with a crystal oscillator, as shown in Fig. 2.

In Fig. 3, the up-down counter will add Mfd and subtract Mfu to obtain MΔf which will be converted to the relay memory. It is then treated to approximate to round figures (such as 10 m^3 or 100 m^3) as the minimum increment on the totalizer. Through the ladder circuit it is also converted into an analog instantaneous flow rate signal. The counting sequence is designed to cancel out the effect of a gradual change in the velocity of sound in water.

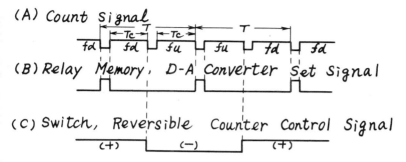

(A) Count Signal

(B) Relay Memory, D-A Converter Set Signal

(C) Switch, Reversible Counter Control Signal

Fig. 3   Time chart of ultrasonic flowmeter

A bi-directional flow measurement can easily be achieved with a polarity identification circuit. This type of arrangement is especially useful in situations subject to tidal variations.

2.3.2 Measurement of flow rate in open channels

The ultrasonic flow meter can be used in open channels built for irrigation purposes or for transport of effluent in a sewerage plant.

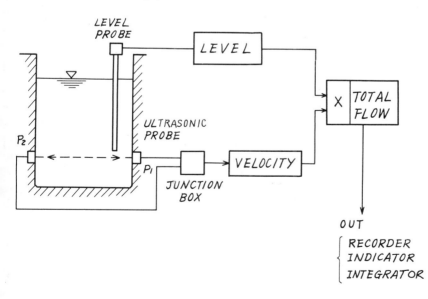

Fig.   4   Schematic diagram of the ultrasonic flowmeter for open channels
(one-channel method)

(1) One path method

The probes are installed at a certain depth; mostly at 0.6 of water level, the vertical distribution characteristics of flow velocity being measured beforehand by sliding a pair of probes from the high water level to the low level and obtaining the velocity at several levels. In multiplying the flow velocity with the signal from the level gauge, the result of the flow distribution measurement given above is used as the correction factor for the level signal (see Fig. 5).

Depending on channel conditions, and when 1:1 relationship is established between level and velocity, the flow rate can be obtained from the flow velocity only, without a level signal, provided the correction factor is combined in the linearizer.

(2) Multi-path method

When the velocity distribution changes or when the direction of flow is changed, a multi-path measurement of flow velocity is recommended. Ordinarily, several sound paths are scanned by switching. The cross sectional area for the respective sound path is multiplied by the velocity and summed. The level gauge is used to determine the cross sectional area for the uppermost sound path. In case the water level changes significantly, provision may have to be made to cut off the upper path.

3   FLOW MEASUREMENT IN LARGE DIAMETER PIPES

3.1 Application to city water supply systems

The first installation of an ultrasonic flowmeter in Japan was made in January 1964 at the pumping station in Itami City near Osaka. At present, more than 400 flowmeters are being used, mostly in city water supply systems. Other applications are in power stations and in open channels and rivers to which reference will be made later. The smallest diameter of flowmeter which has been installed is 25 cm (10 in.), and the largest is 500 cm (200 in.), the latter in a power station.

The newest flowmeter, Model UF-500, is shown in Plates 1 and 2.

In 1964, the ultrasonic flowmeter was a novelty, and tests were conducted several times in order to determine the accuracy and reliability of the meter. Some of the results are shown below. Recently, however, the "sing-around" type of flowmeter has become common among users in Japan, and flow tests are not requested any more. Because of this, all relevant data are no longer available.

3.1.1 Flow test at Magome Reservoir Tower[2]

This test was conducted by Tokyo Metropolitan Water Bureau on 24th May and 8th June of 1965. Two 30 m diameter (100 ft) water towers were used for the measurement. The water level was multiplied by the cross sectional area of the tower and from this the volume of water was calculated. The ultrasonic flowmeter was mounted on a 60 cm diam. (2 ft) cast iron pipe. A venturi meter was installed in the same pipe and operational data were compared. The results proved that at 1 m/s or at higher flow velocities, the ultrasonic flowmeter had an accuracy within ±1% (Table 1 and Fig. 6).

3.1.2 Flow test at Koemon Reservoir Tower[2]

The pipe diameter in the test at Magome Reservoir was comparatively small, and a test was conducted by the same organisation on a larger diameter of pipe. A 43 m diam. (140 ft) water tower was used for the test, and the water pipe diameter was 150 cm (5 ft). As seen in Table 2 and Fig. 7, the overall accuracy was well within ±1%.

3.1.3 Examination

The following are the points to be examined in flow measurement by the ultrasonic flowmeter.

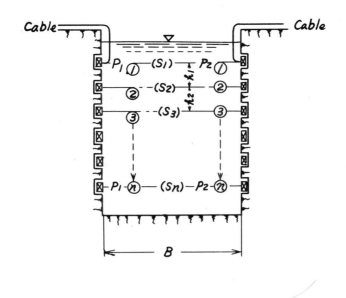

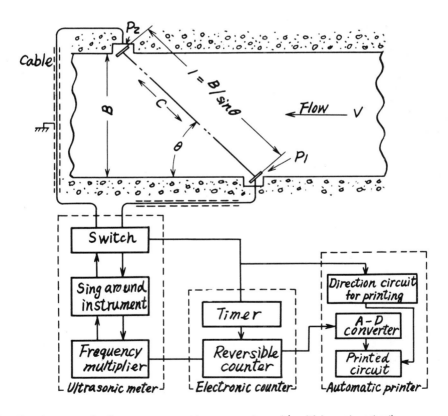

Fig. 5 Apparatus for flow measurement in an open channel (multichannel method)

TABLE 1   TEST DATA (MAGOME)

TEST NO.	TIME (MIN.)	VOLUMETRIC		ULTRASONIC		VENTURI	
		L [m³]	RATE [m³/h]	U [m³]	$\frac{U-L}{L}$ [%]	V [m³]	$\frac{V-L}{L}$ [%]
2 A	40	92.2	138	93	+1.1	47.9	-47.8
2 B	30	229.1	458	218	-4.8	217.5	-4.8
1	120	1940.4	970	1925	-0.77	1953.2	+0.67
2	120	3420.6	1710	3388	-1.0	3361.6	-1.7
TOTAL	310	5682.3	1100	5624	-1.03	5580.2	-1.80

TABLE 2   TEST DATA (KOEMON)

DATE	TEST NO.	PLANNING [m³/h]	TIME		VOLUMETRIC		ULTRASONIC		ERROR
			START-FINISH	TIME (H)	LEVEL [mm]	VOLUME [m³]	COUNT NUMBER	VOLUME [m³]	
9/21	1	1000 (868.4)	0 : 30 ~ 2 : 00	1.30	873.0	1302.7	394	1361.3	+4.49
	2	2000 (2027.9)	2 : 15 ~ 3 : 15	1.00	1359.0	2027.9	590	2038.5	+0.52
	3	4000 (3642.4)	0 : 30 ~ 1 : 00	.30	1220.5	1821.2	536	1851.9	+1.68
9/25	4	7000 (8725.4)	1 : 10 ~ 1 : 40	.30	2253.5	3362.7	987	3410.1	+1.41
	5	10000 (10430.4)	1 : 50 ~ 2 : 10	.20	2330.0	3476.8	998	3448.1	-0.83
	6	3000 (2098.4)	2 : 20 ~ 3 : 00	.40	937.5	1398.9	412	1423.5	+1.75
	TOTAL			4.30	8973.5	13390	3917	13533	+1.07

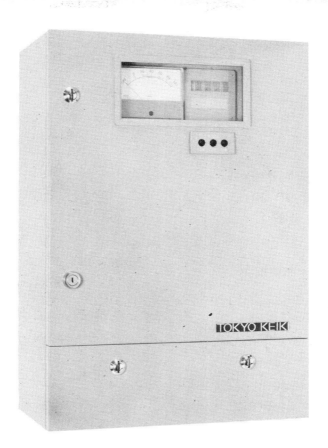

Plate 1

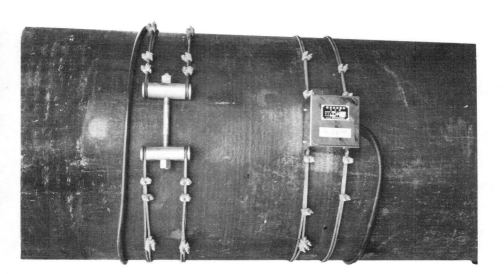

Plate 2

(1) *Confirmation that the sing-around system is formed exactly as calculated, including the probe mounting position:*

This can be achieved by measuring C/f_0

Since $C/f_0 = C t_0 = \ell + \tau C$ ... (13)

and $\ell \gg \tau C$ in the case of large diameter pipes, the sound velocity C can be calculated by measuring the water temperature. The values of f_0 or t_0 can be obtained from the frequency counter which is self-contained in the UF-500 flowmeter. By comparing C/f_0 thus obtained with the design value, the results given by the sing-around system can be examined.

The flow need not be stopped in order to measure f_0, as

$$f_0 = \frac{f_d + f_u}{2}$$ ... (14)

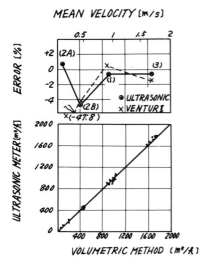

Fig. 6   Test data at magome
Cast iron pipe diameter = 60 cm

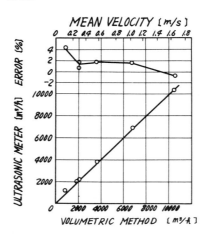

Fig. 7   Test data at Koeman. Steel pipe diameter = 150 cm

(2) Correction coefficient

The results of flow tests other than described above are shown in Table 3. Installations with straight pipes of more than 10D (D = pipe diameter) in the up-stream and 5D in the down-stream direction were selected where the flow distribution was stable.

Figure 8 shows the results compiled with respect to Reynolds number. The theoretical values obtained by calculation are also shown and these indicate fairly good agreement.

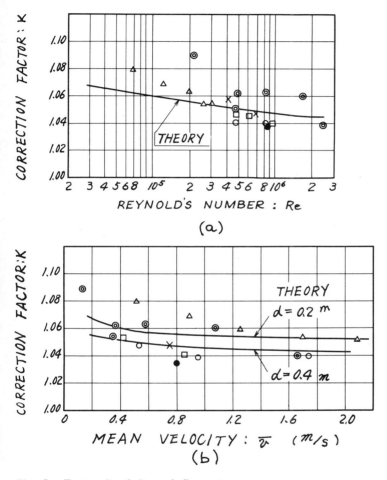

Fig. 8   Test results of ultrasonic flowmeter

    a   Reynolds' number and correction factor

    b   Mean velocity and correction factor

(3) Flow direction along the pipe axis

The flow rate is obtained from the flow velocity in the direction of the pipe axis multiplied by the cross sectional area of the pipe. Therefore, when the direction of the flow is not parallel to the pipe axis, some correction has to be made. A Reflection Method (V type configuration of probe mounting) or Cross Path Method (X type mounting) can be used to provide a solution. We would like to consider the V type mounting here.

In Fig. 9, the ultrasonic pulse is reflected at the opposite pipe wall. The flow velocity V

TABLE 3 Specification — flow test

Test No.	1	2	3	4	5	6
Test Site	Nagoya City Nakagawa West Distribution Stat.	Tokyo Metropol. Measurement Lab.	Tokyo Metropol. Magome Distrib.	Tokyo Metropol. Nagasawa Treat. Plant	Tokyo Metropol. Koemon Distrib. Stat.	Ibaragi Pref. Nakagawa Indust. Water
Test Date	Sept. 11 - 12 '63	Mar. 25 - 26 '65	May 24 & June 8 1965	July 11, '65	Sept. 21, 25 '65	Nov. 17 - 18 '65
Dia. (mm)	1100	200	600	1200	1500	1200
Wall Thickness (mm)	Ductile Cast Iron 18	Probes Exposed to Water	Cast Iron 15.4	Steel 11	Steel 13	Ductile Cast Iron 17
Lining Material Thickness (mm)	Moltar 10	None	None	Coal Tar Enam. 3	Coal Tar Enam. 3	Moltar 10
Water	Clear Water 20 - 21°C	Water 10°C	Clear Water 17°C	Clear Water 18°C	Clear Water 22°C	Industrial W. 15°C
Straight Pipe Up-Stream Down-Stream	More than 20m " 50m	2.4m 2.4m	Approx. 20m " 20m	15m 15m	Approx. 40m " 20m	Approx 30m More than 50m
Vessel for Standard	Distribution Pond 5660m²	Standard Tank 11.5m³	2x30mφ Reservoirs 698.5m² - 8550m³ X 2	Distribution Pond 2339m²	43.6mφ Reservoirs 1492m² 20000m³	Distribution Pond 597m³ Level Change 3m
Level Measurement	1 - Point Scale, 1mm Unit	Scale at the Side of Tank	1 - Point for Each reservoir Scale, 1mm Unit	3 - Points Float Gauge	2 - Point By Wire-Rope with weight	2 - Points Scale

which affects the sound velocity in both down-stream and up-stream directions will be respectively:

$$V_{P_1}R = V_X \cos\theta - V_Y \sin\theta \qquad \qquad \dots (15)$$

$$V_{P_2}R = V_X \cos\theta + V_Y \sin\theta \qquad \qquad \dots (16)$$

Thus, the effect of V_Y can be cancelled out[3].

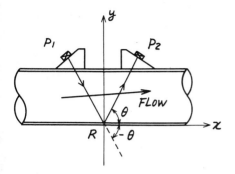

Fig. 9   Reflection method

(4)   *Non-symmetric distribution of flow velocity*

When the distribution of flow velocity is not symmetric in regard to the pipe axis, the multi-sound-path system can be used to obtain a higher accuracy of measurement.

Figure 10 shows the time chart of an installation at a power station. Two pairs of probes are mounted where the sound path for each pair crosses at right angles. The flow rate is measured from the sum of two differential sing-around frequencies on two paths. The effect of slant or circular movement of flow proved to be largely improved by such a method of mounting.

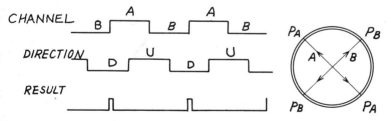

Fig. 10   Two-channel system
D : Down-stream direction
U : Up-stream direction

Further improvement may be expected by increasing the number of sound paths. But not much improvement is seen if the number is increased to more than four paths.

3.2 Application in a power plant

As the ultrasonic method permits the mounting of probes on the outside of a penstock, its use is relatively simple and economic in power plants where pipe diameters are large. The following are

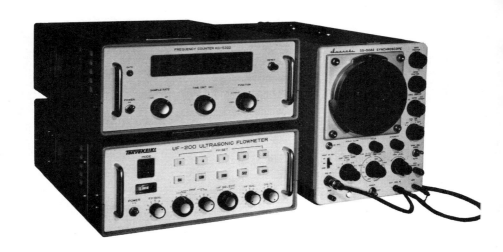

Plate 4

the major applications:

(a) *Permanent installation on a penstock*

To control turbine operation or to measure the rate of discharge of water. A bi-directional model is used in pump storage schemes.

(b) *Measurement of turbine efficiency by a portable model*[4]

The portable model type UF-200 (Plate 3) has been used for this purpose. Probes are mounted on the outside of the penstock by a magnet (Plate 4) which can be used on any pipe diameter. It may be connected to a digitial print-out device, and one person can carry out all the measurements.

Experiments undertaken by the Central Research Institute of the Electric Power Industry, Tokyo are listed in Table 4, and a typical result is shown in Fig. 11. The data are rather obsolete, but it was as a result of this work that the ultrasonic method was specified in a Japanese Electrical Code as a standard means for measuring turbine efficiency, and flow measurement using the UF-200 instrument became a routine operation in major Japanese power stations. The total number of measurements carried out now exceeds 1000.

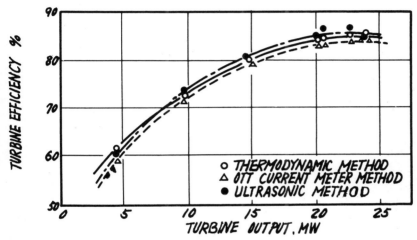

Fig. 11   Test results at Ontake power station (no. 3 turbine)

3.3 Measurements in square-section pipes

One of the features of the ultrasonic flowmeter is that it can be applied to conduits of any shape. Figure 12 shows an application on a conduit with a square cross section. As is usually the case, it is made of concrete. Stainless steel windows for mounting the probes are provided on one of the walls. The reflection plate is located on the opposite wall.

4   FLOW MEASUREMENT IN OPEN CHANNELS

4.1 Irrigation water

In one application, five sets of ultrasonic flowmeters are in operation in the open channels carrying irrigation water from the Tone River. The channels are all trapezoidal-shaped. Because the water level was kept constant by a level control gate, the ultrasonic flowmeter was the only practical approach in this case.

The result of the experiment preceding the permanent installation of the meter, and the outline of the installation is given in Table 5.

Compared with a current meter, the results show agreement within ±5% which was sufficient for the purpose.

TABLE 4   Japanese power stations in which turbine efficiency was measured by the ultrasonic method

Name of Station (No. of Turbine)	Date of Test	Type	Details of Turbine				Difference in Results Compared with other Methods used at the same time*		
			Output	Net Head	Discharge	Speed	A	B	C
			kW	m	m³/S	rpm	%	%	%
Sirakawa :: :: ::	Dec. 1964	V.P-4N	3,100	159.1	2.2	450	+2.0		
Ontake (No.1) ::	Mar. 1965	V.F	25,000	229	11.47	600	0		
Towada (No.3) ::	May 1965	V.F	12,950	180	6.67	600	0.5	-1.5	
Yonekawa :: ::	June 1965	H.P-2N	3,450	246.05	1.68	514	+5.0	+5.0	
Yonekawa :: ::	Aug. 1965	(see above)					+6.0	+6.0	
Kurobegawa No. 2 (No.1) ::	Sept. 1965	V.F	22,000	180.5	17.4	375	+1.5		0
Iori (No.2) :: ::	Sept. 1965	V.F	9,000	240	9.0	600	-4.0	-4.0	
Jintsugawa No. 1 (No.1)	Sept. 1965	V.F	45,000	61.92	80.0	172			
Jintsugawa No. 1 (No.2)	Sept. 1965	V.F	45,000	61.92	80.0	172			
Wadagawa No. 2 (No.1)	Sept. 1965	V.P	68,900	430	16.6	300			
Wadagawa No. 2 (No.2)	Sept. 1965	V.P	68,900	430	16.6	300			
Towada (No. 3) ::	Nov. 1965	(see above)							
Sinkosaka :: ::	Jan. 1966	V.F	32,000	128.9	28.0	360	0		
Ontake (No. 3) ::	Jan. 1966	V.F	25,000	230	12.35	500	-0.5		-2.5
Oigawa (No. 2) ::	Jan. 1966	V.F	23,500	112.73	27.4	360/300			-1.0
Kurokawa No. 1 (No.3) ::	Feb. 1966	A.P.-2R.4N	15,700	244.2	7.35	500	0	-2.5	
Kano :: ::	Mar. 1966	H.F	1,700	93.88	2.30	1,210			(-0.5)†
Oigawa (No. 2) ::	Apr. 1966	(see above)							
Towada (No. 3) ::	May 1966	(see above)							
Yagisawa (No. 2) ::	June 1966	Vertical Pump-Turbine	87,000	111	100	150	+1.0		
Yagisawa (No. 3) ::	June 1966		87,000	111	100	150			
Maki (No. 2) :: ::	June 1966	V.F	16,200	95.4	18.3	360	-2.5	-0.5	
Kurokawa No. 1 (No. 3) Turbine	July 1966	(see above)							
Morozuka Pump ::	July 1966	V.F	54,000	226	27.06	300	0		
		V. Pump	48,000	245.9	18.2	300			
Miyouken (No. 1) ::	July 1966	HF-IRS	1,231	90.0	1.67	720		+4.0	
Miyouken (No. 2) ::	July 1966	HF-IRDS	1,507	37.79	4.175	514		+3.0	
Miyouken (No. 3) ::	July 1966	HF-IRDS	1,529	37.79	4.175	600		-0.5	
Towada (No. 3) ::	Oct. 1966	(see above)							
Numazawanuma (No. 2) ::	Nov. 1966	H.Pump	21,000	211.0	8.6	500	0	0	
Hinobori (No.1) :: ::	Dec. 1966	V.F	5,000	62.79	9.3	450	+4.5	+2.5	

* Shown by "(other)−" (ultrasonic) at highest efficiency value.
A: Difference from thermodynamic method.   B: Difference from Pitot-tube method.   C: Difference from current-meter method.
† This value is not the difference from the current-meter method, but from an electromagnetic flow meter installed in the middle penstock which is one metre in internal diameter.

4.2 Applications in a sewerage treatment plant

Probes were mounted at a fixed level in a square-shaped open channel carrying effluent from a sewerage treatment plant. The change in water level is included in the calculation as a correction factor in the linearizer. It showed agreement within ±5% with a current meter.

Flow measurement at the input to a sewerage treatment plant requires very careful selection of the probe mounting site, since sound will be attenuated by any foreign materials present in the

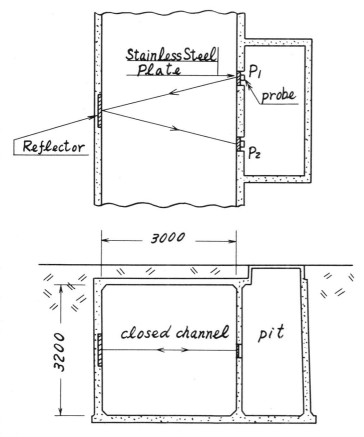

Fig. 12   Reflection Method on the Closed Channel

sewer. If a fairly long distance from the pump can be tolerated, it can be used to homogenise the sewerage and cause heavy materials to settle, and so permit stable operation of the ultrasonic flowmeter.

4.3 Power plant applications[5]

As shown in Table 6, the ultrasonic flowmeter has been used to determine turbine efficiency at power stations where horizontal axis Francis machines, oblique axis turbular units and vertical axis Kaplan turbines are installed.

Ultrasonic flowmeters have been installed in the open discharge channels from the pump, which in most cases are of rectangular section.

TABLE 5 Summary of experiments at open channels

Date	Place	Open Channel Width & Depth
August 10, 1966 September 3, 1966	Saitama Irrigation Water Resources Authority	W. up 5.110 mm W. bottom 3.000 mm D. Approx. 1.9 m

1 Probes – 400 KHz 40 mmϕ Fixed at 655 mm from bottom.
2 Mean flow velocity measurement only, because water level is kept constant by special gate.
3 Compared with current meter – within 2 - 4.5%.

Delivered on order March, 1967	Saitama Irrigation Water Resources Authority	W. up 4970 mm W. bottom 2800 mm D. approx. 1.8 m

1 Probes – 400 KHz 40 mmϕ Fixed at 0.4 H from bottom (H = water level) because of constant level.
2 Compared with current meter – 4% (July 4)

Delivered on order March, 1967	Shibaura Sewage Treatment Plant	W. 2800 mm D. approx. 1.1 m

1 Probes – 400 KHz 40 mmϕ Movable.
2 After measurement of mean flow velocity by changing position of probes, they were fixed at 0.4 m from bottom.
3 Flow rate was obtained by multiplying the obtained flow velocity by water level (capacitance type level device).
4 Accuracy – within ± 5%.

Delivered on order March, 1968	Togoda Water Resources Authority	W. up 5500 bot. 2700 D. approx. 2889 mm
"	Iizumi Water Resources Authority	W. up 2710 bot. 1510 D. approx. 1250 mm
"	Furutone Water Resources	W. up 4250 bot. 2650 D. approx. 1730 mm
"	Sanuki Water Resources Authority	W. up 3650 bot. 2060 D. approx. 1640 mm

1 Probes – 400 KHz 40 mmϕ Fixed at 0.4 H from bottom.
2 Flow rate is obtained by multiplying the obtained value by water level (float type)

In the case of rectangular open channels, the flow velocity is given by:

$$V = \left\{ \frac{\ell}{2B} \left(\frac{C}{f_0} \right)^2 \tan \theta \right\} \, \triangle f$$

In the case of trapezoidal channels, we have the equation,

$$V = \left\{ \frac{\ell}{2(b + 2b_0)} \left(\frac{C}{f_0} \right)^2 \tan \theta \right\} \, \triangle f$$

for which notations are given in Fig. 13.

Typical results of measurements at rectangular and trapezoidal open channels are given in Figs. 14 and 15. The results show sufficiently high accuracy for practical use. A large number of measurements are now carried out using the techniques described here.

TABLE 6 Specification of power stations tested

Name	Test date	Turbine specification				Angle of probe (0°)	Natural frequency of probe	Shape of channel
		Type*	Output (kW)	Discharge (m³/S)	Head (m)			
Kyoken	Mar, '67	HF-IRDS	1500	4.18	37.79	45	400kHz	Rectangular measuring section made by guide wall in trapezoidal channel
Yoshigase	Sep, '66	VR-IRS	11500	10.0	115.7	45	400kHz	Rectangular section
Shingo	Oct, '67	VR-IRS	12500	76.0	21.25	76	400kHz	Rectangular section
Koshi	Aug, '68	ID-IRT	1400	22.0	8.0	45	400kHz	Trapezoidal section

* H = horizontal, V = vertical, I = inclined, F = Francis, K = Kaplan, D = Deriaz, R = runner.

5   CONCLUSION

A modified sing-around type flowmeter has been developed which in practice has been found to operate reliably. One pair of probes only are used and the direction of sound propagation is switched.

The system allows probes to be mounted on the outside of pipe walls. Major installations are in city water supply systems, sewerage systems and in penstocks at hydro-electric power stations.

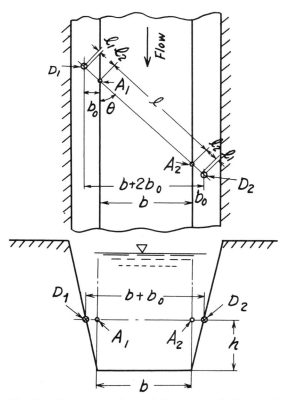

Fig. 13   Probe separation and dimensions of a Trapezoidal Channel

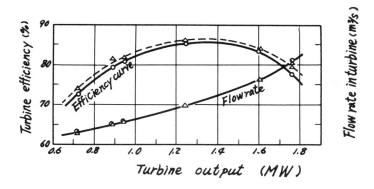

Fig. 14   Turbine efficiency at the Myoken station in March 1967 (rated head = 37.79m)

——o—— ultrasonic meter
— — △ — — current meter (Ott type)

The principal features of the flowmeter are:

1   It is applicable to any large diameter pipe. Measurement is possible from outside the pipe. There is no need to insert a probe into the pipe.

2   There is no pressure loss.

3   Flowmeters can be designed to measure reversing flows.

4   The flowmeter is linear over a wide range.

5   The flow range can be changed easily.

6   Installation costs are small and can be made without stopping the flow: existing pipes can be used.

7   There is no significant difference in the cost of a meter to operate on large or small diameter pipes.

The ultrasonic flowmeter approach is equally suited to flow rate measurement in open channels, and new applications are still being found.

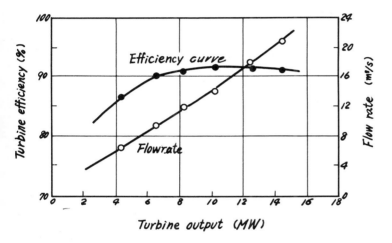

Fig. 15   Turbine efficiency and flow rate in the Koshi station in August 1968 (rated head 8 m)

6   REFERENCES

1   M. YAMAMOTO and K. ITO: 'Ultrasonic Flowmeter for Large Pipes', Journal of the Institute of Electronics and Communication Engineers of Japan, 1965, **48**, 1956-63

2   S. MATSUZAKI and M. YAMAMOTO: 'Flow Test of Ultrasonic Flowmeter for large diameter pipes', Journal of The Instrumentation Control Association Japan, 1967, **16**, pp. 51-60

3   Y. YOSHIDA, Recent applications of Ultrasonic Flowmeter. Journal of the Instrumentation Control Association Japan, 1970, 19, pp. 104-106

4   H. SUZUKI, H. NAKABORI and M. YAMAMOTO: 'Ultrasonic Method of Flow Measurement', Water Power, 1968, July, pp. 266-269

5   H. SUZUKI, H. NAKABORI, T. HOSHIKAWA and T. SATAKE: 'Ultrasonic Method of Flow Measurement in an open channel', Water Power, 1970, May-June, pp. 213-218

DISCUSSION

J.E. Carrington: In the United Kingdom there was interest in the ultrasonic flowmeter in the 1950's, but because of the large amount of space required and the weight of the electronic units at that time attention was transferred to other forms of flowmeter. However, some work has continued, resulting in the use of these versatile instruments for special cases, notably for the measurement of the flow of airborne pulverised fuel by Parkinson of the National Coal Board. In the U.S.A., instruments have been developed and are commercially available for measurements of blood flow pulsations, whilst at Queen's University we are investigating the measurement of pulsating flows for industrial purposes, working with hydraulic mineral oils in circular pipes at pressures up to 1,000 p.s.i.

The authors have taken advantage of the great advances in electronic techniques in the last decade, and are to be congratulated on bringing their flowmeter to such a practical and commercial success. The method of applying the probes is an obvious advantage economically in existing installations.

I should be grateful if the authors would indicate how their instrument compares economically with other forms of meter, say the electro-magnetic type, in new installations.

Also can this method of applying the probes be used for smaller pipes, say, 1.0 m diameter?

Authors' reply: A comparison of the economics of the ultrasonic flowmeter with other types of flowmeter cannot be made easily as the basis for such a discussion is quite different and depends on the country in which installations are considered.

In the case of Japan at present, the ultrasonic meter will be more economic than the electro-magnetic flowmeter at diameters larger than 300 mm, and more economic than the venturi meter at diameters larger than 1200 mm. It will be economic at smaller diameters when the construction of stop valves and bypass loops are involved.

The portable type of flowmeter is a different case, as it has features which are quite unique.

We have experience of applications on a 2 in. diam. pipe in which specially fabricated transducers are used similar to those designed for measurement of cooling water in power stations or for air conditioning purposes in buildings.

However, generally speaking, for smaller pipe diameters, the larger is the error in measurement. In addition, from economic considerations, we recommend application of the ultrasonic meter only in pipes with diameters exceeding 300 mm.

J.E. Carrington: Early workers were inhibited by the apparent difficulties of getting the ultrasonic beam through the pipe walls, due to refraction problems at large angles of inclination of the beam. How did the authors overcome this problem?

Authors' reply: The construction of the transducer is as shown in Fig. 1 in the report. The transducer used by the authors has the incident angle from plastic wedge to pipe wall (steel) of $40°$, the refraction angle into steel being $59.99°$ and the refraction angle into water $23.02°$. An ultrasonic beam of enough energy will be propagated to reach the receiver-transducer.

J.E. Carrington: In approaching an existing installation how important are variations in pipe wall thickness, roughness, and pipe materials, some of which may not be known with precision?

Authors' reply: The accuracy of the meter depends largely on the accuracy of the data on pipe diameter, wall thickness and material (acoustic velocity). The outside pipe diameter can be measured, and the ultrasonic thickness gauge can be used to measure the wall thickness.

According to the "Code for the Measurement of Flow through Hydraulic Turbines by the Ultrasonic Method" established by the Electric Power Technical Committee of the Institute of Electrical Engineers of Japan, it is specified that the error in measurement of the outside diameter of the penstock should be within $\pm 0.1\%$, and wall thickness should be accurate to within $\pm 5\%$. Incidentally, the shear wave velocity in steel is 3200 m/s, but velocities in cast iron or in lining materials should be checked before the measurement is carried out.

The roughness of the inside wall will cause an error in measurement in pipes of small diameter, but the effect can be neglected unless the scale is very thick.

The lower limit of diameter for which Model UF-500 should be used is 300 mm. The instrument also has a built-in frequency counter which enables the sing-around frequency to be checked to find out if the data given for the pipe dimensions is accurate and if there has been build-up of scale during operation.

J.E. Carrington: Do any problems arise from leakage paths for the ultrasonic beams traversing around the pipe wall instead of passing through the fluid.

Authors' reply: It is true that the ultrasonic beam around the pipe wall will cause disturbance to the measurement. But, in the case of large diameter pipes, there exists large time differences between the ultrasonic beams around the pipe and through the fluid, and this is not a cause of error in the measurement of flow rate.

J. Causon: The results presented show considerable variation compared with other test methods reported. Can you account for this?

Authors' reply: There are cases in which considerable variations are shown between the ultrasonic method and other methods. The error can be caused in the ultrasonic measurement, but methods such as the thermodynamic, pitot tube or current meter may also include errors.

Many test results are given in Table 4, but few of these tests were undertaken with the object of making a real comparison of the method of measurement. In most cases, the tests were carried out to obtain data relating to the operation of electric power plant. Therefore, adequate facilities for accurate flow measurement were not always available. A prototype ultrasonic meter was used, and the purpose of the work at that time was mainly to determine if the ultrasonic method was convenient for use rather than to establish the accuracy of measurement.

If we select only the measurements where intercomparison was the first object, and if we pick-up only data obtained at power plants which had satisfactory conditions for accurate flow measurement, the differences are reduced remarkably.

J. Causon: The use of externally located transducers is convenient, but presumably makes accurate determination of the second path length difficult. To what extent would this affect your results?

Authors' reply: The accurate data on pipe material, construction and dimensions are the first requirement for accurate measurement. According to the Code for Flow Measurement mentioned above, the expected error will be ±1.2% of the reading, apart from the error due to the correction coefficient. In practice, as mentioned in section 3.1.3, the given data can easily be compared with the value of c/f_0 obtained by actual measurement.

J. Causon: Using external transducers is it possible to measure accuracy by using more than one set of transducers?

Authors' reply: Yes, generally speaking, if more sets of transducers are used, the accuracy will increase. Especially in cases where the flow velocity profile is not symmetrical in respect to pipe axis the two path method as shown in Fig. 10, is then very useful. This is now the standard method in Japan for the measurement of efficiency of hydraulic turbines. Other methods can be considered such as having acoustic paths not crossing the axis of the pipe. However, as the ultrasonic beam passes through the pipe wall, the incident angle of the beam into the water is limited, and not much improvement in accuracy can be expected.

J. Causon: What is the main purpose of your tests? Is it to determine efficiency or is it to establish a routine check on machine performance?

Authors' reply: The main purpose of tests at hydraulic power stations is to determine the efficiency of the turbine. The ultrasonic flowmeter has its merits in that (1) it is portable and easy to handle, and (2) has good repeatability. It is suited to determining the time when the turbine should be repaired, as well as to comparing the difference in efficiencies before and after repair of the machine. The flowmeter is being used for this purpose in Japan by all the major electric power companies.

B.T. Goldring: I must congratulate the authors on a very interesting paper and would like to know, when measuring water flow in a closed pipe, if their flowmeter is affected by the presence of air bubbles in the water.

Authors' reply: The energy of ultrasonic waves is attenuated by air bubbles in the water. Hence, if many bubbles are included, the ultrasonic beam cannot reach the receiver and a measurement cannot be made.

Air bubbles, by their nature, come up to the upper surface as they float in the water. Therefore, if the probe site is away from the source of bubbles, the acoustic path will be beneath the bubbles and there will be no trouble. But when a volume of bubbles forms a layer of air at the upper side of pipe, an error in measurement is caused as the ultrasonic measurement is based on the assumption that the pipe was full.

B.T. Goldring: Figure 12 shows an acoustic path which goes horizontally across a buried channel. Why do you not have the acoustic path vertical, as this would make the pit unnecessary if the transducers were located in the roof of the channel?

Authors' reply: Ultrasonic waves are reflected from air, and if air exists in the pipe, the ultrasonic beam will hardly pass through. As a rule, air bubbles in water come to the upper surface of the pipe. Therefore, it is not recommended that transducers should be fitted at the upper surface. In addition, in the case of reflection method (V-configuration) where an ultrasonic pulse is reflected back from opposite pipe wall, the possible existence of mud at the bottom of the pipe will considerably absorb the ultrasonic energy. Due to the reasons given above, in order to secure stable measurement, the acoustic path is so designed as to pass horizontally across the buried channel.

ULTRASONICS AS A STANDARD FOR VOLUMETRIC FLOW MEASUREMENT

Seth G. Fisher and Paul G. Spink

Abstract: Leading edge ultrasonic flow measurement systems have demonstrated that volumetric flow, even with badly distorted velocity distributions, can be measured to one percent accuracy without the need of prior calibration. The equipment technique is based on time measurements of sonar pulses propagated in opposite directions across the fluid medium. Measurements taken in several acoustic paths are combined according to a numerical integration technique to compute volumetric flow. Accuracies as precise as 0.1 percent of flow are achievable when the residual error of the integration has been established and corrected. This paper substantiates performance analysis with the supporting evidence of application data. Theoretical accuracies are computed based on mathematical expressions. First presented is the exact mathematical expression for volumetric flow rate. This is then compared with the volumetric flow rate as calculated by a leading edge ultrasonic flow measurement system. An error analysis covering applicable sources of error is performed. The last part of the paper summarizes the laboratory and field comparison evaluations that have been run in support of the accuracy predictions.

1   INTRODUCTION

Ultrasonic flow measurement systems have demonstrated accuracy, reliability, and a general independence of inhomogeniety in the fluid. Use of the leading edge of the received pulse to operate time measuring circuits has proven successful in avoiding problems associated with refraction and multi-path effects in the acoustic propagation[1]. There is clear indication that ultrasonic systems may be successfully employed in a wide variety of fluid flow measurement applications. It is also clear that exploitation of their capabilities requires detailed evaluation of the accuracy which can be achieved in typical flow measurement situations, and an understanding of the effects which may limit the ultimate precision of such systems.

This paper summarizes a detailed analytical evaluation and laboratory confirmation of the accuracy of an unique ultrasonic system applied to the measurement of volumetric flow. It is demonstrated that, with thorough evaluation and application of results of this kind of analysis, accuracy approaching that of standard laboratory calibration facilities, i.e., a weighing tank and associated equipment, may be achieved in field installations.

The system computes volumetric flow by the Gaussian numerical integration technique[2], using measurements taken in parallel acoustic paths. Pulses are transmitted simultaneously from opposite ends of each path, and the flow computation is based on direct measurements of the associated travel times. This approach permits achievement of the best possible accuracy, minimizing the effects of spatial and temporal variations of sound and fluid velocities. This paper considers only a fixed fluid cross-section, i.e., flow in closed conveyances, or open conveyances with limited variations in stage. Slight modifications of the flow computation technique are required to handle the case of varying stage.

Two categories of error arise in the flow measurement: those which are generally independent of conveyance shape and distributions of sound and fluid velocities, and those which are determined by conveyance shape and velocity distributions. The former includes mechanization errors which are governed by accuracies achieved in scaling, computation, and travel time measurement. The latter includes integration and crossflow errors associated with hydraulic conditions and variations in sound velocity in the fluid. The principal point of difference is that mechanization errors may be directly controlled in implementation of acoustic, electronic, and computational system designs, while the influence of integration and crossflow errors is largely a matter of adaptation to environmental conditions of the particular application.

Both types of error require detailed evaluation to insure achievement of the best possible overall accuracy. These evaluations are carried out in this paper for typical installations in pipes of 19.25 in. and 16 ft. diameter. The 19.25 in. pipe application corresponds to the installation employed in recently completed weigh tank tests. The weigh tank tests were designed to confirm the predicted Gaussian integration error for fully-developed flow in a straight pipe and to demonstrate accuracy in an installation involving substantial hydraulic complexity upstream of the measurement section. The 16 ft. pipe application corresponds to the installation for efficiency tests in which this ultrasonic system was used to measure the flow of water through a hydraulic turbine.

The evaluation of integration errors considers a four path Gaussian system, for which accuracy within approximately one percent has been found to result for a wide variety of velocity distributions representative of flow in different hydraulic configurations. In applications in which this order of accuracy is acceptable and only approximate knowledge of the velocity distributions is available, it is ordinarily appropriate to rely upon the inherent accuracy of the Gaussian technique; special calibration of the flow measurement system is unnecessary. On the other hand, achievement of 0.1 percent accuracy, which is within the capability of this type of system, generally requires compensation for the residual error of the Gaussian integration. In order to establish validity of the predicted integration error, it is necessary not only to develop precise techniques for its evaluation, but also to demonstrate a low sensitivity of the error to variations in the velocity distribution. Sensitivity is important in applications where the velocity distributions may not be precisely defined. Accordingly, these subjects are treated in some detail. Techniques for the prediction of integration errors for distributions defined only in terms of measured velocity data are also discussed. A further type of application involves more complex flow conditions, for which analytical prediction of integration and crossflow errors is not readily accomplished and model tests are appropriate. Results of such a test are presented.

2   THEORY OF VOLUMETRIC FLOW COMPUTATION

Evaluation of the accuracy of an ultrasonic system employing several pairs of transducers is facilitated by stating the flow measurement problem as a requirement to measure the net flow of fluid across a plane surface. Figure 1 illustrates this "measurement plane" oriented at an angle θ with respect to the axis of a round pipe. Needless to say, the flow across any oblique plane is the same as the flow across a plane perpendicular to the axis of flow so long as the total flow into the measurement section equals the total flow out. In Fig. 1, an ellipse is defined by the intersection of the measurement plane with the pipe wall, and volumetric flow rate is given by the integral of the normal component of the fluid velocity vector, $V(z, \xi)$, over the area of the ellipse:

$$Q = -\iint_A V(z, \xi) \cdot \hat{\xi} \, d\xi dz = \iint_A \left\{ u(z, \xi) \sin \theta - v(z, \xi) \cos \theta \right\} d\xi dz \qquad \ldots (1)$$

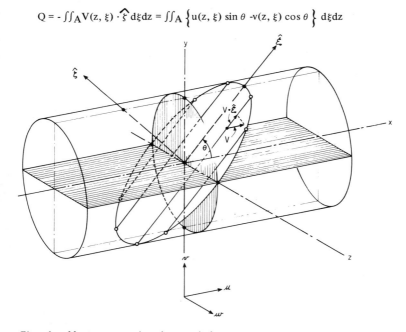

Fig.   1   Measurement plane in round pipe

As shown in Fig. 1, $\hat{\xi}$ represents a unit vector normal to the measurement plane, and u and v are components of the velocity vector in the directions of the x and y coordinate axes. It is noted that w, the z-directed component of fluid velocity in the measurement plane does not contribute to discharge.

In general, the limits of integration are established by intersection of the measurement plane with the boundary of the conveyance. In open channels, the free surface forms an appropriate portion of the boundary.

The ultrasonic flow measurement system considered in this paper employs pairs of transducers installed at the boundary of the conveyance so as to form parallel acoustic paths in the measurement plane. The transducers are employed both as transmitters and receivers, so that a single path is associated with each pair. Fig. 1 shows four acoustic paths.

Discharge is computed by doubly integrating the component of fluid velocity in the direction of the acoustic paths (the direction of the unit vector $\hat{\xi}$), taking into account orientation of the measurement plane with the nominal axis of flow (the x-axis in Fig. 1). The inner integral, the line integration over the acoustic paths, is evaluated directly from measurement os transit times of the acoustic pulses. The outer integral is evaluated numericaly as the sum of weighted outputs of the individual acoustic paths.

As shown in the Appendix, the line integration over path i is given by

$$f_i = \frac{\ell_i^2}{2} \left\{ \frac{\triangle t}{t_1(t_1 + \triangle t)} \right\}_i \tan \theta_i \cong \tan \theta_i \int_{-\ell_i/2}^{\ell_i/2} V(z_i, \xi) \cdot \hat{\xi} \, d\xi$$

$$= \int_{-\ell_i/2}^{\ell_i/2} \left\{ u(z_i, \xi) \sin \theta_i + v(z_i, \xi) \sin \theta_i \tan \theta_i \right\} d\xi \qquad \text{... (2)}$$

where t_1 is the shorter of the transit times for oppositely-directed acoustic pulses in a path, of length ℓ, $\triangle t$ is the difference between the longer and shorter transit times, and the subscript i is used to designate the quantities associated with path i. Comparison of Equations (1) and (2) indicates that the u component of fluid velocity is correctly integrated over an acoustic path, while an erroneous output, or crossflow error, results if the integral of v is not zero. w does not contribute to either discharge or flow measurement system output.

For a system employing N paths, the measured flow rate is

$$Q_m = D \sum_{i=1}^{N} w_i f_i \qquad \text{... (3)}$$

This equation expresses performance of the outer integral by a selected method of numerical quadrature, for which weighting factors, w_i, and path locations, z_i, are uniquely specified, and the scaling constant, D, corresponds to the maximum span of the conveyance across the measurement plane. In practice, the calculation is generally repeated a large number of times to provide a single measurement of the mean flow rate over a selected time interval.

Of various available quadrature methods, that due to Gauss is generally preferred on the basis of its inherent accuracy. For measurements at N locations, it integrates exactly any polynomial of degree 2N-1 or less[3].

The flow computation expressed by Equation 3 is explicitly determined by the physical dimensions and orientations of the acoustic paths, by the time measurements, and by the weighting factors specified for the quadrature integration. In some applications it may be advisable to introduce a small scaling correction to account for predictable error in the quadrature integration, and/ or due to a crossflow effect. A correction of this type generally amounts to one percent or less for a Gaussian system employing four or more acoustic paths.

3   ERRORS INDEPENDENT OF FLUID VELOCITY DISTRIBUTION AND CONVEYANCE SHAPE

Errors not directly determined by spatial distributions of fluid and sound velocity comprise those associated with scaling for the physical dimensions, those determined by the precision of time measurements, and those associated with the precision of computations.

3.1 Dimensional scaling

Dimensions which have critical effect on accuracy of the volumetric flow measurement are those indicated in Equations 2 and 3: the maximum span, D, of the measurement plane; and the length, ℓ_i, and angle, θ_i, of each acoustic path. Minimum resultant error in flow measurement is achieved when as fabricated measured values of these dimensions are used. Typically where the installation is accomplished in the factory D and ℓ_i can be measured to a tolerance of ± 0.002 inches and θ_i can be measured to a tolerance of ± 1 minute of arc. In the field where surveying techniques are used, D and ℓ_i can be measured to a tolerance of $\pm 1/32$ inch and θ_i can be measured to a tolerance of ± 40 seconds of arc, typically. Since metering section size will generally determine the fabrication method used, the tighter tolerances for the linear measurements can be achieved for the smaller sizes resulting in percentage error in flow computation that is small and tends to be independent of size. For example, the error due to measurement tolerances in a 19.25 in. inside diameter pipe tested in the laboratory was ± 0.04 percent, and for a 16 ft pipe where installation took place in the field, the error was ± 0.04 percent.

Operation at elevated temperature, where expansion may be significant, increases the influence of dimensional effects; analysis of a particular heat exchanger application shows an error of approximately ± 0.1 percent.

The flow measurement is less sensitive to errors in lateral positioning of the ultrasonic paths; i.e., placing the sampling points of the numerical integration, than it is to errors in D, θ, and ℓ. Thus, although the errors due to lateral spacing are determined by fabrication, rather than measurement tolerances, they can be made very small, less than 0.01 percent for typical velocity distributions and normal fabrication techniques.

3.2 Time measurements

Inaccuracy in flow measurement due to time measurement error is largely determined by the stability of electronic circuits involved in the measurement of the time difference, $\triangle t$. Required absolute accuracy of measurement of the transit time, t_1, is generally less severe by two or three orders of magnitude. Contributions to error in the $\triangle t$ measurement may be encountered due to delays in propagation of electrical signals through the receivers and in operations of triggering and gating circuits involved in the time interval measurement. Errors are produced by differences in delays in the two received signal channels for a given acoustic path, as opposed to overall delays. They are obviously minimized by employing nominally identical channels. Their variability ordinarily does not appear in measurements taken over a time period of a few minutes, but data taken at intervals of several hours, or days, can be expected to show scatter attributable to this effect.

Noting that sound velocity, c, is given approximately by the ratio of path length to transit time, Equations 2 and 3 can be used to show that a time measurement error, $\delta(\triangle t)$, produces an error in the measured flow rate given by

$$\delta(Q_m) \simeq \frac{D c^2 \tan \theta}{2} \sum_{i=1}^{N} w_i \, \delta(\triangle t)_i \qquad \qquad \text{... (4)}$$

Using this relationship, measured long-term variability in flow rate measurements by a four path system installed at $\theta = 45$ degrees in a 19.25 in. pipe has been analyzed to assess the influence of time measurement errors. Data taken over a period of eleven days show a long-term variability equivalent to a random time measurement error of $\delta(\triangle t) \cong 2.5$ nanoseconds, rms. The system employed time interval measurement circuits with measured stability of one nanosecond, rms, and a 1.6 mHz sonic frequency. Translated into stability of delays in propagation of electrical signals through the receiving circuits, this result is indicative of random phase differences of about three degrees, rms, in each of the four pairs of receivers. The scatter of data points in Fig. 5 illustrates the effect of this inaccuracy in the measurement of Δt for the 19.25 inch pipe. The resultant error in a single measurement of discharge is 0.05 cfs, rms, or approximately 0.15 percent of the maximum flow rate. The corresponding value for installation in a 16 foot pipe with θ = 65 degrees is 1.1 cfs, rms, or 0.014 percent of maximum flow rate.

The travel time measurements are also subject to short-term variation associated with turbulence in the fluid flow. This variation generally can be reduced to any desired degree by employing a sufficient number of individual measurements in each determination of mean flow rate. Typical results are illustrated by measurements taken at a Reynolds number of 3 x 10⁶ in the 19.25 in. pipe. Successive values of mean flow rate calculated over approximately 1200 measurements in each acoustic path showed turbulent variation of 0.07 percent, rms, in the straight pipe, and 0.12 percent, rms, in the heat exchanger model of Fig. 2, where it is reasonable to expect a higher level of turbulence. These tests also showed variability in successive determinations of mean flow rate to be inversely proportional to the square root of the number of measurements, as would be expected for turbulent variations having a time scale which is small compared to the measurement interval. Similar results were obtained when this system was installed in a 16 foot pipe. In this case, the mean flow rate computation encompassed approximately 700 measurements in each acoustic path, and variability among different determinations of mean flow rate under nominally identical conditions was about 0.1 percent, rms.

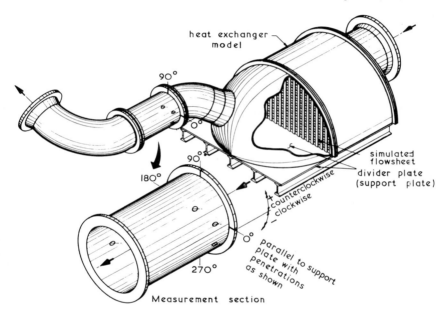

Fig. 2   Measurement section installed in outlet piping of heat exchanger model

3.3 Computational precision

The errors associated with the precision of computation for an all digital system are determined by word byte size and register size. The system can be designed to make the resultant error insignificant. For the flow measurement system evaluated against the weigh tank, computational accuracy was 1 part in 4000 or 0.025 percent.

4   ERRORS DEPENDENT UPON VELOCITY DISTRIBUTION AND CONVEYANCE SHAPE

There are three types of error which are influenced by conveyance shape and spatial distributions of fluid and/or sound velocities over the measurement plane. Two relate to the two integrations involved in the flow computation; the third, designed herein as the crossflow error, is determined by uncertainty in specification of the axis of flow.

4.1 Crossflow

Comparison of Equations 2 and 3 with Equation 1 shows the crossflow error to be determined by the integral of the v component of fluid velocity over the measurement plane. This is obviously zero when there is no secondary flow, as in fully-developed flow in a straight pipe. In installations downstream of simple bends, partially closed gate valves, or devices producing similar profile distortions, it is generally possible to orient the measurement plane, by rotating it about the x-axis, so as to reduce this integral to zero and thus to eliminate the crossflow error. In more complex installations, it is sometimes impossible to establish the axis of flow with precision, so that model tests may be required to verify accuracy of the flow measurement. In such situations, the resultant of crossflow and quadrature integration errors is measured in the model tests, but it is not ordinarily easy, nor is it necessary, to distinguish between the two. An example of this type is illustrated by the heat exchanger of Fig. 2.

4.2 Line integration

As indicated in the Appendix, the line integration, Equation 2, is precise when there is no spatial variation in either fluid or sound velocity. The series expansion outlined therein indicates that the following three significant error terms are encountered when there is systemmatic variation over an acoustic path:

(1)   A relative error which is equal to the ratio of the mean square deviation of fluid velocity, with respect to its mean over the path, to the square of the mean sound velocity;

(2)   A relative error which is equal to the ratio of the mean square deviation of sound velocity, with respect to its mean over the path, to the square of the mean sound velocity;

(3)   A relative error which is equal to twice the negative of the ratio of the mean product of fluid and sound velocity deviations from their mean values, to the product of mean values of sound and fluid velocity over the path.

Knowledge of the distributions makes it possible to calculate errors for each path and to sum their contributions to the flow rate calculation of Equation 3. If fluid velocity approaches sound velocity in magnitude, the first error becomes significant; if severe sound velocity gradients are encountered, the third may be appreciable. Otherwise, the line integration is highly accurate; confirmation may be achieved from calculations based on rather imprecisely defined distributions.

An illustration of typical magnitude of the first error is provided in the measurement of flow of water at ordinary ambient temperatures, where water velocity is generally one percent or less of the sound velocity, so that the relative error in flow measurement is generally smaller than 0.01 percent. Similarly, the second error will ordinarily be less than 0.01 percent, since sound velocity variation along a path is generally substantially less than one percent.

The third error is seen to depend upon spatial correlation of fluid and sound velocities. Thus, if extremes of fluid and sound velocity tend to correspond, i.e., if higher temperature fluid has higher than average velocity, or vice versa, the error in the line integration tends to be greater than would be experienced if this tendency were less pronounced. In the particularly difficult application to a heat exchanger operating at elevated temperature, calculations show an error in flow measurement of about 0.1 percent for the most severe distributions expected to be encountered.

4.3 Integration by Gaussian quadrature

Numerical quadrature techniques are designed to perform accurate integration without any compensatory scaling or functional alterations to account for characteristics of the function being integrated, and to employ in the evaluation no other information than values of the function at a discrete set of points. Successful application to the ultrasonic flow measurement problem requires assessment of the accuracy of the method for the fluid velocity profiles to be encountered. If this can be done with precision, then it is possible to alter the scaling as appropriate to compensate for any resultant integration error. The following summarizes some results of investigations directed toward assessing the accuracy of four sample Gaussian quadrature for fluid velocity distributions for which analytical definitions are available, as well as those having only a graphical definition based upon measured data.

   Figure 3 compares integration accuracies calculated for two different characterizations of fully-developed flow in a straight round pipe. The solid line corresponds to the standard log profile.[4] It was produced by evaluating the closed-form integral of the logarithmic function over each of the four acoustic paths. Also shown in this figure is a point, similarly calculated, corresponding to the parabolic profile for laminar flow. The remaining points represent evaluations for the inverse power profile of Nikuradse,

$$\frac{u(r)}{u} = K \left(1 - \frac{r}{r_0}\right)^{\frac{1}{n}}$$

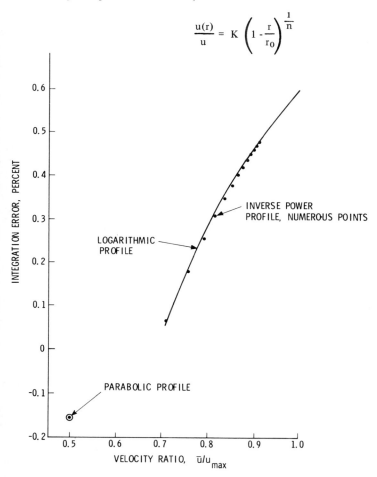

Fig. 3   Four-path Gaussian integration of fully-developed flow in a straight pipe

where r is radial distance from the axis of the pipe, r_0 is its radius, and K and n are constants. Fig. 3 shows evaluations for integral values of n from 4 through 16. In the worst case, these values show a difference of 0.015 percent with respect to integration error for the log profile. This is two orders of magnitude smaller than the maximum difference relative to mean velocity, between fluid velocities at the same point for the two different characterizations of the velocity distribution. This, in itself, is indicative of a low sensitivity of the Gaussian integration to profile variation.

   A further, and perhaps more significant, indication of the low sensitivity of the Gaussian method is provided by the total range of integration error covered by the curve of Fig. 3. It is seen that a variation in error of less than 0.6 percent is realized over profile variation covering velocity ratios from 0.7 to 1.0, the entire range of turbulent flow in a straight pipe.

Figure 4 plots the results of similar calculations corresponding to flow computation from output of a single path placed across the diameter. To illustrate the comparison, it also shows the curve of Fig. 3. The ratio of magnitudes of slopes of these two curves is approximately 27. This demonstrates a great reduction in sensitivity to profile variation when four paths, arranged to perform a Gaussian integration, are employed, rather than a single path calibrated for a specified nominal profile. This general conclusion is also evident in the difference between the logarithmic and inverse-power characterizations, which, for the single path system, approaches 0.7 percent at the low velocity ratios, compared to the maximum difference of 0.015 percent for the four path Gaussian system.

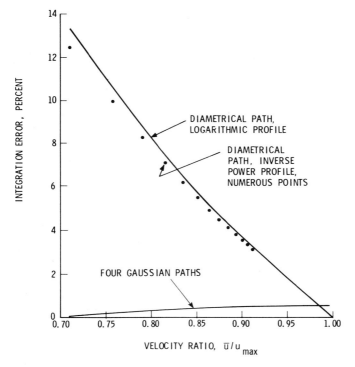

Fig. 4   Integration errors for turbulent flow

Figures 3 and 4 are based upon a particular type of variation in the velocity distribution. The question naturally arises as to ensitivity to other types of variation. Distorted profiles, e.g., as typically encountered downstream of bends, are of special interest, but not amenable to the same concise mathematical analysis as has been accomplished for the symmetrical distributions of fully-developed flow. Because the Gaussian technique is applicable to integration of both odd and even functions, there is no inherent reason to expect that accuracy or sensitivity should be poorer when it is applied to unsymmetrical distributions than has been demonstrated for these symmetrical distributions. It is only necessary to employ other methods to establish the integration error.

Assessment of the accuracy of Gaussian quadrature for velocity distributions defined in accordance with measured data may be readily accomplished by graphical integration. Validity of the assessment is, however, strongly dependent upon the precision with which this process is carried out. One method which has been found to give consistently good results involves a least-squares polynomial fit to a large number (typically 40 or more) of equally-spaced points defining integrals of the velocity profile over parallel lines across the measurement plane. The polynomial is then taken to provide a precise mathematical definition of the function to be integrated by Gaussian quadrature to measure the rate of flow. The evaluation is repeated for polynomials of successively higher degree. Successive calculated values of the Gaussian error generally tend to stabilize and thus to provide an accurate evaluation of integration error for the actual velocity profile.

To test the precision with which this curve-fitting technique is capable of evaluating the Gaussian integration error, it has been applied to the logarithmic profile of fully-developed turbulent flow in a round pipe. Results are presented in Table 1 for a 41 point definition of the integrand, including zero values at the pipe walls. The second column indicates the degree at which the calculated error first falls within the indicated range, where it remains as the degree is increased until computer limitations prevent further improvement in the quality of the polynomial fit to the data points. The excellent agreement with the actual values, within 0.05 percent in the worst case, is believed to justify use of this technique for measured profiles for which there exists no standard of comparison. It is thus possible to realize very precise prediction of integration accuracy in applications in which nominal profiles, or expected ranges of their variation, have been established in model tests, or as a result of hydraulic similarity to installations in which profile measurements have been taken.

TABLE 1   Polynomial approximation for calculating four path Gaussian integration error, logarithmic profile

$\dfrac{\bar{u}}{u_{max}}$	Degree of polynomial where Gaussian error stabilizes	Degree of polynomial giving best fit to 41 data points	Range of calculated Gaussian integration error for polynomial fit, per cent	Gaussian integration error for logarithmic profile, per cent
0.88	6	22	+0.44 to +0.48	+0.43
0.835	16	22	+0.36 to +0.38	+0.36
0.78	14	20	+0.23 to +0.25	+0.25

Calculations of this type have been carried out corresponding to installations in round pipes at various distances downstream of pipe bends of different geometry, partially open gate valves, and other fittings. They show integration by four path Gaussian quadrature to be generally accurate to better than one percent, even for severely distorted profiles; e.g., when the entrance to the measurement section is as close as one-half diameter to the exit of a short-radius bend. For the rectangular duct profile data of Nikuradse[5], the calculated Gaussian error is plus 1.1 percent for path projections parallel to either side of the rectangle.

These results indicate that system accuracy of the order of one percent is realizable in measurement configurations in which profile information is poorly specified, and that better accuracy is obtainable by compensating for the predicted Gaussian error where this information is well defined. When this is done, system accuracy is largely determined by the uncertainty of the prediction. Where crossflow errors can be eliminated by orientation of the measurement plane, it should be possible to achieve accuracy as good as ±0.2 percent by careful analysis of the integration error, even for severely distorted profiles.

In other situations, the combination of a high accuracy requirement and practical difficulty in defining the distributions of fluid velocity magnitude and direction, may dictate the use of model tests in which the flow measurement system is installed to simulate actual operating conditions. An example of this type is discussed in the following paragraph.

5   MODEL TESTS

A series of model tests was conducted during the period from 9 March to 19 March 1971, in which a portable model of the Westinghouse L.E. Flowmeasurement System was programmed to perform a four path Gaussian integration of flow in a 19.25 in. round pipe. The 100,000 pound weighing tank and associated calibration facility of Alden Research Laboratories, Worcester Polytechnical Institute, Holden, Massachusetts, provided the standard against which flow rate measurements were compared.[6]

Purposes of the tests were:

(1)   To confirm the predicted Gaussian integration error for fully developed flow in a straight pipe, and

(2)   To measure the resultant of Gaussian integration and crossflow errors when the measurement section was installed in the output piping of the heat exchanger model of Figure 2.

Before running the straight pipe tests, the velocity distribution at the measurement location was measured using a pitot rake. Analysis of the pitot rake measurements disclosed an excellent fit to a logarithmic profile corresponding to

$$\frac{\overline{u}}{u_{max}} = 0.87$$

for which the analysis indicates a Gaussian integration error of 0.42 percent (see Fig. 3). Results of the weigh tank evaluations are presented in Fig. 5. Each plotted pitot represents the mean, using from 3 to 10 consecutive weigh tank runs, of deviations between flowmeter and weigh tank determinations of the mean rate of flow. A total of 220 runs was made over a period of five days to provide a sample size sufficient to check the variability of measured data against anticipated limitations in performance of the acoustic travel time measurements. These results, as well as similar data taken in the heat exchanger model, support the conclusion that the observed variability may be attributed to this cause. No correlation could be found with other variables, i.e., water temperature, pressure, or mean flow rate; the slight change due to Reynolds number variation predicted from Fig. 3 is obscured by the relatively greater influence of data scatter at the low flow rates.

Plotted on Fig. 5 are a line with 0.42 percent slope, representing the calculated Gaussian integration error as well as a regression line defining the least-squares fit to the measured data. The latter has a slope of 0.48 percent, indicating excellent agreement with the predicted value.

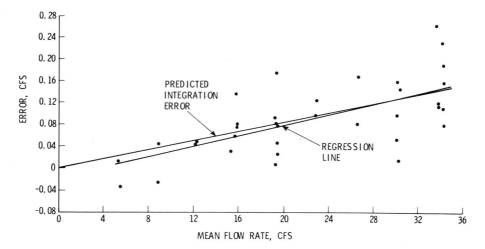

Fig. 5   Weigh tank tests; straight pipe

Figure 6 displays the results of weigh tank evaluations when the measurement section was installed at various orientations in the heat exchanger model of Fig. 2. Figure 6 shows indicated values of the resultant of integration and crossflow errors centered on vertical lines defining 95 percent confidence limits determined by analysis of data variability. The $0°$, $240°$, and $330°$ orientations show narrower confidence limits, reflecting the fact that greater numbers of runs were taken at these orientations.

A degree of speculation is involved in interpreting the variation with angle indicated in Fig. 6. However, there exists an evident positive bias, which is consistent with results usually encountered in Gaussian integration in a round pipe. Also, it is indicated that net flow in the measurement section has an upward and outward (to the left in Fig. 2) tendency. This is influenced in part by the relatively high horizontal velocity of fluid exiting the simulated tubesheet. (Fig. 2 shows the measurement section in the $0°$ orientation. Since the angular scale was fixed to the flange of the measurement section, a positive value of the measured angle corresponds to clockwise rotation of the measurement section).

6 HYDROELECTRIC EFFICIENCY TESTS

Tests were performed from 25 May through 28 May 1971, to measure the efficiency of a hydraulic turbine. Penstock flow was measured according to the method described in this paper. The piping configuration is shown in Fig. 7 which also shows the location of the flow measurement section, just downstream from a converging 10 degree bend. Measurements were made over the full range of gate settings, including speed on load as well as closed, where leakage flow was recorded. The flow range varied from 12 cfs for leakage flow to approximately 7700 cfs for 100 percent gate.

The head on the turbine was near rated value of 500 ft and was essentially constant throughout the test period. The pool elevation increased only 0.5 ft and the plant was operated in such a way as to keep the tailrace level constant within 1.5 ft. For determining efficiency there were 64 runs made at 19 different gate settings made by the use of gate blocks. Each run consisted of the average of 11 readings taken every 30 seconds over a 5 minute interval, and each flow reading represented the average of 64 measurements in each acoustic path.

Consistency of the discharge measurements is shown by the individual run deviations plotted in Fig. 8, where deviations are plotted against the mean value of all runs at each gate setting. Except for a single point at 70 percent gate, which fell approximately one percent below the six other runs at this gate setting, the data show excellent repeatability. As a result, the average of only a few runs at each gate setting establishes mean flow rate well within 0.1 percent. (Cause of the single low discharge measurement was identified as occasional false triggering on electrical

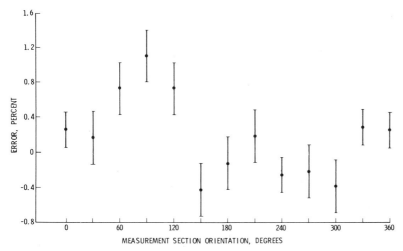

Fig. 6   Weigh tank determination of integration and crossflow errors; heat exchanger model
(Note: 95% confidence limits are shown for each mean value of measured error)

noise due to a fault in the flowmeter circuitry. This occurred during the practice set of runs prior to the start of the official test and was subsequently corrected.) The calculated rms value of run deviations, omitting the low reading and accounting for the effect of a small number of points at each gate setting, is 0.11 percent. The variation is influenced by the repeatability of gate settings (which is considered to have been very good in these tests), and of head corrections, as well as the influence of turbulence in the flow measurements. The indicated contribution of the turbulent variation, i.e., something less than 0.11 percent, rms, over approximately 700 measurements, is comparable to the 0.07 percent value realized over 1200 measurements in the 19.25 in. pipe. Its effect on error in determination of the mean value for several runs is reduced according to the square root of the number of runs. For these tests the flowmeasurement system was installed so as to avoid crossflow errors and scaled to compensate for the predicted value of error in the Gaussian integration. Because the velocity profile was well defined and only moderately distorted, it is expected that accuracy of this compensation was within 0.1 percent.

7   CONCLUSIONS

Table 2 summarizes the error evaluations discussed in detail in this paper for three installations: the 19.25 in. straight pipe, the heat exchanger modelled in Fig. 2, and the 16 ft pipe of the efficiency tests. For the heat exchanger, larger values of dimensional scaling and line integration errors reflect the influences of operation at elevated temperature, while the time measurement

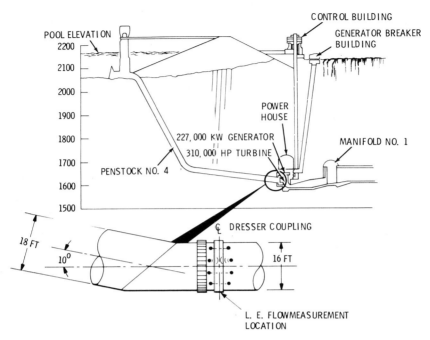

Fig. 7   Hydroelectric plant layout

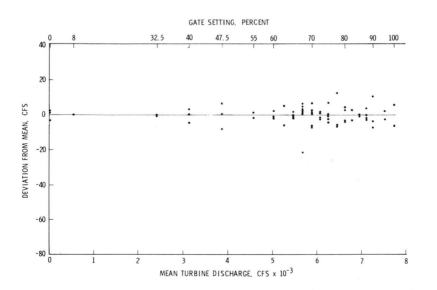

Fig. 8   Turbine discharge measurements; corrected to 500 ft head

error is substantially smaller than in the 19.25 in. pipe, because of the larger flow and the lower sound velocity of the actual application. It is seen that in both full scale applications, overall accuracy is mainly dependent on the precision of determination of the Gaussian integration and crossflow errors.

Generally, experimental results provide excellent confirmation of predicted performance of the flow measurement system. Especially noteworthy is the close agreement between predicted and measured evaluation of the Gaussian error in a straight pipe. A major result of the analysis is demonstration of the extremely low sensitivity to variations in velocity distribution achieved for a four path Gaussian system. It is concluded that 0.1 percent accuracy may be realized in typical field installations.

8   APPENDIX: Evaluation of line integral

The integral of fluid velocity over an acoustic path is related to transit times of acoustic pulses propagated over the path. Thus, transit times for pulses transmitted in opposite directions over path i may be expressed as

$$t_{1i} = \int_{-\ell_{i/2}}^{\ell_{i/2}} \frac{d\xi}{c(z_i, \xi) + V(z_i, \xi) \cdot \hat{\xi}} \qquad \qquad \text{... (A.1)}$$

$$t_{2i} = t_{1i} + \triangle t_1 = \int_{-\ell_{i/2}}^{\ell_{i/2}} \frac{d\xi}{c(z_i, \xi) - V(z_i, \xi) \cdot \hat{\xi}} \qquad \qquad \text{... (A.2)}$$

Here it is assumed that both sound velocity, c, and fluid velocity, V, exhibit spatial variations over the measurement plane. The effects of temporal variations, e.g., due to turbulence in the flow, or time dependency in the temperature of fluid passing through the measurement section, are not included in the statement of these relationships. It is implied that the sampling process, represented by the basing of flow rate computation on the results of a large number of transmissions over each

TABLE 2 Summary of per cent errors (two-sigma) for typical applications of four path Gaussian ultrasonic flow measurement system

Error source	19.25 in diameter straight pipe	Heat exchanger operating at elevated temp.	16 ft diameter pipe
Dimensional scaling	0.04	0.1	0.04
Time measurement	0.3	0.03	0.03
Computational precision	0.025	0.025	0.025
Line integration	< 0.01	0.1	< 0.01
Quadrature integration and crossflow	0.1	0.2	0.1
Overall accuracy (square root of sum of squares of the above)	0.32	0.25	0.11

path, is unbiased, so that time averages correctly reflect the temporal means of c and V. This is, of course, necessary to accurate flow measurement, and in ordinary circumstances easily accomplished when the time measurements are based on the simultaneous transmission of acoustic pulses across the path.

Representing spatial means of sonic and fluid velocities over path i as

$$\overline{c_i} \quad \text{and} \quad \overline{(V \cdot \xi)_i}$$

respectively, and spatial variations with respect to these mean values as

$$\delta_{ci} \quad \text{and} \quad \delta_{vi}$$

then

$$c(z_i, \xi) = \overline{c_i} + \delta_{ci}$$

and

$$V(z_i, \xi) \cdot \hat{\xi} = \overline{(V \cdot \hat{\xi})_i} + \delta_{vi}$$

Using these substitutions, series expansions of Equations A.1 and A.2 may be manipulated to give

$$\frac{\ell_i^2}{2} \left\{ \frac{\Delta t}{t_1 (t_1 + \Delta t)} \right\}_i = \int_{-\ell_i/2}^{\ell_i/2} V(z_i, \xi) \cdot \hat{\xi} \, d\xi \left[1 - \frac{1}{\overline{c_i}} \left\{ \frac{2 \, \overline{\delta_{ci} \, \delta_{vi}}}{\overline{(V \cdot \hat{\xi})_i}} \right\} \right.$$

$$\left. + \frac{1}{\overline{c_i}^2} \left\{ \overline{\delta_{ci}^2} \quad \frac{3 \, \overline{\delta_{ci}^2 \, \delta_{vi}}}{\overline{(V \cdot \hat{\xi})_i}} \quad \overline{\delta_{vi}^2} + \frac{\overline{\delta_{vi}^3}}{\overline{(V \cdot \hat{\xi})_i}} \right\} \right]$$

... (A.3)

which is accurate to the second order in ratios of

$$\overline{(V \cdot \hat{\xi})_i}, \ \delta_{ci}, \text{ and } \delta_{vi}$$

to $\overline{c_i}$ (Equation A.3 uses the horizontal bar over a quantity to designate its mean value over the acoustic path).

Within the accuracy permitted by spatial variations of sound and fluid velocities,

$$\delta_{ci} \text{ and } \delta_{vi}$$

the left side of Equation A.3 correctly evaluates the integral of fluid velocity over an acoustic path, if length scaling is based upon the actual length of the fluid path, and the transit times for acoustic signals propagated over the fluid path are correctly measured. If significant time is spent in other materials than the fluid path, i.e. acoustic windows in front of the transducer faces, then measured transit times must be corrected accordingly.

Equation A.3 shows a single first-order term; this term may be expected to define quite closely the error which results when large sound velocity gradients are encountered. Of the four second-order terms, the first and third,

$$\frac{\overline{\delta_{ci}^2}}{\overline{c_i}^2} \quad \text{and} \quad \frac{\overline{\delta_{vi}^2}}{\overline{c_i}^2}$$

are ordinarily substantially greater in magnitude than the second and fourth.

9   REFERENCES

1   HASTINGS, C.R.: 'The L.E. Acoustic Flowmeter, An Application to Discharge Measurement', presented at the Annual Convention of the New England Water Works Association, Sept. 1969

2   MALONE, J.T., and WHIRLOW, D.K.: U.S. Patent No. 3,564,912

3   LAPIDUS, L.: 'Digital Computation for Chemical Engineers (McGraw-Hill, 1962)

4   SCHLICHTING, H.: 'Boundary-Layer Theory' (McGraw-Hill, 1968)

5   NIKURADSE, J.: 'Untersuchungen üben die Geschwindigkeitsverteilung in turbulenten Strömungen', Heft 281, 1926

6   HOOPER, L.J.: 'Report of The Evaluation of The Westinghouse Leading Edge Acoustic Flow Measurement System at the Alden Research Laboratories', ARL-M-71-85, March 1971

DISCUSSION

F.C. Kinghorn: This is a very interesting paper, and in particular the evaluations of Gauss' quadrature method of integration, as applied to flow measurement, is most useful.

To my mind the most important source of error in the technique described would appear to be the effect of swirl or "cross-flow". How do the authors ensure that the detectors are aligned in such a way that swirl has no effect on the results? In complex flows is it necessary to build a model in order to determine the nature of the swirl by using point velocity measurements?

W.R. Loosemore: Can the authors comment on the sensitivity of the method to changes in the velocity profile over the length of pipe between the two measuring transducers. Although independent of any general asymmetry in profile, does not the technique assume an invariance in the profile over the measuring region during the time of measurement?

How is the zero set before a measurement? Is it necessary to assume that the fluid is stationary?

G. Causon: It would appear that the high accuracy claimed for the power station measurements is due in part to the fairly close location of the test section to a small bend and convergence.

How would you proceed in the more usual case where measurement would necessarily be several diameters downstream of a bend? In such a case the flow has a rotational component which may continue for as much as 30-40 diameters.

Frequently pipes pulsate, changing shape from truly circular. Have you any comment on the effect of this on accuracy?

How much simplification would be possible if you set out to work to an accuracy of ±0.25% rather than ±0.1%? For example could you use externally mounted transducers?

Authors' reply: A further explanation of the effects of crossflow and swirl is appropriate in view of the comments of the last three speakers.

Let us first define terms. We use the term "crossflow" to describe any net transverse mass velocity vector in the flow measurement pipe section. Such transverse flow occurs downstream of a bend, for example, where the distorted velocity profile resulting from the acceleration forces in the bend gradually becomes symmetrical, due to the viscous forces in the straight section following.

We define swirl as any rotational component to the mass velocity. Such swirl will exist downstream of two tightly coupled non-planar bends.

Crossflow and swirl can exist separately or concurrently in actual fluid systems. In the high Reynolds regime where our ultrasonic system finds its most frequent applications, crossflow effects are typically insignificant 10 diameters or more downstream of the disturbances which caused them. On the other hand, rotational effects, once induced, will persist 50 to 100 diameters downstream of the disturbances which caused them.

Our ultrasonic system is insensitive to the rotation of velocity distributions. That is, swirl alone produces no error. Any tangential velocity vector due to fluid rotation which projects onto one ultrasonic path is exactly compensated by an equal and opposite tangential velocity vector projecting onto the symmetrically opposite ultrasonic path. This insensitivity to fluid rotation is in marked contrast to that of energy-type fluid meters such as flow nozzles. Here the change in rotational energy as the fluid proceeds through the nozzle will reflect itself in an increased differential pressure reading for the nozzle for a given flow – a change in calibration. Because a long straight section is required to eliminate rotational effects at high Reynolds numbers, we believe that such effects are contributing significant, though undetected, errors in many current flow nozzle installations.

With respect to crossflow, the error, if any in our system, is determined by the orientation of the ultrasonic paths relative to this crossflow. Crossflow produces an error only if the vector representing the transverse mass velocity projects itself onto the ultrasonic path. In this situation, it adds to or subtracts from the axial velocity vector and can result in error. The magnitude of the error clearly depends on the magnitude of the crossflow. On the other hand, if the ultrasonic paths are oriented at right angles to the crossflow then there is no effect on the velocity vector along the ultrasonic path. So it becomes important to understand the direction of the crossflow, if any, produced by disturbances – bends, valves, etc., in fluid systems.

We have performed numerous model and full scale tests as well as theoretical analyses to establish the fluid velocity profiles and crossflow effects for many disturbances such as bends, valves and more complex geometries. Consequently, we are able to orient our meter to eliminate crossflow errors in most applications. Model tests can be provided if the geometry is not one which we have examined. In most instances, it is sufficient to orient the meter such that the projections of the ultrasonic paths onto the pipe cross section are perpendicular to the plane of the nearest upstream bend. Crossflow produced by such bends lies in the plane of the bend, unless the system geometry just upstream of the bend is complicated by other components.

To provide you with some feeling for the total uncertainty or error, with various fluid velocity distributions we have attached Table A. The errors in the table include errors due to the uncertainty in the direction of the crossflow as well as the errors resulting from the four path Gaussian quadrature integration procedure.

TABLE A: Summary of accuracy capability (Two-Sigma) for various flow conditions

		FLOW CONDITIONS		
		Well behaved ● Fully developed ● Axis symmetrical ● Slight distortion	Distorted, but known ● Simple bends ● Gate valves	Distorted and unkown
ACCURACY (2 SIGMA)	BEST	(Requires compensation for known Gaussian error)	(Requires compensation for known Gaussian error plus proper known angular orientation to eliminate crossflow error)	(Requires model test, compensation and proper angular orientation)
		±0.1%	±0.2%	±0.2%
	NOR-MAL	±0.5%	±1.0%	±1.5%

Zero set, questioned by Mr. Loosemore, is accomplished in either of two ways, each of which is capable of system alignment well within the stated accuracy capability. This is a one-time setting for a given metering section and set of electronics. Either method requires only the use of a dual trace oscilloscope and involves adjusting the gain of each of the four receiver pairs, the only adjustment the system has.

Where the fluid flow can be stopped, each pair of receivers for each ultrasonic path is first adjusted to obtain the proper signal to noise ratio and then to obtain identical leading edges of the received pulses. The system is then in a balanced or zeroed condition.

Where the fluid flow cannot be stopped, each pair of receivers for each ultrasonic path is again first adjusted to obtain the proper signal to noise ratio and then to obtain identical amplitudes of the first half cycle of the received pulses. To ensure proper balance in this case, the transducer leads for each pair are then reversed at a given flow and if the negative flow measurement agrees with the positive flow measurement, then the system is truly balanced or zeroed.

M.J.A. Woodley: Can you give a brief description of the computational hardware which was used?

What is the approximate cost of the system you have described?

Authors' reply: A description of the hardware, involves mainly the electronics portion of the system; the rest consisting of transducers and transducer cables. The electronics are divided into two basic units — an analog ultrasonic unit and a digital processing unit. The ultrasonic portion consists of a common transmitter, a pair of passive "switching" networks, and a pair of receiver/ comparators for each ultrasonic path.

The digital unit contains a network that quantizes time for the t_1 and $\triangle t$ measurements to the required degree of precision. Timing and control circuits, non-volatile memory circuits, scaling and programming circuits, and a simple delicated serial arithmetic computer make up the remaining portion of the processing system. Buffering and a digital display are provided for the output, and D.C. power supplies are included. Circuits are included to reduce false alarm rates, validate data, and to perform an online self test for each averaging period. Scaling and averaging periods are adjustable by means of changeable plug-in circuitry. All the electronic hardware is of a solid state design utlizing integrated circuits.

J.C. Schuster: In applying ultrasonic flow meters to large diameter, relatively thin-walled pipes, does the movement of the wall affect the velocity measurement in both the "swing around" and direct timing methods? What order of magnitude (that is: length between transducers and transducer velocity) is the error?

Authors' reply: Analysis of the effect of changing pipe dimensions, as requested by Schuster and Causon, discloses a further example of the inherent insensitivity to changes of the Gaussian integration technique. We have evaluated the difference in flowmeter output corresponding to the change from elliptical to circular cross section. The result is a change in output which is less than 0.015 of the relative difference between major and minor axes of the ellipse. For instance, if the pipe is out of round by one percent, the maximum error in the flow measurement due to the pipe changing shape under pressure is 0.015 percent. In an installation in which the shape is pulsating, corresponding time variations in measured flow would be experienced, further reducing the mean error. In practice, where out of roundness is not expected to exceed one percent in the worst case (Reference: Section 8 — Division 1 of ASME Boiler and Pressure Vessel Code) the effect on accuracy is negligible.

T. Agar: How is the accuracy dependent on the angle θ, and by how much is this angle varied due to scattering by gas bubbles, for example?

R.K. Turton: Being only vaguely familiar with the theory, and having no knowledge of the application, can I ask if there are any limitations to the ultrasonic method imposed by fluid properties.

Is there a lower limit to the flow rate at which linearity can still be achieved?

R. Theenhaus: Do you have any experience of the influence of the effect of solid particles and impurities within the fluid on the accuracy of measurement, including the measurement of water flow in a river?

What is the price for one unit (1000 mm diam. pipe size)?

Authors' reply: The last three speakers have prompted further discussion on problems associated with ultrasonic propagation and angle.

First, accuracy is directly related to any error made in the tangent of the angle. Since this function is least sensitive to error at 45° this nominal angle is preferred, if no other constraints are involved. However, each path is scaled separately so slight differences in measured angles can be accounted for, causing no net error in the flow measurement due to this source.

Ultrasonic propagation has two properties which must be considered. First is amplitude attenuation which is made up of absorption loss, scattering loss or both, and spreading loss. The latter is the well-known inverse square function of the ultrasonic path length. The former are functions of impurities, their composition and their size. Small sizes do not permanently block the wave front but of course do attenuate its amplitude. Large sizes will block the wave front and if their attenuation is great enough then a blind spot or shadow will appear behind the object. Amplitude fluctuations are simply handled by hardware design considerations of ultrasonic propagation frequency in transducer design, transmitter power output, and receiver sensitivity to produce the required threshold to noise ratio and signal to noise ratio to cope with the extremes of conditions to be encountered. Total signal loss, if it is momentary, is of no consequence since to obtain averages the actual number of received measurements is used, not the number of transmissions made. Receiving no pulses in a given measurement interval, of course, produces no new data and one can only rely on the system memory of the last measured interval which for steady flow conditions will still be valid. Since transmissions are simultaneous from each end of the same ultrasonic path the principle of reciprocity applies, and what happens in one direction also happens in the opposite direction. This means that any "jitter" in the leading edges occurs to both pulses. Thus, no error is produced in the critical time measurement $\triangle$t. This has been amply demonstrated in independent tests of this technique conducted in a towing tank facility. Varying concentrations of air bubbles, up to four percent by volume, were injected directly across the ultrasonic paths without any detectable error in the performance of the system. Figure A is a photograph showing the conditions for a concentration of one percent.

Fig. A

The second ultrasonic property relates to thermal gradients, more specifically sound velocity gradients, which can cause the ultrasonic beam to bend. Depending on the narrowness of the beam, the ultrasonic path length, and the distance from the path to the wall or the water surface, beam bending can cause total signal loss. Because of the relatively short path lengths and the turbulent mixing which reduces temperature gradients, this is not a problem in pipe or closed conduit flow and man-made channel flow but must be considered very seriously for large, relatively shallow rivers.

Our experience includes not only the fresh water pipe case but also man-made canals, up to 200-feet wide and 30-feet deep carrying sediment loads, raw sewage, and industrial wastes. In natural rivers we have successfully covered widths up to 1600 feet and depths as deep as 100 feet and as shallow as 20 feet.

M. Pappalardo: How do you compensate for a variation in accuracy due to a change in the temperature of the fluid?

Authors' reply: How changes in sound velocity (i.e., as caused by changes in fluid temperature) are handled, as requested by Mr. Papalardo, is best covered by reviewing the mathematics involved. The principle is defined by two equations involving the transit times of pulses in the medium, t_1 and t_2; the path length, ℓ; the speed of sound, c; and the fluid velocity, V. By transmitting pulses simultaneously, the two equations are essentially simultaneous, permitting independent solutions for the two unknown variables, c and V. Physically, the process is such that as variations in temperature (sound velocity) occur, the transit times t_1 and t_2 change thereby taking the variation into account.

$$t_1 = \frac{\ell}{c + V} \qquad\qquad t_2 = \frac{\ell}{c - V}$$

$$c + V = \frac{\ell}{t_1}$$

$$c - V = \frac{\ell}{t_2}$$

$$t_2 = t_1 + \triangle t$$

$$V = \frac{\ell \triangle t}{2t_1 (t_1 + \triangle t)}$$

C.A.E. Clay: We critically examined the Westinghouse L.E. flowmeter about 2 years ago, and were impressed with its potential in the field of large water flow measurement. In our report we put forward an approximate correction factor of 0.995 to correct for errors in the Gaussian integration method for certain flow measurement situations, and it is gratifying to note that the authors' more rigorous assessment produced similar values.

There appear to be two main problems with this flowmeter, one being the price. The second is the difficulty of convincing second parties that the displayed results are accurate. Would the authors briefly describe the inbuilt checking facilities and any simple way of demonstrating the accuracy of the overall system?

Authors' reply: The demonstration of accuracy consists of two distinct steps. First, clearly, all parties should understand the physical principles of the system, which are subject to analysis independent of the particular mechanization or "black box" involved. These physical principles have been amply verified by independent evaluation using the weigh tank and associated equipment as the standard of comparison. It is to this end that our paper addressed itself.

The second step is the demonstration or proof that the hardware or "black box" is performing the measurements and the computations required to the proper degree of precision and with the proper scaling for the particular metering section.

To verify the precision of the time measurements, laboratory tests of the time measuring circuitry are performed using suitable certified precision time generating instruments (e.g., precision pulse generators, fixed precision delay lines). It is usually required to perform this test only once to verify proper manufacture. The time measuring circuits are tested for precision as a part of the final factory test on each L.E. Flow Measurement System built.

Scaling and computational precision are checked routinely by a built-in, on-line self test feature for each measurement period. Results are displayed on a two color display-green for pass, red for fault. In addition, it is possible with a special test set to stop the automatic machine sequence in a manual test mode. Then verification of the proper test inputs can be made using the same test set. With these inputs one can hand calculate the proper flow for the given metering section involved and then verify the accuracy of the flow measurement system or "black box" by reading its displayed output.

A final on-line check is made to determine the number of ultrasonic pulses received for each measurement period to ensure meeting statistical averaging requirements. A green display is used to signify compliance.

Questions of cost, as raised by Messrs. Woodley, Causon, Clay, and Theenhaus, are best discussed by stating that a basic system is not only capable of metering one location and includes eight transducers with mounts, a complete electronics package with digital display and on-line test features, but also has expansion capabilities to time share three other metering locations. The price for this basic system is approximately £22,000. This price is unaffected by pipe size. Significant simplification of hardware at the expense of accuracy is not possible within the scope of the unique ultrasonic technique.

F.A. Inkley: I would like to ask the authors of this very interesting paper whether their technique has been applied in the field of bulk oil measurement. At Oil Terminals very large flowrates of crude oil are encountered and flow measurements are made using positive displacement meters or, more recently, using turbine flowmeters. In either case frequent meter calibrations are required and this is done using very large, very expensive, volumetric line provers.

Do the authors believe that thoroughly reliable ultrasonic flow measuring systems of comparable accuracy could be designed either to replace the metering system (also very expensive) or the line proving system, or the combination of both? In the former case, would they recommend replacing the metering system or the line proving system?

Authors' reply: An evaluation of the ultrasonic technique to the measurement of large flows of oil has been successfully accomplished in the laboratory. This evaluation covered a wide range of types of oil from crude to refined, and included various water-oil emulsions. Bunker C Oil and Silicone Oil have also been successfully measured over a wide range of temperatures.

As presently designed, our system will achieve highly reliable performance with the accuracies described when applied to the measurement of bulk oil flow rates. Simply stated, it would replace both the present metering system as well as the line proving system.

L.A. Salami: How do you know ehen there is crossflow in the pipe when this same method is a line integration process?

When models are used to study complicated flows how do you know that these complicated flow patterns have been correctly reproduced in the model?

How were the locations of the chords determined?

Have other numbers of paths been used apart from 4 paths and how do these affect the result and accuracy of this method?

What difficulties are there if any, when the paths of integration are along 4 diameters instead of along 4 chords. If these were possible I am of the opinion that this procedure will give better results than using 4 chords. Simple addition of the line integrations obtained using the four diameters will give the flowrate. The paper by Salami (see page   ) shows that fairly accurate results can each be obtained using this procedure i.e. four diameters.

On the basis of the work carried out by Salami (see page 381), I do not share in the authors' opinion that the fairly good results obtained can be carried over to difficult asymmetric flow.

Authors' reply: The numerous questions raised by Salami are best treated by grouping them by subject matter.

The general question of chord location and number is handled in the following way. Location is specified by the numeric integration technique used; namely, Gauss' Quadrature, and for four paths is 0.06943D, 0.33001D, 0.66999D, and 0.93057D, where D is the diameter or maximum width of the conveyance perpendicular to the four parallel paths or chords. In general, the accuracy improves with the number of measurements (paths). For example, two paths have produced accuracies of the order of 5%, three paths 2%, and so on. Five paths would give better results than four paths for the accuracy shown in the NORMAL row in TABLE A of this discussion. Although six paths and above would continue to improve the integration error, it would not improve the overall accuracy because of other error sources. Our choice of four paths is a practical one. It takes into account the increase in accuracy provided by a greater number of paths while recognizing the increase in transducer mounting complexity and additional electronic equipment required. These type results are quite similar to those obtained in Mr. Salami's own paper.

The general questions of crossflow and the use of models are both answered earlier in the discussion. It is felt however, that re-emphasis is in order to clarify certain points. Knowledge of crossflow direction only is required to achieve maximum accuracy. As far as accuracy of model is concerned, the integration techniques employed tend to achieve accuracies independent of flow conditions. As can be seen from TABLE A, the best accuracy achievable does require knowledge that the model is valid; otherwise normal accuracies listed apply. These conditions are not unlike those found in Mr. Salami's own paper in the discussion where knowledge of crossflow, as well as swirl, is required to prevent error in the pitot readings. The pitot tube is subject to error at flow angles of greater than $5°$ with respect to its own axis. The maximum achievable accuracy is also a function of the knowledge of the flow conditions and the integration method used. There is one significant difference between the prior work reviewed by Mr. Salami and ours in that the evaluation of our method, conducted by others, has been against a primary standard — the weigh tank, where the sum of all errors (technique plus hardware) is measured.

In the ultrasonic method, the integration of fluid velocity over a path is provided as an output; its velocity distribution along the path is not. Methods analogous to those treated by Mr. Salami are therefore not appropriate. The use of multiple diametrical paths is generally inferior to the Gaussian Method, or other methods using parallel paths for an ultrasonic system. The diametrical arrangement has the evident disadvantage of weighing too heavily the velocity near the center of the pipe using the line integration technique. This has been confirmed by our field and laboratory experience with both arrangements, chordal and diametrical, evaluated against other flow measurement standards.

We chose to analize in detail symmetrical distributions for which standard mathematical definitions are available. It was our purpose to illustrate by precise calculation how the Gaussian Method provides an accurate integration over a wide range of variation in profile. We noted that there is nothing inherent in the method which suggests accuracy should be better for symmetrical distributions than for unsymmetrical ones, but analytical demonstration for the latter is not so easy. We have, however, evaluated unsymmetrical profiles numerically and in the laboratory against the weigh tank with the good results as previously stated. We were impressed with the approach taken by Mr. Salami in his paper and believe it would be interesting to evaluate the Gaussian Method for the unsymmetrical profiles he defined. We have confidence that the results would be similar to those we obtained for the symmetrical distributions.

4·1

THE PERFORMANCE OF LONG BORE ORIFICES AT LOW REYNOLDS NUMBERS

T. Cousins

Abstract: There is a need to measure small flow rates confidently without the necessity for individual calibration of the device. Orifices could fulfil this requirement. In view of the difficulty of manufacturing standard orifices of small diameter, tests have been carried out on parallel bore orifices whose length equals the diameter. Experiments have been successful in predicting the calibration of 1.6 mm to 5.0 mm diameter orifices at Reynolds numbers (Rd) down to 1000 to a standard deviation of ±2.5%. Two major experimental problems arise from these tests, blockage of the orifice and leakage in both the test system and around the orifice. Due to the small size of these orifices these effects are very critical. The accurate manufacture of these orifices is difficult, requiring microscopic examination of the faces and edges. The tests have shown that very stringent tolerances are required for the upstream edge, the upstream face and orifice diameter.

1   INTRODUCTION

Small flows, of about 1 g/sec, such as those found in pilot plant operation or dosing, may be measured by an integral orifice differential pressure cell, rather than using the conventional orifice carrier with tappings connected to a differential pressure cell. An orifice is installed at the exit of the high pressure chamber in a manifold, which takes a tapping downstream of the orifice directly to the low pressure chamber. The fluid flows through the high pressure chamber and a differential is caused by the orifice between the high and low pressure chambers. A series of orifices of less than 5 mm. diameter were required to be used as integral orifices, to measure clean fluids, both gaseous and liquid, with an uncertainty of ±2½%.

The manufacture of standard orifices is very difficult for diameters below 5 mm., for example a quarter circle orifice is not recommended below 15 mm. diameter, and a conical entrance not below 6.35 mm. diameter (1). A quarter circle orifice of 2.0 mm. diam. would require a radius of 0.2 mm. diameter to be accurately ground on the upstream edge and a conical entrance orifice of the same diameter would be 0.2 mm. thick coming down to 0.042 mm. Thus not only are such orifices difficult to make at such small diameters, but they are also becoming mechanically weak.

An investigation was therefore initiated into the possibility of using square edge, thick plate, parallel bore orifices. The reasons for choosing this type were:

(a) The strength of such orifices against temperature, pressure and flow force effects.

(b) A square edge would be easier to produce than either a shaped entry or a sharp edge.

The orifices were designed to have the same thickness as their diameter for the following reasons:

(a) The value of the coefficient of discharge changes slowly with 1/d.

(b) This length or bore was considered reasonable for manufacture, any longer and it would have been difficult to produce a parallel bore.

(c) These orifices are good for low Reynolds number measurement.

Other investigators of l/d = 1 thick plate orifices tend to be contradictory in the assessment of their performance. The coefficient of discharge of a thick plate orifice depends on the degree of contraction of the fluid passing through the orifice. Some investigators suggest that they are very sensitive to minor secondary factors, such as pressure pulsations, dissolved air in water, cleanliness or the surface condition of the throat. For low Reynolds numbers, they conclude, these orifices may not be used when calibrated, let alone uncalibrated (3).

2 TEST EQUIPMENT

2.1 Description of orifices

The orifices were parallel bore, the length being equal to the orifice diameter. Both the upstream and downstream edges were nominally square edged. Series A were first constructed and were machined from stainless steel, but series B and C were made from mild steel. The shape of the orifices was determined by the differential pressure cell for which they were designed (Fig. 1). Series A and B contained five orifices of nominal bores 5.0, 2.8, 1.6, 0.9 and 0.5 mm diam., series C contained three sets of nominally 5.0, 2.8, 1.6 mm diam. orifices. The details of these orifices are shown in Table 1.

Two larger orifices were built to confirm the trend of results of the other orifices. These were of 38.1 and 12.7 mm diam. and were machined from perspex.

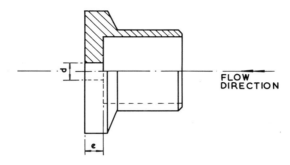

Fig. 1   Parallel bore orifice

2.2 Differential pressure cell

The upstream pipe was connected into a chamber on the high pressure side of a capsule. A manifold clamped the orifice in position at the exit of the chamber. Drilled through the manifold, downstream of the orifice, was a tapping to the low pressure chamber. The capsule was connected to either a pneumatic or electronic force balance system which is normally used to indicate the differential across the orifice (Fig. 2). For all experiments the differential was obtained by connecting a manometer to tappings in the high and low pressure chambers. Upstream of the orifice was a large chamber reducing to a short length of threaded pipe just before the orifice. The downstream tapping to the low pressure chamber was 3 mm from the rear face of the orifice. Teflon rings were used to seal the orifice and low pressure connection. A new seal was used for each test.

2.3 The orifice carrier

After the results of series 'A' had been so scattered, an orifice carrier was constructed to check that the unusual conditions upstream of the orifice were not the cause of the bad results. The carrier was machined from mild steel and designed to have a smooth constant bore of 15 mm diam. upstream of the orifice. Sealing was by two soft 'O' rings. The tappings were at D and D/2 (Fig. 3).

2.4 Air test equipment

Air was supplied from an airbell at a given volumetric flowrate. A scale along the length of the airbell was graduated directly in cubic feet. Air was supplied to the airbell, after filtering, from a compressed air supply. The airbell was connected to the integral orifice or orifice carrier by 15 mm diam. steel pipe via an air filter. The flowrate was controlled by a needle valve at the outlet. The differential head across the orifice was given by a Betz manometer and the airbell pressure and orifice upstream pressure by single limb oil manometers (Fig. 4). The maximum error was ±0.6% at the lowest flowrate, and the average error ±0.3% (Appendix 10.1).

2.5 Water test equipment

The initial water tests using the series 'A' orifices used water supplied directly from the mains

TABLE 1

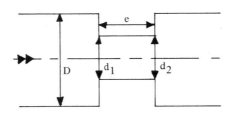

Orifice Number	Nominal d & e	d_1 (mm)	d_2 (mm)	e (mm)	e/d	d/D	r (mm)	Maximum Upstream Roughness (Micrometers)	Bore Roughness	Graph Numbers
A10	5.0	5.047	5.030	4.990	0.990	0.335				7a
A20	2.8	2.797	2.784	2.820	1.009	0.186				7b
A30	1.6	1.623	1.595	1.600	0.986	0.107				7c
A40	0.9	0.895	0.884	0.870	0.973	0.060				7d
A50	0.5	0.506	0.502	0.500	0.986	0.034				7e
B10	5.0	5.050	5.045	5.003	0.991	0.335				9f
B11	5.0	5.055	5.052			0.335				9f
B12	5.0	5.025	5.030			0.335				9a
B20	2.8	2.860	2.850	2.855	0.999	0.186				9b
B30	1.6	1.610	1.610	1.606	0.998	0.107				9c
B40	0.9	0.928	0.925	0.880	0.950	0.060				9d
B50	0.5	0.514	0.512	0.470	0.514	0.033				9e
C10	5.0	4.970	4.967	4.956	0.999	0.335	.0090	.30	2.0	11a 12a
C20	2.8	2.796	2.779	2.804	1.006	0.186	.0045	.15	1.6	11b 12b
C30	1.6	1.601	1.586	1.578	0.985	0.107	.0080	.15		11c 12c
C11	5.0	4.969	4.964	5.006	1.009	0.335	.0090	.20	2.5	11a 12a
C21	2.8	2.790	2.790	2.816	1.009	0.186	.0050	.25	1.2	11b 12b
C31	1.6	1.568	1.548	1.590	1.014	0.107	.0055	.20		11c 12c
C12	5.0	4.978	4.965	4.982	1.001	0.335	.0110	.25	2.5	11a 12a
C22	2.8	2.790	2.787	2.787	1.000	0.186	.0050	.15	1.6	11b 12b
C32	1.6	1.587	1.582	1.596	1.009	0.107	.0040	.15		11c 12c
D10	12.7					0.357				10
E10	38.0					0.266				10

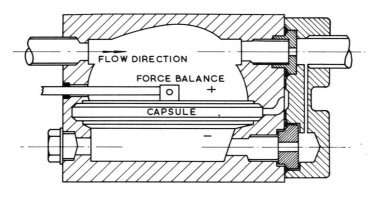

Fig. 2   Differential pressure cell

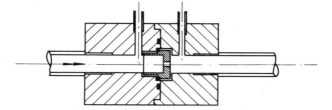

Fig. 3   Carrier for orifices

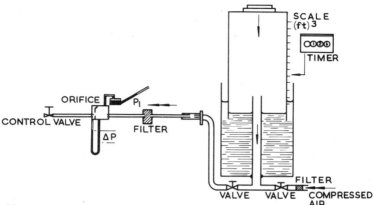

Fig. 4   Air test equipment

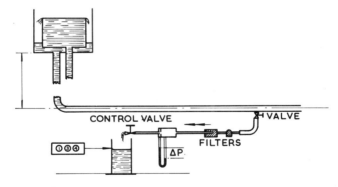

Fig. 5   Water test equipment

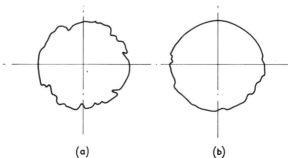

Fig. 6   Shadowgraph of dirt blockage in 1.6 mm diameter orifice

    a   Dirt distorted orifice
    b   Cleaned orifice

water supply through a pressure regulator. Fears that this may be aerated, which later proved unfounded, led to the use of water supplied from a constant head tank (Fig. 5). Two filters were used upstream of the orifice, one commercial filter and one made of mesh and fine gravel. A needle valve at the outlet was used to control the flowrate. The flowrate was determined by timing the filling of calibrated tanks. A water on mercury U-tube manometer, and an air on water manometer, were used to measure respectively, the differential at the higher and lower flowrates. The maximum error was ±0.5% and the average ±0.25% (Appendix 10.1).

2.6 Supplementary test equipment

2.6.1 Air test of 38 mm diameter orifice

The 38 mm orifice was calibrated in a 135 mm diam. airline against a water calibrated thin plate orifice. The differential heads were obtained from Betz manometers and the upstream pressures from single limb oil manometers.

2.6.2 Water test of 12.7 mm diameter orifice

The 12.7 mm diam. orifice was calibrated in a 38 mm diam. water line against a volumetric measuring tank. The differential head was obtained from a water on mercury manometer and for smaller flows an air on water manometer.

3   FLOW EQUATIONS

$$Q = C_d E_a \sqrt{2 \triangle P \rho_A}$$

from hydrostatics $\triangle P = \rho_L g h$

$$Q = C_d E_a \sqrt{2 g \rho_L h \rho_A}$$

Let $K = C_d E_a \sqrt{2 g \rho_L}$

now as the upstream velocity was very small.

$E = 1$

Therefore $K = \dfrac{\pi C_d d^2}{4} \sqrt{2 g \rho_L}$

Therefore, for a given orifice, $K = \dfrac{Q}{\sqrt{h \rho_A}}$

The experimental results have been plotted as K/d^2 as d is the only variation between the orifices. K/d^2 is therefore a "universal" coefficient for all the orifices.

The coefficient has been plotted against the orifice bore Reynolds number defined by

$$R_d = 12.7 \frac{Q}{\mu d}$$

4   EXPERIMENTAL INVESTIGATION

4.1 Series 'A'

The initial tests showed a considerable spread of data and no correlation between the air and water tests. There was also no apparent relationship between the individual orifices. The two

major sources of experimental error were found to be dirt blockage and leakage both externally
and internally around the orifice. When dealing with orifices of such small diameters the most
minute dirt particles become large when compared with the orifice. Fig. 6, a shadowgraph tracing,
demonstrates the effect of dirt blockage experienced in a 1.6 mm diam. orifice. Leakage is the
same in scale, for example, a "pinprick" of 0.1 mm diam. represents a 4% effect on the discharge
coefficient of a 5.0 mm diam. orifice. For all future experiments the filtering was improved, the
pipework thoroughly cleaned, and each orifice cleaned and degreased by soaking in white spirit.
The apparatus was thoroughly tested for leaks before each series of tests. Aeration of the mains
water was suspected of effecting the calibration, future tests were therefore carried out using
water from a constant head tank. Later results showed that the calibrations derived from the
mains water tests were correct, but the tests were continued on the constant head tank supply.

The results of the tests on series 'A', when repeated, showed a considerable improvement in
repeatability and matching of air and water results (Fig. 7a, b, c, d, e). There was still however
a large difference between the results of the individual orifices (Fig. 8).

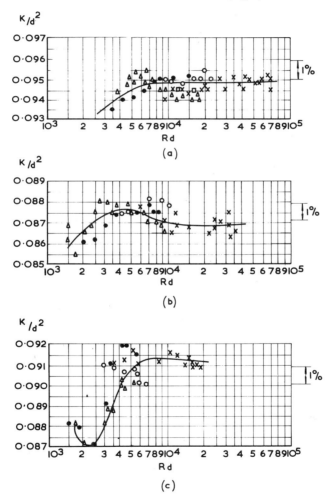

Fig. 7   Individual orifice calibrations of Series A

x——x	mains water tests — in the Deltapi	a 5.0 mm dia. orifice
o——o	air tests — in the Deltapi	b 2.8 mm dia. orifice
△——△	constant head water supply — orifice carrier	c 1.6 mm dia. orifice
o——o □——□ }	air tests	

As the approach velocity was so small compared with that through the orifice, it was hoped that the unusual upstream conditions presented to the orifice by the differential pressure cell would not effect the calibration. To confirm this an orifice carrier with smooth continuous upstream pipe was constructed. The results (Fig. 7 a, b, c, d, e) are very similar to those obtained from the integral orifice.

The orifices when inspected under a magnifying glass appeared to be reasonably well made. However, an investigation of them under a microscope revealed that they were very poorly manufactured. The most noticeable faults were the rounding of the upstream edge, burrs around the edge and a very rough surface finish. In one particular case the orifice had been drilled off centre and the machine grooves formed a jagged upstream edge. Fig. 9a shows some of the irregularities in the series 'A' orifices.

4.2 Series 'B'

A new set of orifices were constructed with much tighter tolerances. The 5.0 mm diam. orifice was made first and after three attempts a satisfactory technique was found. Fig. 9b shows how much better Series 'B' orifices were compared with series 'A'. Fig. 9c shows the first attempt 5.0 mm diam. orifice, the light reflection showing the rounding of the upstream edge.

The orifices were tested in air and water and the curves shown in Fig. 11 a, b, c, d, e. When plotted against the throat Reynolds number these formed a more coherent curve (Fig. 12) than did the series 'A' orifices. The three 5.0 mm diam. orifices when plotted on the same curve showed that the orifice with the rounded upstream edge had a higher calibration constant (Fig. 10f).

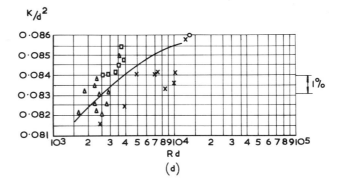

(d)

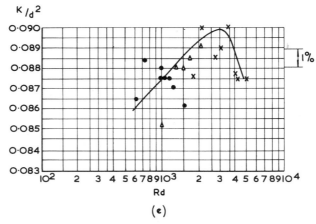

(e)

Fig. 7 (continued)

    d 0.9 mm dia. orifice
    e 0.5 mm dia. orifice

4.3 Series 'C'

To confirm that the orifices could be reproduced, and to obtain an idea of the tolerances required, a set of nine orifices were constructed. It was decided for the time being the 0.9 and 0.5 mm diam. orifices would be too difficult to manufacture to the required tolerances.

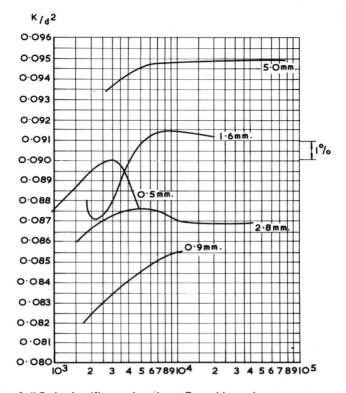

Fig. 8   Plot of all Series A orifices against throat Reynolds number

The results (Fig. 13 a, b, c), using the nominal bores, fall within $\pm 2\frac{1}{2}\%$ of the values obtained from the series B orifices. One orifice, a 1.6 mm diameter orifice, gave a low discharge coefficient. When the true bores were used (Fig. 14 a, b, c), this orifice came into line with the others.

From the results and detailed measurement of these orifices a set of tolerances was established, Table 2.

4.4 Series 'D' and 'E'

The two larger orifices were made to confirm the trend of the results obtained from the smaller orifices. As can be seen from Fig. 10 they do confirm that the curves do not become distorted at the higher Reynolds numbers.

5   COMMENTS

Small flows can be measured in many ways by using different types of calibrated meters. Such meters include filters, through which the fluid resistance is measured, capilliary tubes and some variation on an orifice. They need not be linear or square root law meters, as long as they are repeatable after calibration. In many cases the calibrated uncertainty of these meters is $\pm 5\%$, which is adequate if the requirement is only for a repeatable meter and an indication of a change in flow. It is a different matter, however, to produce a meter to measure small flows with an *uncalibrated* uncertainty of $\pm 2.5\%$.

Two major sources of error, when dealing with low flow measurement, which must be eliminated before obtaining any satisfactory results are dirt blockage and leakage. With such small dimensions and flows any meter is liable to dirt blockage, for even the minutest particles are large compared with the meter dimensions. Long bore orifices have been critisized as being sensitive to dirt blockage, but they are much better than filters and capilliary tubes and only slightly worse than other orifice configurations. Leakage is a major problem, especially gas leakage. Liquid leakage is relatively easy to detect but the detection of gas leakage requires very careful observation. Water aeration was also suspected of influencing the water test results, but later tests showed that this was not the case. Other investigators have found, however, that long bore orifices are very sensitive to aeration.

The orifice discharge coefficients have been plotted against the throat Reynolds number. The throat Reynolds number has been chosen for two reasons, the first is that the differential pressure cell has no genuine upstream pipework, and the second reason is the upstream pipework is large enough compared to the orifice not to effect the mechanics of the flow. As other investigators have found the discharge coefficient does not vary uniformly with Reynolds number, but individual orifices do keep within a 5% band. The results of series B in the orifice carrier have however been plotted against the pipe diameter Reynolds number (Fig. 11b).

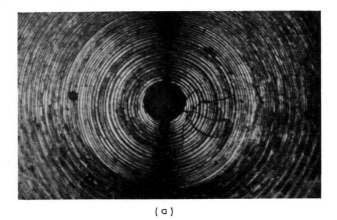

(a)

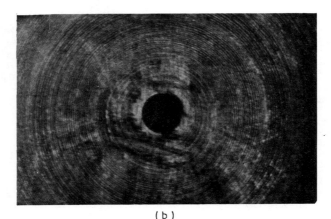

(b)

Fig. 9   a   Series A, 0.9 mm dia. orifice magnified 20 times
          b   Series B, 0.9 mm dia. orifice magnified 20 times

The most important result of these tests is that the measurement of small flows using long bore orifices is not really a flow but an engineering construction problem. Long bore orifices may be used to measure small flows providing they are constructed to strict tolerances. The series A orifices were found, by inspection under a microscope, to be poorly constructed. Fig. 9a shows that the 0.9 mm diam. orifice, which was one of the worst of this series, has a very rough upstream surface and a jagged entry. When all the series A orifices are plotted against Reynolds number (Fig. 8) there is no correlation between the individual discharge coefficients. The series B orifices are much better constructed typified in the photograph of the 0.9 mm orifice (Fig. 9b). Fig. 9c and Fig. 9d show two of the three series B 5.0 mm diam. orifices, the first of which has a very rounded entry. When the two are compared with their discharge coefficients, (Fig. 10f), the orifice with the rounded entry has a higher discharge coefficient. This is consistant with the contraction area, which increases with an increase in upstream curvature, becoming like a nozzle. The roughness of the upstream surface is important in that it defines the entry curvature, it is thus important that the tolerance on the upstream roughness be more stringent than the entry curvature. Also the flow is accelerating along this face and thus the roughness will have an effect on the mechanics of the flow.

The series C orifices were made to find what tolerances could be expected in manufacture and whether such tolerances would be sufficient to give discharge coefficients within a 5% band. The discharge coefficients all fell within the required tolerance when the true measured diameter was used (Fig. 13), although using the nominal diameter one of the 1.6 mm diam. orifices was outside the tolerance (Fig. 12). Table 2 shows the final tolerances resulting from these experiments. All of

(c)

Fig. 9 (continued)   c   Orifice B10, 5.0 mm dia. orifice magnified 20 times (see Fig. 10f)

them are tighter than those achieved in the construction of Series C but are possible with careful manufacture. For the present the 0.9 and 0.5 mm diam. orifices have not been recommended to be used as uncalibrated meters, the required tolerances making them very difficult to make.

The larger orifices were built to show that at higher Reynolds numbers no unusual phenomena would be encountered. They were also easier to make, the tolerances not being so critical, thus they were a useful confirmation of the results of the smaller orifices.

The discharge coefficients of series B, D, and E still do not follow a continuous curve when plotted against Reynolds number (Fig. 11a). This is the result of a number of factors which are difficult to separate. At the lower Reynolds numbers the flow in the orifice bore is becoming laminar, which may account for the distortion of the 0.5 and 0.9 mm diam. curves. The orifices are also not exactly geometrically similar, that is the l/d ratio varies and the upstream entry curvature may vary slightly. A major factor would appear to be the variation of discharge coefficient with β, the ratio of the orifice diameter to the upstream pipe diameter. Fig. 14 shows a smooth, almost linear, increase in mean discharge coefficient with increase in β. The linearity of the curve is probably a coincidence, as there is a mismatch in Reynolds numbers between the orifices, and the values of the mean discharge coefficient are very approximate but it does indicate that there is a substantial β effect. It also helps to explain the difference in the orifices D and E.

An interesting comparison of tolerances required for small bore orifices is shown in

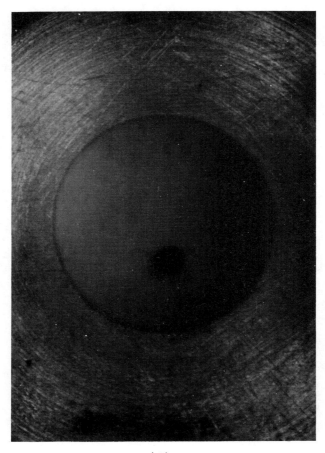

(d)

Fig. 9 (continued)   d Orifice B12, 5.0mm dia. orifice magnified 20 times (see Fig. 10f)

Appendix 10.2. Voss [4] formulated his tolerances after testing a large number of orifices. As can be seen the two sets of tolerances compare favourably especially for the large 5.0 mm orifice. A comparison of discharge coefficients for $l/d = 1$ orifices is shown in Fig. 15. The values obtained in this paper are compared with the results obtained by Koennecke [7] and as can be seen they are of a similar magnitude. His plots however have a much greater curvature, also for the two orifices

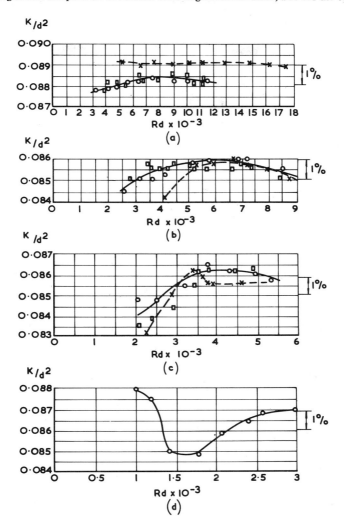

Fig. 10   Individual orifice calibrations of Series B

x———x   mains water tests — in the Deltapi
o———o   air tests — in the Deltapi
△———△   constant head water supply — orifice carrier
o———o⎫
           ⎬ air tests
□———□⎭

a   5 mm dia. orifice
b   2.8 mm dia. orifice
c   1.6 mm dia. orifice
d   0.9 mm dia. orifice

plotted the discharge coefficient goes in the opposite direction with β. Grace and Lapple, (3), were unable to obtain a curve for their $l/d = 1$ orifices because of the huge scatter of results. This appears to be mainly due to the poor manufacture of the orifices, as they were only 0.0794 mm in diameter. In particular they allowed a very large tolerance on the length of the orifice bore which can cause a large variation in discharge coefficient. Lichtarowicz and Markland (2) show a curve of ultimate discharge coefficient against l/d obtained from the results of other investigators. While it is difficult to see from the results the ultimate coefficients of the orifices they would appear to be similar to the discharge coefficients for $l/d = 1$ orifices given by this curve.

6   CONCLUSIONS

This paper shows that down to 1.6 mm diameter, thick plate orifices with a length to diameter ratio of unity may be reproduced within an uncalibrated uncertainty of $\pm2.5\%$ in discharge coefficient. The tolerances on such small orifices are necessarily stringent, but down to 1.6 mm diameter can be achieved.

7   ACKNOWLEDGEMENTS

The author wishes to thank K.J. Zanker for his help and advice in supervising the work described in this paper and H. Kendall for the excellent test work he carried out.

I would also like to thank Kent Instruments Ltd. for permission to publish these results.

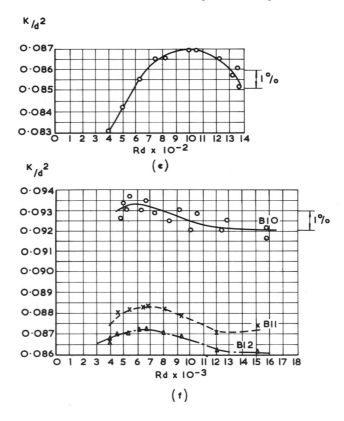

Fig. 10 (continued)   e   0.5 mm dia. orifice
                       f   B10, B11, B12; 5.0 mm dia. orifices (see Figs. 9c and 9d)

8 REFERENCES

1   'Methods for the Measurement of Fluid Flow in Pipes'. Part 1 orifice plates, nozzles and venturi tubes, British Standard (1042 : Part 1 : 1964)

2   KASTNER, L.J., McVEIGH, J.C., 'A Reassessment of Metering Orifices for Low Reynolds Numbers'. The Institution of Mechanical Engineers, proceedings 1965-66, **180**, Pt. 1, No. 13

3   GRACE, H.P., LAPPLE, C.E.: 'Discharge Coefficients of Small-Diameter Orifices and Flow Nozzles'. Trans. A.S.M.E., 73, 1951, p. 639

4   LAVERNE R. VOSS: 'The Design of Small Multiple Orifice Meters for 1% Air Flow Measurements', Research Lab., General Motors Corp., presented at the 'Symposium on Flow—Its Measurement and Control in Science and Industry', May 10-14, 1971, Pittsburgh, Pennsylvania

5   FILBAN, T.J., GRIFFIN, W.A., 'Small Diameter Orifice Metering', Journal of Basic Engineering, Sept. 1960, p. 735

6   HANSEN, M.: "Düsen und Blenden bei Kleinen Reynoldsschen Zahlena, Fursch. Ingnes", 1933-34, 64.

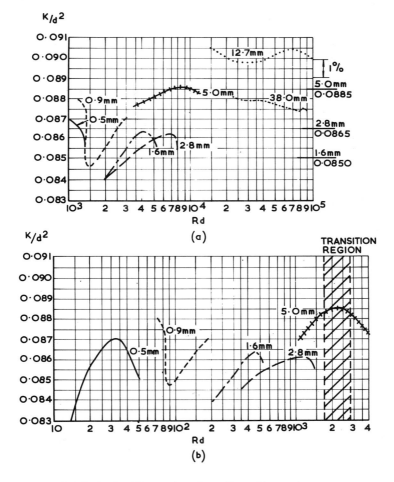

Fig. 11   a   Plot of all Series B orifices against throat Reynolds number

   b   Plot of all Series B orifices against upstream pipe Reynolds number

7 KOENNECKE, W.: 'Neve Düsenformen für Kleinere und Mittlere Reynoldszahlon', Forschung 9.Bd./Heft 3, 1938, pp. 109-125

8 FILBAN, T.J., GRIFFIN, W.A.: 'Small-Diameter Orifice Metering', Trans. A.S.M.E., Series D, 82, 1960, pp. 625-638

9 DEKKER, B.E.L., and CHANG, Y.F.: 'An Investigation of Steady Compressible Flow Through Thick Orifices'. Proc. Instn. Mech. Engrs., 1965-66, 180, Pt. 3

10 LANDSTRA, J.A.: 'Quarter-Circle Orifices', Trans. Instn. Chem. Engrs., 38, 1960

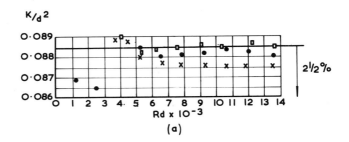

(a)

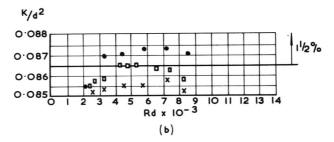

(b)

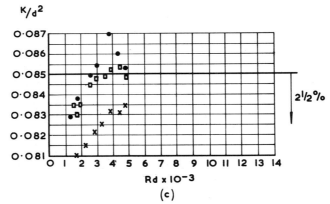

(c)

Fig. 12   Individual calibrations of Series C orifices using their nominal bore

Series C
□———□ 0
o———o 1
x———x 2

a 5.0 mm dia. orifices
b 2.8 mm dia. orifices
c 1.6 mm dia. orifices

9 DEFINITIONS OF SYMBOLS

L = Length of Orifice bore, mm

d = Diameter of Orifice bore, mm

a = Orifice area, mm^2

D = Diameter of upstream pipe, mm

Q = Flowrate, g/sec

C_d = Discharge coefficient

ΔP = Pressure difference, g/cm^2

E = $\sqrt{1 - (d/D)^2}$

h = Differential head, mm H$_2$O

g = Gravity, cm/sec^2

ρ_A = Density of air, g/cm^3

ρ_w = Density of water, g/cm^3

μ = Absolute viscosity, poises

ρ_L = Density of manometer fluid, g/cm^2

R_d = Throat Reynolds no.

R_D = Pipe Reynolds no.

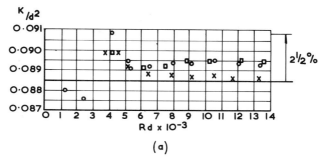

(a)

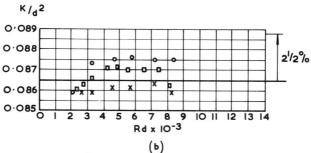

(b)

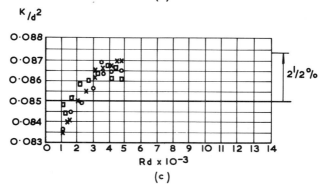

(c)

Fig. 13   Individual calibrations of Series C orifices using their actual bore

Series C

□————□   0

o————o   1

x————x   2

TABLE 2   Tolerances for parallel bore orifices

Item for tolerance	Orifice bore (mm)		
	1.6	2.8	5.0
Bore	1.596 mm 1.604	2.793 mm 2.807	4.9875 mm 5.0125
Thickness	1.584 mm 1.616	2.772 mm 2.828	4.950 mm 5.050
Surface Finish (Faces)	0.4	0.4	0.4 micrometers
" " (Bore)	1.5	1.5	1.5 "
Radius at Bore Entry and Exit	0.004 mm	0.004 mm	0.004 mm

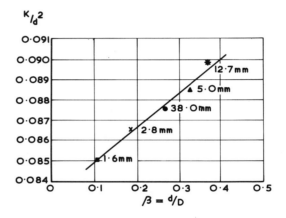

Fig. 14   Discharge coefficient plotted against β

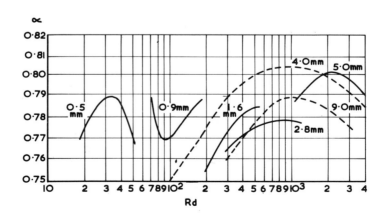

Fig. 15   Comparison of discharge coefficient with the results obtained by Koennecke[7]

x — — — x   Koennecke's results

10   APPENDIX

10.1 Tolerances on experimental investigation

(a)   Air Tests

Readability of Differential $= \pm 0.05$ mm H_2O
      "   "   Pressure $= \pm 0.3$ mm Hg
      "   "   Temperature $= \pm 0.25^{\circ}C$
      "   "   Air Volume $= \pm 36$ cm^3
      "   "   Time $= \pm 0.1$ sec

Maximum % error $\pm 0.60\%$ at lowest flow
Average % error $\pm 0.30\%$
Leakage rate $= 3 \times 10^{-2}$ cm^3/sec

(b)   Water Tests

Readability of differential
(i)   Air on water manometer $= \pm 0.25$ mm H_2O
(ii)   Water on mercury   " $= \pm 0.25$ mm Hg
     Water volume $= \pm 8$ cc
     Time $= \pm 0.05$ sec
     Maximum % error $= \pm 0.50\%$
     Average   " $= \pm 0.25\%$

The orifice dimensions were obtained by using very accurate measuring equipment, such as a shadowgraph.

10.2 Comparison of orifice tolerances with those obtained by L.R. Voss[4]

Voss has obtained tolerances very similar to those obtained in this paper by testing a large number of orifices. He varied bit by bit the essential dimensions of small bore, square edge thin plate orifices.

TOLERANCES	COUSINS	VOSS
Orifice dia.	1·6mm 2·8mm ±0·25% 5·0mm	3·05mm ±0·4% 4·76mm ±0·27% 6·35mm ±0·2%
Hole inlet edge	Maximum radius to be 0·0004mm and free from burrs	Must be square and sharp, free from either burrs or rounding— so that it does not visibly reflect light from 10 X magnification
Surface finish	Maximum surface roughness 0·4 micrometers	Maximum surface roughness 0·4 micrometers
	5·0mm : flat to 0·05mm 1·6mm : flat to 0·017mm	Surface must be, flat to within 0·0254mm and parallel

DISCUSSION

D.L. Smith: Have you considered using jewelled orifices to obtain the necessary high finish.

B.T. Goldring: Further to the suggestion that jewelled orifices could be used, I had a small number of synthetic sapphire orifices made a few years ago. These orifices were quite cheap, and the variations between individual calibrations were negligible. The bore was approximately 0.5 mm, the L/d ratio was 2.0 and the inlet was rounded.

A.N. Crossley: Further to the last speakers comments regarding jewelled orifices it may be possible to obtain wire drawing dies. Alternatively, it should be possible to make a brass wire lap impregnated with suitable material and polish the bore. Then each face may be lapped in the normal manner, this should produce a square edged entrance and exit.

E.A. Spencer: The problem of accurate manufacture to which the author of this paper and Okafor and Turton in a later paper refer is not limited to the orifice plates of very small diameter which they used, although the difficulties are undoubtedly increased greatly with decreasing size. In a paper published in BWK in 1970 (A) the writer presented results from a long study of a series of orifice plates in an 8-inch pipe. A special technique was developed at the National Engineering Laboratory (B) for examining the edge sharpness of the orifice bore. Plastic replicas can be made which can be sliced, polished and studied at high magnification. The effect of rounding off the edge increases with decreasing area ratio. It is probable not only that many plates do not conform at the time of manufacture, as Harning found, but that many lose their predicted discharge co-efficient by edge erosion when in service in dirty fluids. Regular inspection is important there-fore, and the National Engineering Laboratory technique enables permanent physical records to be kept for comparison.

References:

A   SPENCER, E.A., CALAME, H. and SINGER, J.: 'Edge Sharpness and Pipe Roughness Effects on Orifice Plate Discharge Coefficients'. Brennst.-Warme-Kraft 1970, **22**, (2), pp. 56-62. and NEL Report No. 427, East Kilbride, Glasgow: National Engineering Laboratory, August 1969

B   GALLACHER, G.R.: 'Measuring edge sharpness of orifice plates', Engineer, Lond., 1968, **225**, (5860), pp. 783-785

D. Wolfman: The author mentioned in his paper troubles arising from leakages around the orifices and from the atmosphere. We overcame similar problems by producing light alloy or copper orifices and fastening them between two stainless steel knife-edge flanges known in the vacuum industry as Conflat or Hall flanges. This ultimately solved the problem of leakages.

Do not you think that in order to correlate similar flows one has to guard a fixed ratio between d (orifice diameter) and D (tube diameter)?

Author's reply: In reply to Mr. Smith and Mr. Goldring, the high precision finish obtainable from jewelled orifices would have been most suitable, but our enquiries showed that these could only be obtained in quantities far in excess of the numbers likely to be required. The three larger orifices can be made cheaply within the required tolerances and so for the present the idea of using jewelled orifices has been abandoned. The major criterion for the bore of the orifice is that the diameter should be uniform rather than smooth. The suggestion made by Mr. Crossley is con-cerned more with the finish of the bore and would be useful rather than necessary. Dr. Spencer's comments on measuring the sharpness of an edge are very useful. It is very difficult and costly to define an edge for the purposes of inspection and, while I suspect the method he discusses would be too costly for our particular application, it appears to be a very useful technique.

Previous investigations into the effect of dissolved air in water on the discharge coefficient of separated flow devices, such as Dall Tubes, made me very conscious of the large errors this would have on long bore orifices. In my experiments however I do not think that there was any problem as the water was not aerated. Regarding Mr. Wolfman's comments about internal leakage, the problem was solved by using compressible teflon seals, with a new one for each orifice tested. In the paper I have mentioned there appears to be some effect, but it is very difficult to separate this from the other effects, such as variations in Reynolds' number.

A.T.J. Hayward: The author has emphasised the need for scrupulous cleanliness when calibrating small-bore orifices. This accords with our experience at the National Engineering Laboratory in calibrating carburettor jets; unless we employ very elaborate degreasing and filtration techniques, errors of 20 or 30 per cent can easily occur. I should like to ask what happens to your carefully calibrated small-bore orifice meters when they are installed in an industrial system of normal industrial standard of cleanliness?

Author's reply: While I agree that these orifices are susceptible to blockage by dirt, they are relatively no worse than most other low flow devices. Because such small sizes have to be used, blockage will always be a problem, but it is hoped that industrial users will recognise this problem and takes adequate precautions against it.

S.C. Okafor: Is further investigation on the pressure gradient along the orifice bore length possible? I personally believe that the future of long orifices is bright as a solution against erosion effects on sharp thin orifices. An analytical evaluation of effect of the geometric ratio, L/d is necessary to permit the prediction of discharge coefficient at the design stage.

Author's reply: I do not intend to do any further investigations into long bore orifices, but many people have been and are working on these types of orifice and obtaining some interesting results. From the paper by Dekker and Chang, Ref. 9, there appears to be some grounds for believing that hysteresis occurs near the threshold of choking, as Dr. Brain points out. In my work I did not come across this hysteresis as I did not go to a high enough pressure ratio. The orifices I have described are not intended to be used in or near critical flow.

P. Jepson: Could this type of orifice plate ever be used to obtain flow measurement accuracies better than the ±2.5% quoted?

I would think not, since it would appear impractical to manufacture orifice plates of such small diameters with upstream edges which conform to the BS 1042 specifications on edge sharpness; e.g. edge radius to BS 1042 $<$ 0.0004d which for d = 1.6 mm gives the edge radius to be less than 0.00064 mm

Could this factor account for the relatively poor accuracies quoted and could the author suggest any method of measuring such small edge radii?

Author's reply: I did not envisage using these orifices for accuracies within ±2.5% : to do so would require high manufacturing tolerances resulting in uneconomical production costs. However, an uncalibrated uncertainty of ±2.5% is by no means bad for the measurement of such small flows, especially as a large proportion of that market requires only repeatable devices.

As Dr. Jepson points out, the rounding of the inlet was found to be one of the major causes of the scatter of results.

As has been said in the report, several methods have been used to measure the inlet curvature. L.R. Voss (Ref. 4) suggests that the carvature be measured visually be examining the incident light reflection from the edges under a microscope at a given magnification. This method I felt to be too arbitrary as it depends on the opinion of the person inspecting the orifice.

F.W. St. Clair: I would like to mention that my company has manufactured an integral orifice assembly fitted to a differential pressure cell for at least 10 years. Standard orifices machined from selected materials are available in seven bore sizes ranging from .350 ins. to .020 in. In answer to comments from other delegates regarding jewelled orifices I would like to state that these are also available mounted in a stainless steel holder in bore sizes down to 0.002 in. I would confirm the authors comments regarding surface finish as undoubtedly this has a marked influence on the range of Reynolds numbers which can be covered. Also that satisfactory operation presumes the measured fluid is clean and free from all suspended material which would rapidly build up on, or plug these small orifices.

4·2

NON-NEWTONIAN FLOW MEASUREMENT USING CONVENTIONAL PRESSURE DIFFERENCE METERS

J. Harris and A.N. Magnall

Abstract: This work is an extension of previously published results[1] relating to the flow of non-Newtonian solutions through British Standard venturi and orifice meters. It was found in the prior work that the discharge coefficient did not correlate well on a generalised Reynolds number basis and an alternative more effective correlation was presented. The departure of experimental data from standard curves is found in the present work to be even more exaggerated in the case of suspensions and again the alternative correlation of data appears more efficient.

1   INTRODUCTION

Many of the fluids encountered in industry possess complex rheological properties. The performance of standard orifice and venturi meters with rheologically complex non-Newtonian fluids is not well understood at the present time.

There are standards available in several countries of the world relating to the above types of meters, but the calibrations invariably are intended to refer to Newtonian fluids which may be characterised in isothermal flow by a single viscosity and a simple form of Reynolds number[2]. In contrast it is much more difficult to characterise non-Newtonian fluids; ideally a Reynolds number would be found which would enable adaption of the calibration such as those found in the British Standard BS 1042[2] to be made to the non-Newtonian case, but no such dimensionless group or groups has been found up to the present. An alternative technique has therefore been sought and the present work records further applications for the prior technique[1].

2   PRIOR WORK

A review of the various factors affecting the performance of orifice meters has been given in a previous publication by Harris and Magnall[1]. In this publication it was shown that for several aqueous polymer solutions the discharge coefficients of one venturi meter and several orifice plates would correlate on a basis of the generalised Reynolds number for generalised Reynolds numbers greater than about 10^4, and furthermore at these higher Reynolds numbers the experimental points approximated to the British Standard curves BS 1046[2].

In the region of Reynolds numbers less than about 10^4 no well defined relation of the form,

$$C_D = f(N_R) \qquad \qquad \text{... (1)}$$

appears possible.

In equation (1), C_D is the discharge coefficient defined in the usual way (see for example Kay[3]) and N_R is the so-called generalised Reynolds number, defined by,

$$N_R = \frac{\rho U_m^{2-n} D^n}{K 8^{n-1}} \qquad \qquad \text{... (2)}$$

This definition of Reynolds number was introduced by Dodge[4] and by Metzner[5] and is evaluated from *steady laminar flow data.*

The apparent success of the generalised Reynolds number basis of correlation above $N_R \sim 10^4$ should not be overestimated since the discharge coefficient is practically constant then and the pressure drop in a venturi or orifice meter is mainly an inertia dominated phenomenon and hence the viscous properties, whether Newtonian or non-Newtonian are largely irrelevant. Practically any formulation of Reynolds number would suffice. In the lower ranges of Reynolds numbers, where viscous effects are important, it was found that the generalised Reynolds number became ineffective for correlating the results.

Prior experience had shown that a formula due to Bowen[6], namely

$$D^X \tau_W = B U_m^W \qquad \qquad \ldots (3)$$

was very effective in correlating pipe friction data. This has been checked for its effectiveness by Harris[7] and Harris and Quader[8].

The effective correlation of pipe friction data for equation (3) leads to the consideration of possible application in the case of venturi and orifice meters. A development of the formula given by

$$(m^{1/2} D)^X \triangle P = B U_m^W \qquad \qquad \ldots (4)$$

was used as a correlating basis for the prior work on solutions[1] and just as in the case of pipe friction data, considerable improvement was recorded in correlating experimental data over that achieved by a generalised Reynolds number approach.

Similar results are now available for suspensions, and these are treated in the current work.

3   CURRENT RESULTS

The experimental results quoted in this work were obtained with the same experimental apparatus as those given in the paper by Harris and Magnall[1]. The meters were installed singly in a 0.0508m (2 in.) diameter pipe line in accordance with BS 1042. The principal dimensions are:

TABLE 1

Meter	Dia (meters)	Area Ratio (a/A)
Venturi (Throat)	0.03175	0.391
Orificies	0.0304	0.360
	0.03175	0.391
	0.0349	0.474
	0.0381	0.563

Five concentrations of china clay suspensions were used in the experiments. The powder was supplied by English China Clays Ltd., St. Austell, Cornwall, U.K. The concentrations and densities are given in the following table:

TABLE 2

% Solids	Density (Kg/m^3)	n	K(S.I.)
20	1109	0.386	3.7
27	1138	0.116	66.0
29.5	1148	0.138	110.0
31.5	1155	0.141	154.0
35.5	1175	0.152	300.0

The values of n and k are those in the power-law rheological formula;

$$\tau = K \dot{\gamma}^n \qquad \qquad ... (5)$$

and were obtained using a Haake-Rotovisco variable speed viscometer.

3.1 Comparison with standard curves

Following the prior form of presenting results, experimental results have been prepared on the basis of equation (1). These are displayed on Figs. 1 to 5 and compared with standard curves. Several factors are apparent in these graphs: a) There is considerable scatter in the experimental data for both the venturi meter and all the orifice meters. b) The experimental points appear to cross the standard curve in the case of the venturi meter results in Fig. 1, but are almost invariably found below the standard curves in the case of the orifice meters. c) There is no correlation for the various concentrations in a single meter. d) At higher Reynolds numbers the experimental values of discharge coefficient tend to constant values which are below those of the standard curves.

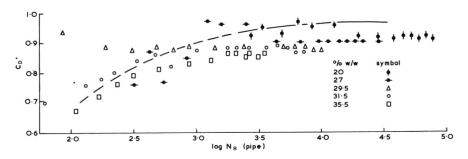

Fig. 1   Comparison of standard curves of discharge coefficient C_D against Log N_R (pipe) with experimental data for Venturi meter (throat dia. = 0.03175 m) (area ratio, m = 0.391) China clay slurry
———BS 1042[8]
– – – Ref. (1)

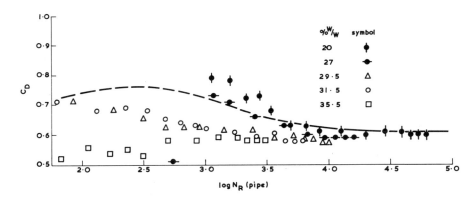

Fig. 2   Comparison of standard curve of discharge coefficient C_D against Log N_R (pipe) with experimental data for orifice meter (dia. = 0.0304 m) (area ratio, m = 0.36) China clay slurry
———BS 1042[8]
– – – Ref. (1)

182   Harris and Magnall

3.2 The alternative correlation scheme

On the basis of Figs. 1 to 5 it appears at the present time as though correlation of discharge

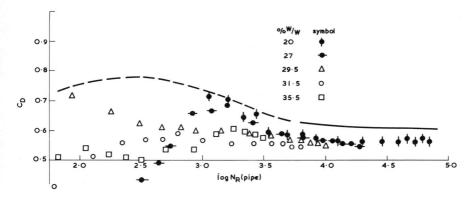

Fig. 3 Comparison of standard curve of discharge coefficient (C_D) against Log N_R (pipe) with experimental data for orifice meter (dia. = 0.03175 metres) (area ratio, m = 0.391) China clay slurry
———BS 1042[8]
— — — Ref. (1)

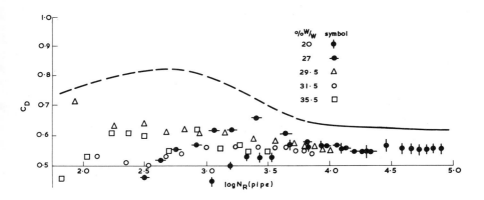

Fig. 4 Comparison of standard curve of discharge coefficient C_D against Log N_R (pipe) with experimental data for orifice meter (dia. = 0.0349 m) (area ratio, m = 0.474) China clay slurry
———BS 1042[8]
— — — Ref. (1)

coefficient against generalised Reynolds number is very inefficient as a means of presenting data. As an alternative, the same data was prepared on the basis of equation (3) and this is presented in Fig. 6 for the Venturi meter and Figs. 7-11 for the orifice meters.

3.2.1 Venturi meter

It is clear from Fig. 6 that a greater semblance of order is achieved for this type of presentation as compared with Fig. 1. All but one of the concentrations (29.5%) appear to possess two branches each a straight line on a double logarithmic plot. The 29.5% results appear to fall on a single straight line and this cannot be explained at present.

Considering the upper limb of the curves, values of x, w and B are plotted against concentration

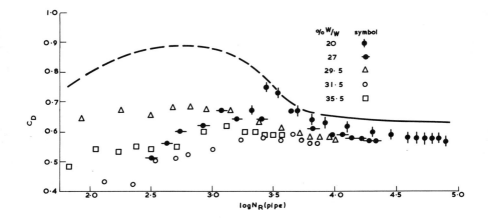

Fig. 5 Comparison of standard curve of discharge coefficient C_D against Log N_R (pipe) with experimental data for orifice meter (dia. = 0.381 m) (area ratio, m = 0.563) China clay slurry
———BS 1042[8]
— — — Ref. (1)

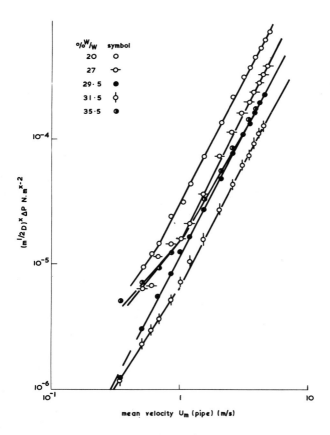

Fig. 6 Log–log plot of $(m^{1/2}D)^X \Delta P$ against U_m for Venturi meter (area ratio m = 0.391)

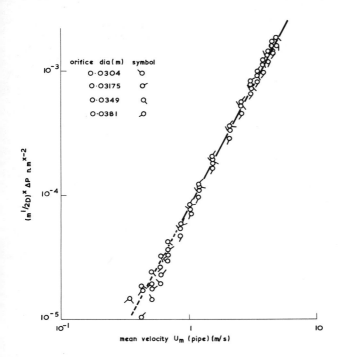

Fig. 7 Correlation of orifice data for 20% china clay slurry
$x = 4.0, w = 2.1, B = 7.6 \times 10^{-5}$

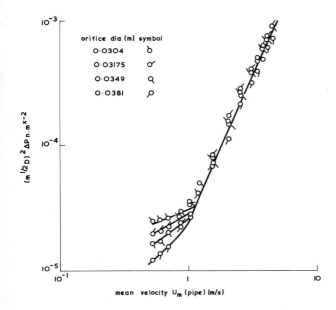

Fig. 8 Correlation of orifice data for 27% china clay slurry
$x = 4.2, w = 2.28, B = 2.65 \times 10^{-5}$

of solids in Fig. 12. Further work is required to establish these trends and to establish comparable results for other slurries. Since only one throat diameter was used the values of x have been arbitrarily chosen to be the same as for the orifice meters. This only affects the vertical position of the curves in Fig. 6 and enables a comparison to be made of the shape of the curves of B against concentration in Fig. 12a.

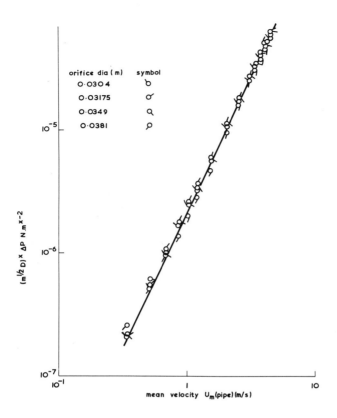

Fig. 9   Correlation of orifice data for 29.5% china clay slurry
x = 5.0, w = 2.22, B = 2.1 x 10^{-6}

3.2.2 Orifice meters

There is added interest in the orifice results Figs. 7 to 11 in that four orifice diameters were used for each concentration. Overall there is considerably more effective ordering of the results over that achieved on a generalised Reynolds number basis, Figs. 2 to 5 inclusive.

Values of x, w and B are plotted for the upper limb of each curve in Figs. 7 to 11 in Fig. 12. There appears to be a definite similarity in shape between the w and B curves for the orifice and venturi meters. Nothing can be stated about the x curve in this way since only one venturi throat diameter was used. Values of n for the venturi were allocated from the orifice results for the reasons stated above.

It may be noted that at higher flow rates correlation for the various diameter orifices is achieved in Figs. 7 to 11. Divergences tend to occur at lower flow rates especially notable in Figs. 8 and 10. All the results require to be extended to lower flow rates to investigate this phenomenon.

3.3 Tabulated results

Tabulated results for x, w and B are given below:

TABLE 3

Concn. %	Venturi			Orifice		
	x	w	B	x	w	B
20	4.0	2.08	3.1×10^{-5}	4.0	2.1	7.6×10^{-5}
27	4.2	2.12	1.5×10^{-5}	4.2	2.28	2.65×10^{-5}
29.5	5.0	2.04	1.15×10^{-6}	5.0	2.22	2.1×10^{-6}
31.5	4.5	2.0	6.60×10^{-6}	4.5	2.21	1.2×10^{-5}
35.5	5.0	1.87	1.44×10^{-6}	5.0	1.9	3.0×10^{-6}

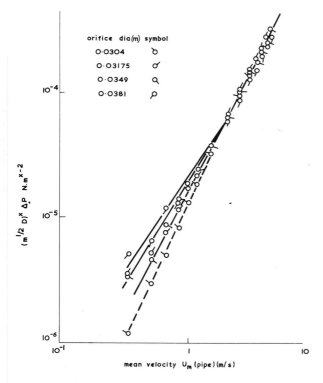

orifice dia(m)	symbol
0·0304	ƀ
0·03175	ơ
0·0349	৭
0·0381	ρ

Fig. 10 Correlation of orifice data for 31.5% china clay slurry
$x = 4.5$, $w = 2.21$, $B = 1.2 \times 10^{-5}$

4 CONCLUSION

It appears from the current work that a broadly similar result is obtained for the slurries used here as was obtained for the solutions treated in a prior publication[1] namely; venturi meter and orifice meter calibrations cannot be very effectively presented on a basis of discharge coefficient against generalised Reynolds number. It is found however that an alternative presentation of results based upon a modified Bowen formula is much more efficient in correlating results.

Divergence of the experimental data at lower flow rates was found for various orifice diameters and this distinguishes the graphs in Figs. 7 to 11 from those for solutions presented in an earlier publication.

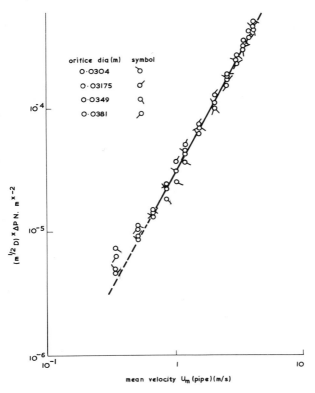

Fig. 11   Correlation of orifice data for 35.5% china clay slurry
$x = 5.0$, $w = 1.9$, $B = 3.0 \times 10^{-6}$

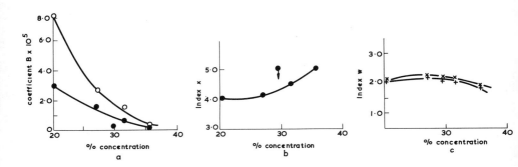

Fig. 12   a   The effect on coefficient B (eq. 4) of solids concentration
○ orifices
● Venturi

b   The effect on index x (eq. 4) of solids concentration (orifices)

c   The effect of index w (eq. 4) of solids concentration
x orifices
+ Venturi (estimated)

It is considered that further experimental work using a range of pipeline diameters is required to clarify the reasons for the above behaviour of slurries at lower flow rates and also to check on the effectiveness of the correlations presented.

It cannot be recommended on the basis of the present work that venturi and orifice meters be used on the basis of standard calibration curves for flow in which the generalised Reynolds number is less than 10^4 whilst above this value errors of the order of 20% may be encountered.

For generalised Reynolds numbers above about 2000, calibrations of the type shown in Figs. 7 to 11 may be used. The linearity in the higher flow rate ranges for single orifice diameters is generally quite accurate and interpolation of the results should be possible with accurate results. It is considered that improved correlations of the higher flow rate data could be achieved by slight adjustments to the value of x, w and B.

5   SYMBOLS

a	=	Throat cross-sectional area
A	=	Duct cross-sectional area
B	=	Bowen coefficient, eqn. (3)
D	=	Duct diameter
K	=	Power-law coefficient, eqn. (5)
m	=	a/A area ratio
n	=	Power-law index, eqn. (5)
N_R	=	Generalised Reynolds number, eqn. (2)
U_m	=	Mean fluid velocity
x		
	=	Indices in the Bowen eqn. (3)
w		
$\triangle P$	=	Meter differential pressure

6   ACKNOWLEDGEMENTS

The authors wish to acknowledge the generosity of English China Clays Ltd., St. Austell, Cornwall, U.K. in supplying the china clay powder used in the experimental work.

7   REFERENCES

1   HARRIS, J. and MAGNALL, A.N.: 'The Use of Orifice Plates and Venturi Meters with non-Newtonian Fluids', Symp. on the Flow Measurement of Difficult Fluids, by I. Chem. Engrs. at Bradford University, 15-16 April 1971

2   BS 1042: Pt. 1: 1964: 'Methods for the Measurements of Fluid Flow in Pipes', Pt. 1. Orifice Plates, Nozzles and Venturi Tubes

3   KAY, J.M.: 'Fluid Mechanics and Heat Transfer', (Camb. Univ. Press, 1957)

4   DODGE, D.W., and METZNER, A.B.: Rheol. Acta, 1958, 1,205

5   METZNER, A.B.: in 'Handbook of Fluid Dynamics' 1961, Vol. 1, Ed.: STREETER, M. (McGraw-Hill)

6   COULSON, J.M., and RICHARDSON, J.E.: Chem. Eng., Vol. 1 (Pergamon)

7   HARRIS, J.: Rheol. Acta, 1968, 7, 228

8   HARRIS, J., and QUADER, A.K.M.A.: Brit. Chem. Engr., 1971, 16, 307

DISCUSSION

A.T.J. Hayward: The expression $m^{1/2}D$ in equation 4 is equal, by definition, to the orifice diameter, d. What is the purpose of expressing it in the more complicated form of $m^{1/2}D$, instead of in the simple form, d?

Authors' reply: Mainly convenience.

K.J. Zanker: What is the effect of concentration on the correlation with Reynolds' number in the laminar-transitional-turbulent range of $10^2 < Re < 10^5$? I would not expect a correlation with Reynolds' number.

Authors' reply: The effect of concentration is displayed in figures 1-5 (inclusive). It is mentioned in the text that no well-defined relationship on a Reynold's number basis (equation (2)) appears possible especially at lower values of the Reynolds' number. This may well stem from the fact that the Reynolds' number (equation (2)) is a weak characteristic of the flow.

From a study of the published literature there does not appear to be any apriori reason why a correlation on a Reynolds number basis should be definitely excluded. In view of the fact that such a correlation is published for Newtonian flow (refs. 1 and 8) there is encouragement to the corresponding situation in relation to non-Newtonian flow. The discharge coefficient is a function of a number of parameters as outlined in ref. 5.

S.B. Au: I would like to know what accuracy for the discharge coefficient of the venturi meter the authors obtained in their tests on non-Newtonian fluids. Is there any difficulty in measuring the pressure differential as far as the non-Newtonian fluid is concerned.

Authors' reply: The pressure differences were measured on manometer tubes and the accuracy of measurement is that normally associated with this type of equipment. Providing a purging water supply is available for use to ensure freedom from blockage, no further practical difficulty should be encountered.

S.B. Au: Have the authors been able to develop any theory or method to predict the discharge coefficient of the venturi meter in measuring non-Newtonian fluids?

Authors' reply: Work has been carried out along these lines and will be published later.

A DIFFERENTIAL PRESSURE FLOWMETER WITH LINEAR RESPONSE

D. Turner

Abstract: Differential pressure flowmeters have hitherto been classified as either (a) constant orifice area, variable differential pressure, or (b) constant differential pressure, variable orifice area. Neither type yields an instrument with a dynamic range exceeding about 10:1. This paper describes a hybrid instrument in which the orifice area is a function of the differential pressure. Proper selection of the functional dependence results in an instrument giving a differential pressure directly proportional to flowrate over a range of 100:1. Alternative arrangements produce exponentially varying differential pressures, which may be processed electronically to measure flowrates over a range of 1000:1. A commercially available instrument based on the above concept is described. Test results are presented showing the dynamic range and linearity achieved in practice, and the effects of variations in the temperature, specific gravity, and viscosity of the liquid being metered are described.

1   THE NEED FOR LINEAR RESPONSE

All pressure differential flowmeters rely on the insertion of an obstruction — the measuring orifice — in the flow path, and depend ultimately on the well-known relationship

$$Q = kAh^{1/2}$$

where 'Q' is the mass flowrate of liquid, 'A' is the area of the measuring orifice, 'h' is the pressure differential across the orifice, and 'k' is a proportionality constant. The simplest and most often used primary elements based on this relationship are the orifice plate and the Venturi tube. With these devices, and any others in which the area 'A' is fixed, the flowrate must be deduced from the square root of the signal 'h'. If measurements are required over a large range of flowrates this places impossible demands on the instrument responsible for determining 'h', for at one tenth of maximum flowrate the pressure differential has already dropped to one hundredth of that at maximum flowrate. Since a significant fraction of 'h' represents power loss, the value at maximum flowrate is normally kept as low as possible, and the signal at small flowrates is therefore liable to be very small indeed. For this reason it is usual to restrict the measuring range of pressure differential installations to about 5:1.

The use of a primary element giving a differential pressure linearly proportional to the flowrate has the immediate effect of squaring the range of flowrates corresponding to a given range of differential pressures. The range over which meaningful measurements of flowrate may be made is increased by a larger factor still, because the slope of the response curve (dh/dQ) at the lower limit is also increased. This increase in the range of measurement is the principal benefit from the use of a linear primary element, and it is important to realise that the addition of a square root extractor to an orifice plate installation does not amount to the same thing. Although this certainly produces a linear response in the end, the readings at low flowrates are just as much in doubt as they were without the square root extractor, and for the same reason that they are based on the measurement of an infinitesimal pressure differential at a point on the response curve where the slope is almost zero.

There are many applications in existing industrial and pilot plants where flowrate measurements over a wide range are necessary or desirable. As an example, consider the case where a quantity of 'out of spec.' product is to be salvaged by blending it with a suitable quantity of 'super spec.' material to produce a blend which meets its specification by a comfortable but not unprofitable margin. To ensure good mixing, and to enable the final blend to be checked as the run proceeds, the two materials may be pumped simultaneously through separate metered lines to the same tank inlet. The proportions of the two materials required are however likely to vary widely on different occasions, therefore wide range flowrate meters are desirable.

In the design of new plant the engineer has at his disposal three main variables; temperature, pressure, and flowrate. Of these, temperature and pressure have been accurately measurable over very wide ranges for many years, but flowrates only over small ranges. As a result plant designers have been constrained to keep flowrates as nearly constant as possible, and to use temperature and pressure as control variables. Reliable wide range flowrate measurements make available a new degree of freedom in plant design.

2   A PRIMARY ELEMENT WITH LINEAR RESPONSE

If a primary element is constructed so that the area 'A' of the measuring orifice varies as the square root of the differential pressure 'h' so that

$$A = k_1 h^{1/2}$$

then this value of 'A' may be substituted in the expression

$$Q = k_2 A h^{1/2}$$

to give

$$Q = k_2 k_1 h^{1/2} h^{1/2}$$

which leads to

$$Q = Kh$$

where k_1 k_2 and K are proportionality constants.

This is achieved in Gervase Instruments Ltd. linear response primary elements* by the construction shown diagrammatically in Fig. 1. The measuring orifice is carried on the free end of a metal bellows, the spring rate of which is known. The other end of the bellows is sealed to the end plate of the unit. Also fixed to the end plate is the shaft of a trumpet-shaped control member, which passes through the measuring orifice and is arranged so that, with the bellows relaxed, the flared end of the trumpet just closes the measuring orifice. The inlet of the unit is on the outside of the bellows, and the outlet is on the inside.

When liquid flows into the inlet chamber, the measuring orifice through which it must flow to reach the outlet is closed. The pressure in the inlet chamber therefore increases and the bellows is compressed. The measuring orifice rides on the end of the bellows towards the narrower stem of the control member, which consequently no longer completely closes the orifice and allows liquid to flow through to the outlet. Thus the differential pressure and the orifice area increase together until equilibrium is reached. The equilibrium position is very stable indeed, and is reached rapidly without any tendency to 'hunt'.

If the bellows spring characteristic is linear, the axial movement of the measuring orifice along the control member is strictly proportional to the differential pressure. It is then straightforward (though tedious) to calculate the shape of control member which will maintain the required constant ratio between the orifice area and the square root of the differential pressure.

In fact the spring rate of metal bellows is seldom constant over the working range, and it is not possible with any shape of control member to maintain proportionality between the orifice area and the square root of the differential pressure right down to zero. However the linearity of response which can be achieved in practice is shown by the graph of Fig. 2, which is the response curve of a unit from current normal production. The entire curve lies between two parallel straight lines drawn at a distance apart corresponding to ±1% of maximum flowrate, enabling meaningful measurements to be made over a flowrate range of about 30:1.

3   A WIDE RANGE PRIMARY ELEMENT

If a simple conical control member is substituted for the trumpet shaped one in Fig. 1, the relationship between the orifice area 'A' and the differential pressure 'h' is changed to (to a first

* Linear 'GILFLO' units

approximation)

$$A = k_1 h$$

Substituting, as before, in the equation

$$Q = k_2 A h^{\frac{1}{2}}$$

leads to the relationship

$$Q = K h^{1\frac{1}{2}}$$

A primary element constructed in this way* may be used with a simple differential manometer to measure flowrates over a range of 250:1, since the flowrate scale expands progressively towards the zero. This is illustrated by Fig. 3, which shows the response curve of an actual instrument of this type together with the mercurvy manometer scale derived from it.

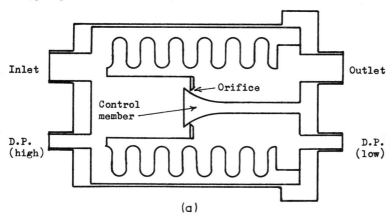

(a)

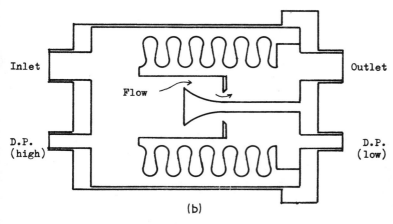

(b)

Fig. 1   Simplified diagrammatic section of linear response primary element

a   Zero flowrate
b   Maximum flowrate

(Note: This diagram is not to scale and is not intended to show details of actual construction)

* Wide-range 'GILFLO' unit

The wide range type of element is considerably more accurate over the whole measurement range than the linear element because (a) small variations in the axial positioning of the control member and orifice have less effect on the signal, and (b) because the conical shape can be machined far more precisely than the complex trumpet of the linear unit.

4   A WIDE RANGE LINEAR RESPONSE FLOWRATE MEASURING SYSTEM

The full dynamic range and accuracy of wide range primary elements can be exploited without relinquishing the benefit of a linear relationship between the flowrate and the signal. The first step is to convert the differential pressure signal to a direct current analogue by means of a

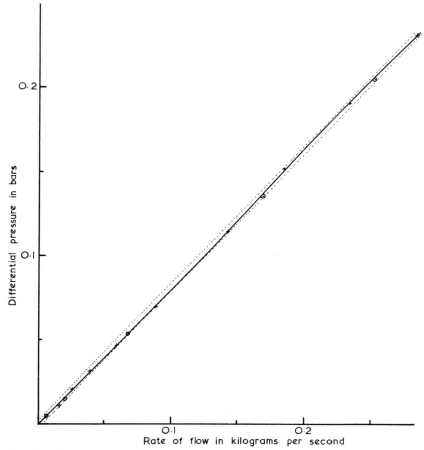

Fig. 2   Response curve of linear primary element
Test fluid: kerosene, S.G. 0.794 at 17° C
The two broken lines represent a tolerance of ±1% of maximum flow
+ flowrate increasing
0 flowrate decreasing

standard electronic differential pressure transmitter. The electrical analogue is fed into a 'black box' * from which it emerges as a standard 0 - 10mA d.c. signal directly proportional to the flowrate. A block diagram of this system is shown in Fig. 4.

Once this step has been taken, an almost infinite range of options is presented. The linear

* 'GILFLO'

194   **Turner**

electrical signal may be displayed (via range switching) on a calibrated meter, or directly, across a suitable scaling resistor, on a digital voltmeter. It may be fed into a data acquisition system, integrated over a period to give total flow, recorded graphically, and printed out, all at the same time if required.

It is not always necessary to use a separate 'black box' for each primary element. A switch may be used to select any one of a number of signals for processing and display. Alternatively the

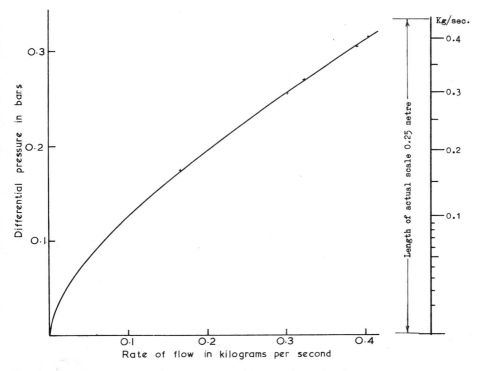

Fig. 3 Response curve of wide range primary element and associated mercury manometer scale
Test fluid: water; temperature, 20°C

Note: Of the 34 experimental points, all but the five shown fell within the thickness of the line

outputs from a number of primary elements may be multiplexed through one 'black box' to an equal number of display or other devices, updating each reading every half second or so.

The flowrate measurement range achieved with this system exceeds 1000:1 and brings with it a problem — the difficulty of finding valves which shut off tight enough to give a zero flowrate indication.

5   CORRECTIONS FOR PHYSICAL PROPERTIES OF THE METERED LIQUID

5.1 Specific gravity

A change from D_1 to D_2 in the specific gravity of the liquid being metered causes a change from h_1 to h_2 in the differential pressure so that:—

(i)   for a constant volumetric flowrate

$$h_2 = h_1 \frac{\sqrt{D_2}}{\sqrt{D_1}}$$

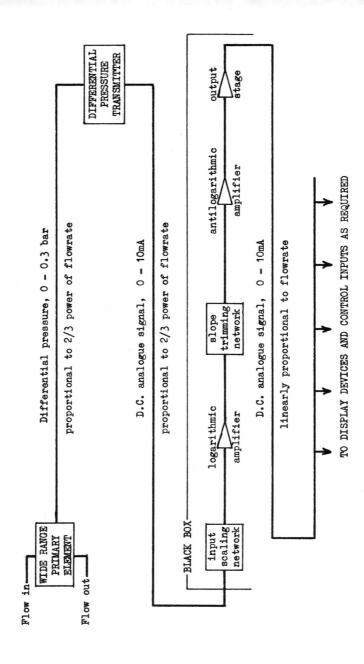

Fig. 4   Block diagram of wide range linear response flowrate measuring system

(ii)   and for a constant mass flowrate

$$h_2 = h_1 \frac{\sqrt{D_1}}{\sqrt{D_2}}$$

The effect of small variations in the specific gravity of the liquid on the indicated (volumetric) flowrate is therefore only half that on a conventional fixed area differential pressure system.

5.2 Temperature

Temperature variations affect the units mainly by changing the spring rate of the bellows, an increase in temperature reducing the spring rate. With stainless steel bellows the temperature coefficient, determined experimentally, for the reduction in differential pressure so caused is $8 \times 10^{-4}/^{\circ}C$.

It is interesting to note that the coefficient of expansion (cubical) of most hydrocarbon liquids is of the order of $12 \times 10^{-4}/^{\circ}C$. For any given mass flowrate the resulting decrease in specific gravity due to a temperature rise therefore causes an increase in the differential pressure of $6 \times 10^{-4}/^{\circ}C$, and goes some way towards restoring the loss caused by the change in spring rate of the bellows.

5.3 Viscosity

The effect on the differential pressures produced due to the viscosity of the liquid being metered is not definable as a simple correction factor, and is therefore described here in general terms only.

For liquids with viscosities below about ten centistokes the effect may be ignored altogether. As the viscosity rises, the differential pressure begins to increase, at first for the lowest flowrates only, and a small hump appears at the bottom end of the response curve. Further increase causes the hump to extend further up the curve, until eventually a linear response is distorted into a good imitation of that normally obtained from a wide-range unit.

If the unit is required for continuous duty on a viscous liquid, and the viscosity under operating conditions is well defined, it is possible to produce specially adjusted primary units to perform well in these conditions only. In this way, heavy fuel oil to furnace burners has been successfully metered over a range of 20:1.

In other cases where wide-range metering of viscous liquids is required the unit may be calibrated under working conditions and the response curve so obtained used as the basis of flowrate measurement.

6   PERFORMANCE ON GAS STREAMS

Within certain limitations, the units described in this paper may be used for measuring the flowrate of gases, and produce pressure differentials that are substantial and easily handled. Response curves may be derived from the normal water calibrations by applying the specific gravity corrections described in section 5.1. The specific gravity of the gas (relative to water) at the temperature and pressure on the outlet side of the unit should be used in the calculation. This pressure must be constant under operating conditions to avoid measurement errors caused by changes in specific gravity.

The main difficulty which arises when using the units on gases is vibration of the bellows, excited at a critical flowrate within the measuring range of the instrument. If this is allowed to build up damage may be caused by hammering of the edge of the orifice on the control member. It is therefore necessary either to work at flowrates below the critical one — which varies from one installation to another — or to pass rapidly through the critical range and work above it. It is expected that future research work will eliminate this problem, but at this date a certain amount of experimental work must be anticipated if the unit is applied to the measurement of gas flowrates.

DISCUSSION

D.N. Ahad: This question concerns the sensitivity of the instrument to pressure and other disturbances, since there are no damping forces, other than small frictional forces. The instrument is understood to be more sensitive to disturbances when used on air.

If due to some disturbance the bellows are made to oscillate, it will do so with a frequency depending on a combination of natural frequencies, of bellows, valve seat and air cushion. These oscillations would significantly affect the calibration of the instrument and would be extremely difficult to eliminate.

Author's reply: When the instrument is used with gases it is indeed possible to run into trouble with bellows oscillations at certain flowrates. We do therefore have some reservations about this application. Nevertheless, many people have put them into service on gas streams with remarkable success, and have shown astoundingly linear response curves.

With liquids, the damping is very high, and the problem does not arise. We have been told that the response time of our instrument is 12 milliseconds, but I do not know how one makes step changes in liquid flowrates sufficiently fast to be able to check it. The steepest transients we have been able to produce have always failed to produce oscillations.

J. Agar: Could you please show the conventional error versus Reynolds' number curve.

Author's reply: With this instrument, because of its complex internal geometry, the Reynolds' number at the measuring orifice is not sufficiently well defined for the curve you mention to be very helpful.

The significant point here is that the Reynolds number at the actual orifice is high, even at low flowrates, because the orifice is then small. This, I think, is the reason there are no discontinuities in the response curves. It also gives the instrument a useful insensitivity to upstream conditions. Recently we checked the response curve of one flowmeter, first with an approach section of 100 diameters of straight smooth pipe, and then with an inlet connection made up of two consecutive elbows in different planes at right angles. There were no differences within the nominal range of the instrument greater than one per cent of span, and we were only able to produce a difference of 2% by increasing the flowrate to 1.3 times the nominal maximum.

J.W. Webb: Could you please give me some indication of the shape of the control member used in the wide range ($Q = Kh^{3/2}$) meter which you described.

Author's reply: It is approximately conical. The sides are not quite straight, but slightly convex, otherwise an addition (and unwanted) exponential 5/3 term comes into the response curve.

A.T.J. Hayward: How does the calibration of your instrument change with time, over a long period of use?

What is the effect of viscosity variation on the calibration of your instrument?

Author's reply: Just before this conference we borrowed from one of our customers a wide-range instrument which had been in service for over two years. We reran its calibration test, and then compared the new response curve with the one we obtained in 1969. There were differences of the order of one per cent of span at the highest flowrates, and smaller differences at the lowest. Over the middle two-thirds of the range, when the graphs were superimposed, the two curves merged into one. The owner of the instrument told us that he had recalibrated it twice, and produced two graphs identical with the original.

We are running a long-term test, with two instruments cycling from zero to maximum flowrate every two minutes, and hope to have no significant differences in results during the next few years.

The half-inch size flowmeter remains within its ±1% tolerance on linearity for fluids with viscosities up to at least 15 centistokes. In general terms, the effect of high viscosity is to produce a hump in the response curve at the low end, which extends up the curve as the viscosity increases.

This distortion is less marked in the larger sizes, and is continually being reduced by refinements in design. All these factors combine to make it difficult to give a satisfactory quantitative answer to your question.

J.E. Roughton: This is a most ingenious method of overcoming one of the fundamental limitations of differential pressure devices but one wonders whether the benefits of linearity and wide range have not been obtained, at the expense of reliability. For example, has the author any information on the life expectancy of the bellows and on calibration drift?

In addition, I was particularly interested to learn of the use of the instrument with heavy fuel oil and would like to have information on the working temperature, maximum flow-rate and accuracy achieved, if known.

Author's reply: Heavy fuel oil is metered with this instrument at a temperature of approximately 90°C. The half-inch size has been used at flowrates up to 300 gallons per hour. Larger sizes are made, and I would expect them to be less sensitive to the difficulties this fluid presents.

We have not been able to discover the accuracy achieved in service. Our calibrations, with oil at a lower temperature and at the same viscosity as the fuel oil under working conditions, gave an accuracy of ±1%. That is at least 90% of the experimental data fell between two parallel straight lines spaced 1% of span on either side of a line through the origin.

On the subject of reliability, I know of no failures so far not caused by misadventure. We had several cases where the flow was connected backwards with disastrous results — since then the design has been modified to allow for this. One failure was caused when heavy fuel oil was allowed to solidify in the instrument, and the flow restarted without first warming it. The resulting over-compression of the bellows caused a spectacular zero shift. The design now includes over-travel stops to prevent damage from this cause. The instruments are designed to use only 60% of the allowed bellows compression and only about 5% of the allowed pressure differential, so the bellows life should be very long indeed.

We are doing our best to wear one out, but don't expect to succeed for several years yet.

4·4

INFERENTIAL FLOWMETERS WITH PHOTO-ELECTRONIC TRANSDUCER SYSTEMS

J.A. Ryland

Abstract: Three types of flow meter are discussed which are based on simple mechanical principles and which use a photoelectric cell as the transducer to obtain an electrical output signal. The Micro Flowmeter operates over the range 0.5 to 5 x 10^{-2} ml/min and is based on the movement of a fluid in a capillary by-pass loop. The Small and Medium Flow meters use conventional variable area devices and operate at flow rates of several hundred litres/min. The design of analogue systems for automatic operation is briefly discussed. Previous reluctance to use photoelectric devices, due to a temperature sensitivity of the photocell and instability in the light source, is now claimed to have been overcome.

1   INTRODUCTION

The purpose of this paper is to describe three types of flowmeter based upon easily understood mechanical principles and having a means of transmitting an analogue signal proportional to flow. The transmission system involves the use of photocells, which, in effect, simulates the human eye when making a visual observation. Generally speaking, the devices have a relatively low cost and are used in situations usually dealt with by more expensive apparatus.

2   MICRO METER

The micro flowmeter is shown in Fig. 1 and consists basically of an inclined transparent capillary tube connected to a reservoir chamber all in the by-pass pipe around a shut-off valve in the main flow line. If the valve is closed when a flow is passing the level in the reservoir is depressed and brings about a movement of the reservoir fluid in the capillary tube. The fluid in the capillary tube is known as a thread, and the rate of travel of this thread is directly related to the incoming flow to the flowmeter. The rate of flow can, therefore, be determined by measuring the time taken for the thread to move over a known distance.

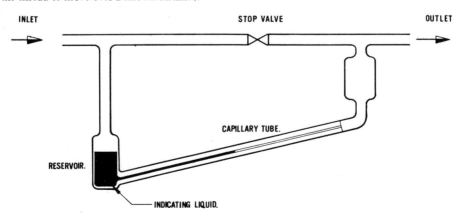

Fig.   1   Basic micro-flow meter

The flow through the meter is not interrupted because, although the shut-off valve is closed, the capillary thread continues to displace the measured fluid from within the capillary tube, pushing it through the outlet port. When the capillary thread reaches the end of the capillary tube, the valve must be opened so that the thread may return to its starting position under the action of gravity, thus restoring the original reservoir level. A practical range of capillary tube bore (d) is 0.5 - 3 mm with an effective minimum length (L) of 5 mm. If a maximum time (t) of 12 seconds, i.e. one-fifth of a minute is considered to be a reasonable maximum time, then the minimum flow rate which can be handled is 4.9×10^{-3} cc per minute. This figure is based on a measured length of 5 mm.

Assuming that the time for the maximum flow rate over this length is 1.2 seconds then the range of the instrument would be approximately $0.5 - 5 \times 10^{-2}$ cc per minute. By increasing the measuring length to the full 100 mm, this maximum flow can be increased by a factor of 3 and a full scale reading of 10 cc/min can be achieved. The apparatus thus described can be used manually with simple equipment. However, if a signal is required which is proportional to the flow rate, it is necessary to have some automatic sequence timing and an analogue output unit attached to the device. This takes the form as shown in Fig. 2.

The shut-off valve in this case is an electrically operated normally open solenoid valve. A synchronous motor drives the sequence cam plate which operates the necessary micro switches and at the same time drives a variable lift circular cam. This cam operates the core of a transformer fed with constant A.C. primary voltage, thus the position of the core brings about a variation in the secondary voltage. A photocell is arranged along the capillary tube so that when the cycle is started

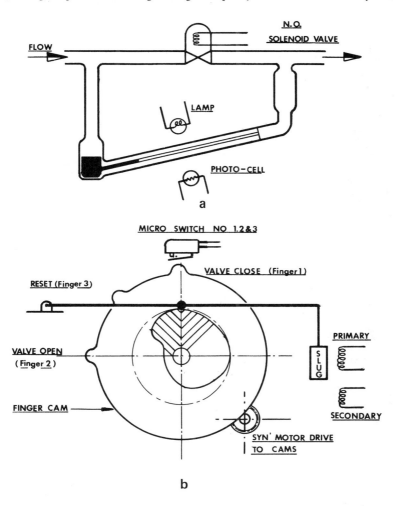

Fig. 2   Transmitter mechanism diagram:
         micro-flow meter

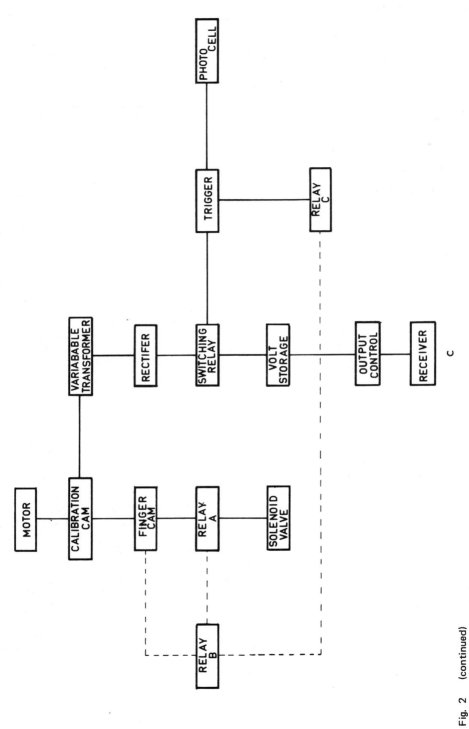

Fig. 2   (continued)
Transmitter block diagram

by closing the valve, the time taken before the light beam is interrupted is a measure of the distance that the cam travels in moving the variable transformer. At the instant the photocell beam is interrupted two things happen:

(1) The instantaneous secondary voltage is applied for a brief moment to a storage capacitor, and

(2) The shut-off solenoid valve is caused to open — in doing so the thread returns to the starting position and awaits the commencement of the next cycle.

A simple logic circuit is arranged so that should the flow be so small that the thread does not reach the photocell, then at the end of the cam stroke a micro switch automatically opens the solenoid valve and impresses on the capacitor the secondary voltage equivalent to that position. The voltage held in the capacitor is used to drive the analogue current output of the electronic unit. This unit is similar to what is described in the second flowmeter and can be of two forms, one a cathode follower circuit using a triode valve whereby the voltage from the condenser is applied to the grid, or secondly by controlling the current of a field effect transistor by applying the capacitor voltage to its gate. Between each cycle, the output remains sensibly constant, and therefore for wildly changing rates of flow within the frequency of the cycle of operation, no change in output is apparent. This is thought to be reasonable because most micro flow systems do not respond very quickly. However, should the flow change from near maximum at one reading and immediately fall to zero before the next cycle, the capacitor system will follow this and within one cycle will change the output from maximum to minimum. Although it can be seen that this meter makes a positive measurement, it has been placed under the category of an inferential device because it does not continually measure the flow, but merely samples for about fifty per cent of its operating time. The indicated flow is therefore inferred from this sampling.

3   SMALL FLOWMETER

The second example of a photo-electric detection method is shown in Fig. 3 where the primary measurement is made with the use of a variable area flowmeter of conventional form consisting of a tapered bore glass tube with a cylindrical float riding on the fluid flow. In principle the weight of the float acts upon its area bringing about a pressure drop across the float and, because flow is a function of pressure drop and escape area at high flows, the float needs to move to the top of the tube where the annular clearance is large and vice versa at low flows. In the small sizes the forces acting are extremely small. For example, a flowmeter with a full scale reading of 50 cc per minute may have a float weighing only 10 mg. Any constraint in the way of a force acting upon the float, for example a force of 1 mg, would bring about a change in the pressure drop of 10 per cent and a change in the flow indication of approximately 5 per cent. It is therefore very difficult with the small sizes to detect the position of this float other than by photo-electric means. A block diagram showing the mode of operation of the device is shown in Fig. 3. Basically the tube is continuously scanned by a photo-electric head, and on receiving an impulse brought about by the float position at a certain fluid flow, the variable transformer secondary output is connected to a capacitor, as with the micro flowmeter. This capacitor charge is held until the next impulse.

The variable area flowmeter is fairly fast in response to small changes in flow, and may dither in certain circumstances, so rather than have a flow sensing head which could follow the float about which would involve a fair amount of electronics in sensing which direction to go, for example, the system here has a steady motion and produces an output independent of any float dither. The variable area flowmeter is generally described as a linear scale device. This is not true for a number of reasons, the main one being the change in discharge coefficient of the float at different annular proportions. This discharge coefficient can also be affected by the Reynolds number and therefore in many cases the calibration can be non-linear by a fair amount. By having a cam which can have a suitable profile cut on it the secondary voltage can be made proportional to the position that the photo-electric head detects on the flowmeter in respect of the flow rate. For this reason the detection system is applied also to large variable area flowmeters where a transparent construction is satisfactory for the application. Figs. 4 and 5 show the thermionic or the transistor circuit diagrams.

4   MEDIUM FLOWMETER

The third example of photo-electric detection is shown in Fig. 6 and indicates a flowmeter of an all metal construction having a variable area principle, or a target metering discs principle. Both

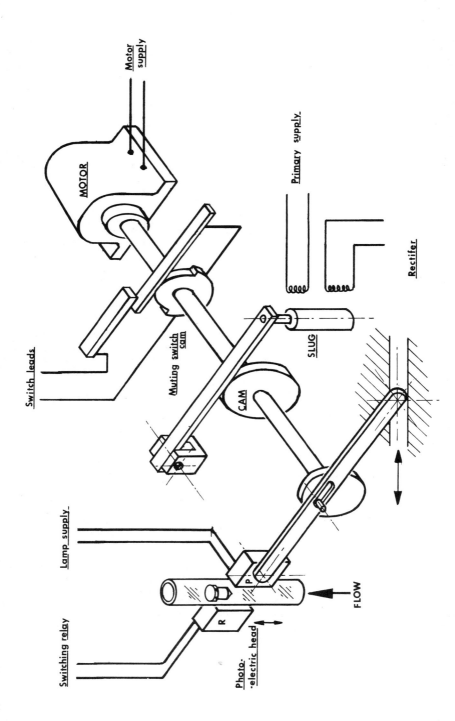

Fig. 3   Variable area flow meter transmitter diagram

a   Diagrammatic sketch

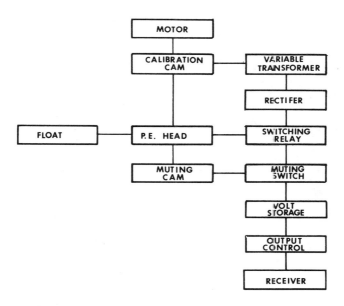

Fig. 3   (continued)
         b Block diagram   b

designs can use gravitational forces to oppose the flow forces acting through the measuring discs, or, alternatively, a spring may be used. This enables the flowmeter to be used in any orientation. The use of the spring enables the device to be made relatively small in that for normal pipe velocities the complete device can be within a housing no bigger than the pipe bore in which it has to fit. In a variable area form on liquids it can deal with velocities of up to 10 feet per second and in the target metering form up to 30 feet per second. In both cases, and more so with the target, the output of the magnetic following mechanism will not produce an exactly linear relationship to flow. With the target system the force on the disc is proportional to the square of the flow velocity. Once again a calibration cam is incorporated but this time it is a mask on a transparent quadrant which moves over a photocell. The photocell is illuminated from a small lamp. The photocell resistance changes according to the amount of obscuration. The relationship between resistance of the photocell and the flow rate through the flowmeter can therefore be determined by the characteristic or shape of the mask. The photocell is used in a Wheatstone bridge circuit with a second photocell in the dummy arm but also illuminated from the same lamp. In this way, changes in ambient light condition affect both cells and do not affect the particular imbalance of the bridge which is a function of flow rate.

5   CONCLUSION

None of the three flowmeters possess any great novelty, although the design may have some different or unusual presentation. The use of a photo-electric detection system provides the user of these various kinds of flowmeter with the facility of being able to check that the system is operating correctly and that the output function is responding correctly to the sequence of operation. If a fault should occur, the sequence of operation can be simulated with a suitable piece of cardboard by noting the effect of obscuring the photocell.

In the past there has been some reluctance towards using photo-electric devices, mainly for two reasons. First the cost of stable devices with respect to varying working temperatures: but this situation has changed remarkably over the past few years. Photocells which operate satisfactorily in ambient conditions of 100° centigrade are now quite normal.

The second criticism concerns the light source and lamp reliability. However, in the devices described the length of the beam is very short and the amount of power required is small. For this

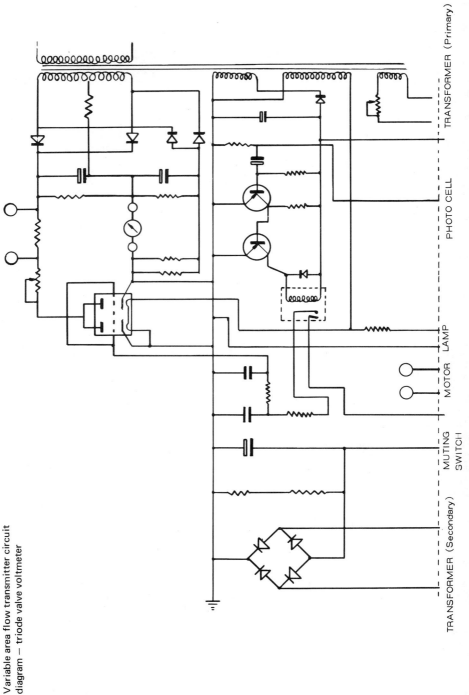

Fig. 4   Variable area flow transmitter circuit
diagram — triode valve voltmeter

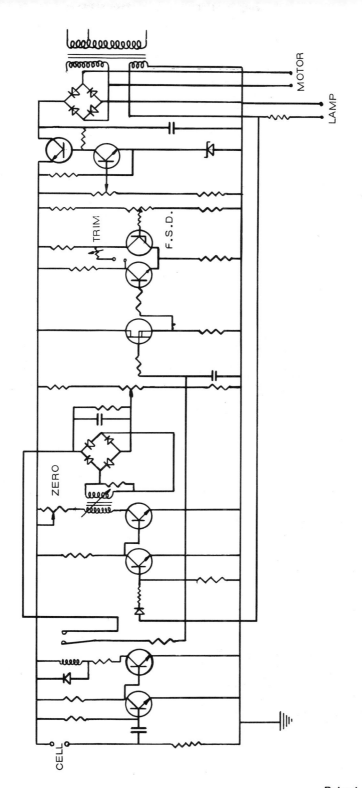

Fig. 5   Variable area flow transmitter circuit diagram (f.e.t. voltmeter)

reason lamps can be operated at voltages far below their rating [about 70 per cent] resulting in 100,000 hours of satisfactory operation. Even after this time there is no perceptible change in light emission. At low voltage operation the filament is not white hot, and therefore, remains mechanically strong. Any tendency to 'black-up' at low voltages is virtually non-existent.

DISCUSSION

A.N. Crossley: Regarding the micro flowmeter, I would be interested to hear the opinion of the speaker with regard to the leakage across the solenoid valve in the closed position.

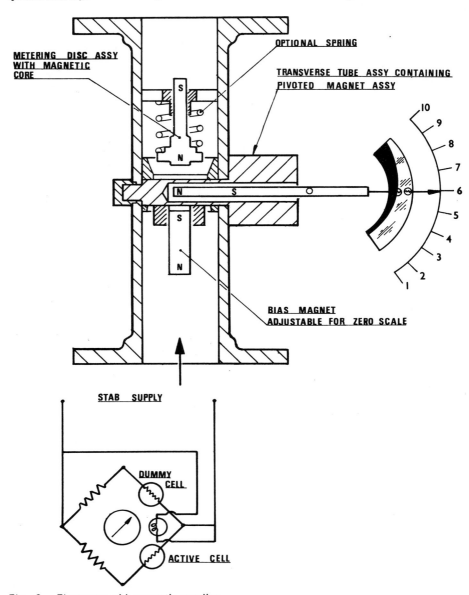

Fig. 6   Flowmeter with magnetic coupling

It has been my experience that solenoid-operated valves (S.O.V.'s) with soft seats do leak and it would be of interest to know if the speaker has experienced trouble due to this cause.

If the liquid is mercury, may I suggest a tungsten wire permanently located in the tube. The circuit is then completed as the level rises. It is possible to position a number of such wires and hence select electrically the point at which the remainder of the system starts operating.

This would have the advantage of removing the photoelectric cell with its attendant electronics.

Author's reply: We have also experienced leakage in seats of S.O.V.'s. In our opinion this is due to the relatively low force available from the solenoid. Our current practice with this micro flowmeter is to use the S.O.V. as a pilot operating a compressed air diaphragm shut-off valve.

We prefer a photo-cell for the capillary thread detection as this system allows the measured fluid to be either a gas or a liquid.

4·5

SMALL DIAMETER ORIFICES IN SERIES

S.C. Okafor and R.K. Turton

Abstract: Data is presented on the flow characteristics of small orifices placed in series. It was obtained as part of a research contract sponsored by the United Kingdom Atomic Energy Authority. A short survey of the literature on multiple orifices and non-sharp edged orifices is included, and the conclusion is drawn that very little reliable evidence for small orifices of diameter ratio less than 0.3 is available. Experimental data obtained from a rig using water passing through orifices placed in a 1.5 in. internal diameter pipe is presented. Conical, quadrant, and sharp edged concentric orifices of diameter ratio 0.065 and 0.2 are described, as are two eccentric arrangements of 0.065 diameter ratio orifices. The effect of L/d ratio and orifice spacing is shown for a range of Reynolds numbers based on orifice diameter between 70,000 and 240,000. Tentative conclusions based on the evidence presented are made.

1   INTRODUCTION

There is a need, as part of the continuing programme of development at the United Kingdom Atomic Energy Authority, Dounreay, for flow metering devices that will fit in a tube approximately 0.75 internal diameter, have a predictable calibration when passing a mixture of liquid sodium-potassium (NaK), maintain this calibration for a long period, and in extreme cases accommodate suspended small particles of material. As part of a research contract sponsored by the U.K.A.E.A. a survey of possible devices was made, as a result of which work was undertaken on orifices in series.

A paper by Sherman and others[1] describing devices using multi-orifice or multiple slot plates arranged in series that had been developed for a similar application in a water moderated reactor suggested that multiple orifice plates were a possible solution. They demonstrated that a wide range of pre-set metered flows was possible by adjusting the alignment of the holes or slots, but the duct was 12 in. in diameter and thus allowed many holes to be provided.

An indication of the possibility of using two simple orifice plates in series may be found in a design type article by Bloom[2] who summarised experimental data for orifices in series, and gives data which indicate that C_D is constant over a range of almost 10:1 in R_e, based on orifice diameter, up to a value of 4,000. He does not however give details of spacing or size, though an example quotes an orifice diameter of 0.12 in.

A systematic study, reported by Agar[3] gives information obtained from two orifices in series in a 1.5 in. diam. pipe, using sharp-edged orifices, and varying the spacing eccentricity, and relative alignment of the holes. Diameter ratios varied from 0.166 to 0.416, spacings from 0.125 in. to 2 in., and various combinations of hole alignment were used. He showed among other things that C_D was effectively constant over a Reynolds number range from 0.9 to 2.3 x 10^{-5} for the orifice assemblies used, that at spacings above about 1.5 in. the orifices acted independently, and that when the orifices were axially in line, whether eccentric or concentric with the pipe centre line the value of C_D was highest, and was higher than that for a single orifice of the same diameter ratio.

Sharp-edged orifices are subjected to erosion and consequent loss of calibration, so other designs were considered for use in a double orifice assembly. Work by Kreith and Eisenstadt[4] and Lichtarowicz and others[5] on capillary tube meters indicated that C_D was linear, for a given length to diameter ratio, over a wide range of Reynolds numbers, with limitations imposed by geometry, and that reliable prediction of C_D and pressure drop should be possible. Ziemke and McCallie[6] in a design type article discussed a device provided with a cone inlet to a parallel tube which discharged through another cone into the pipe, they proposed both cones should be 82° included angle, and the tube be 2.5 times as long as the bore and claimed reliability for this very simple device.

The eccentric orifice data of Agar[3] has been recently extended to a smaller size of pipe by

Hinz and others[7]. They tested sharp edged orifices with diameter ratios of 0.3, 0.4, 0.5 and 0.6 in a 1 in. diam. pipe, and with eccentricity of orifice from concentric to the point where the pipe and the orifice had a common tangent, and used water between 75°F and 85°F. They found flow instability occurred with holes half way between the concentric and fully eccentric positions and proposed a formula to predict the flow coefficient of an eccentric orifice from the calibration of a concentric orifice of the same diameter ratio, and showed that C_D slowly decreased in a linear manner for an eccentric orifice as R_e increased.

A consideration of the duty, particularly the pressure drops proposed, indicated orifice diameters were needed that gave diameter ratios that were much less than 0.3. Similitude considerations, noting that the application involves liquid sodium at temperatures between 200°C and 400°C, suggested water at 70°C to be suitable, and a 1.5 inch internal pipe diameter, with diameter ratios from 0.065 to 0.2 and a range of Reynolds number (based on orifice diameter) from about 7,000 to 240,000. A number of orifices were investigated, some of which are reported here.

2   DESCRIPTION OF EXPERIMENTAL APPARATUS AND METHOD

2.1 The rig

The flow circuit is shown in Fig. 1. The main tank is lagged and provided with immersion heaters to maintain the correct temperature. Water is pumped through one of three rotameters into the orifice assembly, from which it flows through the back pressure control valve to the main tank. The control valve is provided with a small by-pass valve to permit fine trimming control.

Mercury in glass thermometers are fitted in strategic positions in the main section, and air bleed points have been provided where initial experiments indicated a need.

The main section consists of two 30 in. long 1.5 in. inside diameter perspex tubes between which is clamped the orifice assembly, Figs. 2 and 3. As can be seen, two casing tubes, an inner and an outer, are used. The orifices, separated by a spacer, fit inside the outer tube, and are clamped by the inner tube being tightened up by the external clamping bolts. The design allows for all the permutations of geometry.

Pressure tappings are provided as shown in Fig. 1, the pressures being measured on a mercury in glass manometer that can be set at any desired angle. It is also possible to measure the pressure in the space between the plates.

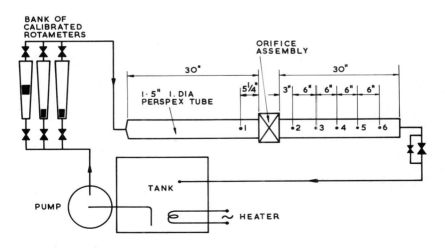

Fig.   1   Orifice system

2.2 Orifice details

The geometry and dimensions of the orifices whose performance is reported in section 4 are given in Figs. 4, 5, 6 and 7. Figure 8 defines the term 'spacing' for the double orifice assembly.

As Figs. 2 and 3 indicate, all parts were manufactured from perspex, and carefully finished.

Fig. 2   General view of the orifice assembly

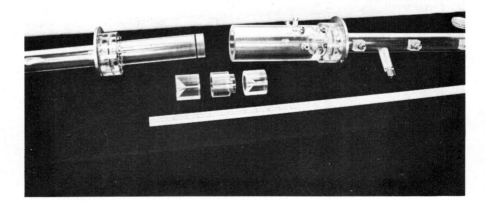

Fig. 3   A view of the component parts of the orifice assembly

2.3 Experimental method

Hot water is pumped round the system at approximately the correct rate until temperatures settle and then the flow is adjusted to the correct value and pressure readings at stations 1 to 6 (Fig. 1) noted. For each flow rate setting stable conditions are ensured before readings are taken.

It has been noted that there is a drop of less than $1^\circ C$ through the main section, so no lagging has been provided so far.

The rotameters have been calibrated by weighing, and checks have not revealed any variation in operation.

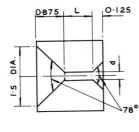

d	L/d	β
0·0975	2·051	0·065
	4·103	
	6·154	
	10·256	
0·302	1·532	0·2

Fig. 4   Dimensions of the conical orifices (inches)

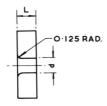

d	L/d	β
0·0975	1·282	0·065
0·302	0·828	0·2

Fig. 5   Dimensions of the quadrant edge orifices (inches)

d	L/d	β
0·0975	2·564	0·065
0·302	0·828	0·2

Fig. 6   Dimensions of the sharp edge orifices (inches)

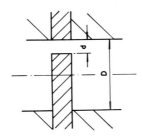

Fig. 7   The fully eccentric orifice

3   EXPERIMENTAL RESULTS

Values for overall coefficient of discharge for several orifice systems are presented in Figs. 9 to 14 on a base of Reynolds number.

Discharge coefficient and Reynolds number are defined in the List of Symbols where it can be seen that C_D is based upon overall pressure drop, and R_e is based upon the orifice diameter.

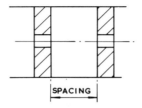

Fig. 8   The spacing length

3.1 Effect of L/d ratio on a single conical orifice

In the investigations so far completed exhaustive tests of the effect of L/d have not been possible, but the general trend is illustrated in Fig. 9. Here the variation of C_D with R_e for 4 values of L/d is shown. It can be seen that the general shape of the curves is the same, but the value of C_D falls as the length to diameter ratio increases, following the trend described by Lichtarowicz and others[5] for sharp edged cylindrical orifices. The maximum value of C_D for an L/d of 2.05 is 0.88 compared with their quoted value of 0.81 at Re values above about 50,000 for an L/d ratio of 2.

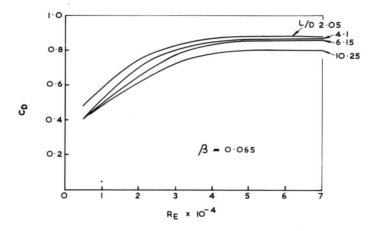

Fig.   9   The effect of *L/d* ratio on the conical orifice C_d

3.2 Concentric orifices in series

Figure 10 presents data for two diameter ratios relating C_D variation and spacing. As can be seen, C_D tends to be a constant for all spacings above a Reynolds number of 50,000.

There appears to be a considerable reduction in C_D from the single orifice for the small diameter ratio orifice, and very little change with spacing, suggesting that the orifices are almost independent in their action even with a spacing of 0.125 in. The graphs for the larger diameter ratio orifice show a wider spread.

214   Okafor and Turton

The conventional hump in the C_D curve reported in the literature for sharp edged orifices does not appear in either single orifice curve, C_D rising in each case to a maximum value.

3.3 Concentric quadrant-edged orifices in series

The effect of spacing on C_D for the different diameter ratio quadrant-edged orifices is shown in Fig. 11. There are differences between the two but both show steadily rising curves of C_D to maximum values, and a marked fall off in C_D as the spacing increases.

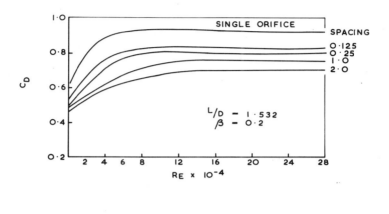

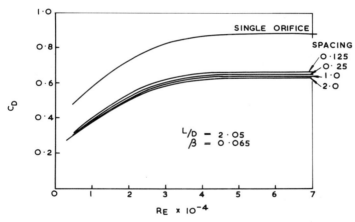

Fig. 10   The effect of spacing on the concentric conical orifices in series

3.4 Concentric sharp-edged orifices in series

Figure 12 illustrates the variation of C_D for sharp edged orifices in series as spacing increases. The small diameter ratio graphs follow the same trend as the other types of orifice though C_D values are lower.

The larger diameter ratio graphs show the familiar hump both for the single and the series orifices. It is interesting to see that close spacing gives higher values of C_D than the single orifice, due perhaps to interaction of the two plates improving the discharged jet and hence the pressure recovery.

3.5 Fully eccentric in line sharp-edged orifices

Figure 13 shows the effect on C_D of placing the orifices so that they have a common tangent with the pipe. Comparison with the concentric orifice performance shown in Fig. 12 for the same area ratio, but smaller L/d indicates very little difference in characteristic shape of maximum value of C_D fully eccentric cone and quadrant edge orifices in line have also indicated little difference in the C_D characteristic when compared with the equivalent concentric orifice performance data.

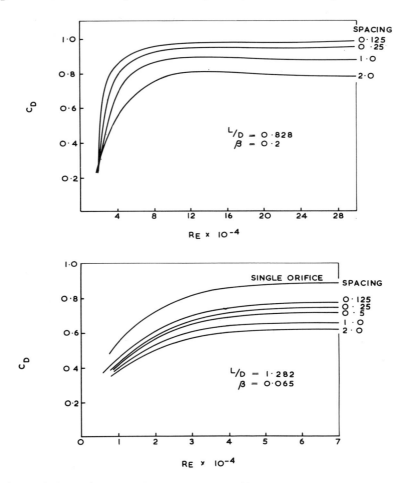

Fig. 11   The effect of spacing on the quadrant edge orifices in series

3.6 Comments on other eccentric combinations

Several arrangements other than in-line combinations of the same type of orifice have been investigated. For example, Fig. 14 shows how high C_D is for a single eccentric orifice with no sharp edge at inlet, and the considerable reduction in C_D which results when a second identical orifice is inserted downstream but at the opposite end of the pipe diameter. Also shown is the observed small effect of spacing on C_D for a pair of orifices arranged in this way.

4   CONCLUSIONS

Orifices of diameter ratio 0.065 and 0.2 have been tested in a 1.5 in. diam. duct, and though extended investigation in the areas covered is still needed, the following tentative conclusions may be drawn:

4.1 The small diameter ratio concentric conical orifice follows the same trend of C_D variation as larger cylindrical orifices, and tends to a constant value at the higher Reynolds numbers tested.

4.2 The conical and quadrant-edge type orifices arranged concentrically in series have a constant C_D over a wide range of Reynolds number, and show no signs of flow instability in the flow range used.

4.3 The small diameter ratio concentric sharp-edged orifice in series also shows a good discharge coefficient curve, but the larger ratio orifices exhibit the same flow instability effect on C_D as the single orifice plate.

4.4 The effect of spacing is quite marked for all the concentric arrangements except that using the conical design.

4.5 The fully eccentric sharp-edged orifices of diameter ratio 0.065 arranged with holes in line showed almost the same variation of C_D as the concentric sharp-edged orifice of the same size, indicating there is little difference in flow behaviour.

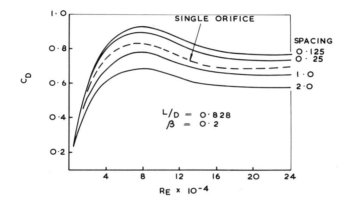

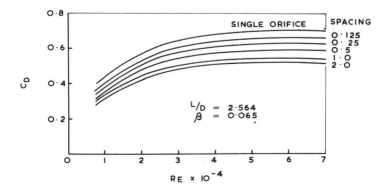

Fig. 12   The effect of spacing on the sharp-edged orifice in series

4.6   The single eccentric orifice with a quadrant inlet edge has a very high value of C_D, and when placed in series with an identical plate but with the hole at the opposite end of a diameter is very little affected by spacing.

4.7   The results reported indicate that small orifices, of diameter ratio 0.065 have a stable discharge characteristic over a range of Reynolds number based on orifice diameter of 7000 to 70,000. The trends suggested by other workers are shown to be valid for small orifices in series.

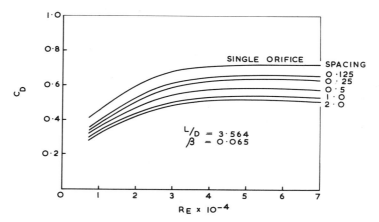

Fig. 13   The effect of spacing on the fully eccentric sharp-edged orifices in series

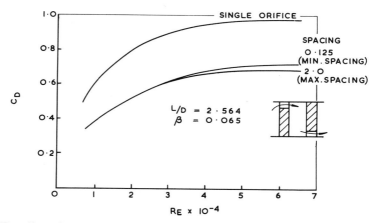

Fig. 14   The effect of spacing on opposed eccentric quadrant-edge orifices

5   ACKNOWLEDGEMENTS

The authors acknowledge both the permission to publish and the financial support given by the Reactor Group of the United Kingdom Atomic Energy Authority for this work.

6   REFERENCES

1   SHERMAN, J. et al.: 'Variable Flow Resistance with Adjustable Multi-Hole Orifice Plates in Series', Trans. A.S.M.E. Jnl of Basic Eng., Sept. 1960, pp. 645-653

2   BLOOM, G.: 'Errorless Orifices', Product Engineering, Oct. 25, 1965, pp. 61-64

3   AGAR, J.D.: 'Pressure Drop Characteristics of Two Eccentrically Positioned Orifices in Series'. M.Sc. Thesis, University of Washington, 1965

4   KREITH, F., and EISENSTADT, R.: 'Pressure Drop and Flow Characteristics of Short Capillary Tubes at Low Reynolds Numbers', Trans. A.S.M.E., 79, July 1957, pp. 1070-1078

5   LICHTAROWICZ, A. et al.: 'Discharge Coefficients for Incompressible Non-Cavitating Flow through Long Orifices', Jnl. of Mech. Eng. Sci., 7:2:1965, pp. 210-219

6   ZIEMKE, M.C., and McCALLIE, B.G.: 'Design of Orifice Type Flow Reducers' Chemical Eng., Sept. 14, 1964, pp. 195-198

7   HINZ, G.A. et al.: 'Empirical Evaluation of the Eccentric Orifice in Small Pipes'. A.S.M.E. Paper 67-PET-21

7   LIST OF SYMBOLS

d   =   Orifice diameter

D   =   Pipe internal diameter

L   =   Orifice length (defined in Figs. 4, 5 and 6)

p_1 =   Pressure at Station 1

p_2 =   Pressure at Station 2

V   =   Orifice mean velocity

β =   Diameter ratio d/D

ρ =   Fluid density

μ =   Fluid absolute viscosity

C_D =   DISCHARGE COEFFICIENT

   = $\dfrac{\text{Actual mass flow rate}}{\dfrac{\pi d^2}{4}\sqrt{\left\{2\rho\,(p_1 - p_2)\right\}}}$

R_e =   REYNOLDS NUMBER

   = $\dfrac{V \times d \times \rho}{\mu}$

8   APPENDIX

Relation between total discharge coefficient and the discharge coefficient of the first orifice

It is proposed that the total pressure drop $\triangle p_T$ of a group of orifices may be expressed with the pressure drop due to the first orifice $\triangle p_1$ as

$$\triangle p_T = \triangle p_1 + \alpha p_T$$

where α is the proposed correlation factor.

This may also be written as

$$\alpha = \frac{\triangle p_T - \triangle p_1}{\triangle p_T}$$

or

$$\triangle p_T = \frac{\triangle p_1}{(1 - \alpha)}$$

Curve fitting gave polynomial equations up to the fifth power of the spacing s, examples of which in order for the sharp-edged orifices in series (suffix s) and the quadrant edged orifices (suffix Q) are

$$\alpha_s = 0.04534 + 0.91907s - 1.49478s^2 + 1.73209s^3 - 0.97238s^4 + 0.97238s^5$$

$$\alpha_Q = 0.09253 + 0.46842s + 0.97238s^2 - 0.49521s^3 + 0.32143s^4 - 0.064205s^5$$

It is also suggested that discharge coefficients may be related as follows, (using the same identities as suffices)

$$C_T = C_{D_1}\sqrt{1-\alpha}$$

Figs. 15 and 16 give typical plots of the variation of α with β and s for the sharp-sharp and quadrant-quadrant combinations.

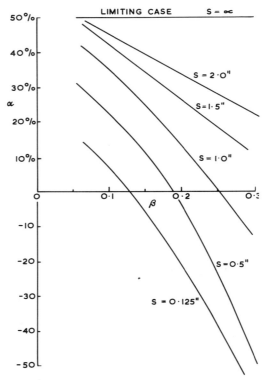

Fig. 15   Effect of S on sharp/sharp orifices in series

DISCUSSION

P. George: I would like to bring the authors attention to work on labyrinth seals, where a spacing factor α is used extensively by designers.

Authors' reply: We are grateful for the comment of Mr. George that labyrinth seal data may be relevant and this will be examined.

J.C. Schuster: Will pressure studies lead to a device for measurement and pressure reduction in one instrument?

Authors' reply: Yes, if the device is stable in operation and gives repeatable results.

V.R. Withers: How accurate was the agreement between the statistical correlations and the experimental data for the coefficient α.

Authors' reply: This point is interesting but cannot be dealt with satisfactorily at the moment and much more experimental data needs to be obtained before statistical statements can be made with any confidence. Corrleations of the type presented are therefore only tentative at this point in time.

A. Lichtarowicz: I was pleased to see that orifice flow problems still attract considerable attention. Since the paper refers in several places to some of my work in this field, I would like to make a few comments.

The paper is of an experimental nature so that actual experimental points must be given on graphs in addition to the smoothly drawn curves. Having performed many tests in this field, I am

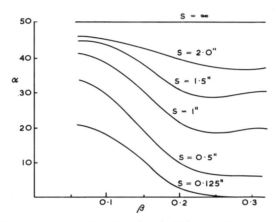

Fig. 16 Effect of *S* on quadrant-edge/quadrant-edge orifices in series

certain that the variation of C_d with Reynolds number is not as well behaved as it appears from the authors' curves.

It is well known that the condition of the upstream edge of the orifice greatly affects its performance as was stressed in another paper at this Conference. The actual sharpness of the corner becomes progressively more important as the orifice diameter is reduced. In view of these remarks, I would like to hear from the Authors why their test orifices were made from Perspex, a relatively soft material with which it is difficult to produce sufficiently sharp corners to give repeatable results?

In section 3.1 of their paper, the Authors compare their results given in Fig. 9 with data given in my paper. Their value of $C_d = 0.88$ for an orifice having $L/d = 2.05$ is contrasted with $C_d = 0.81$ at Re greater than 50000 for $L/d = 2$. This comparison is inadmissible since the Authors' orifice had a conical entry (Fig. 9) and my results refer to a sharp edge orifice with no bevelling. Thus the data presented in my paper can only be compared with the Authors' data given in Fig. 12, where $C_d = 0.7$ for $L/d = 2.564$. This value is considerably lower than the more generally accepted value of about 0.8.

I would like to ask the Authors to specify more precisely thann on page 214 what pressure difference they have used in the evaluation of the discharge coefficients quoted? This may explain the low values of the discharge coefficient.

The effect of spacing of sharp edge orifices in series having $L/d = 0.828$ is different from all other orifices tested. In all cases shown except this one the placement of an orifice downstream reduces the effective C_d for the system. In this case, however, at small spacings the effect is opposite. C_d is increased and only at large spacings (greater than about 10 diameters) C_d falls below the value for a single orifice. I wonder if the Authors could comment on this aspect?

Finally I would like to add a warning to the conclusion 4.7 of the paper. At high Reynolds numbers one must ensure that cavitation does not take place otherwise 'stable discharge characteristics' will be affected and the discharge coefficient will be reduced (see Reference given below).

Authors' reply: Dr Lichtarowicz has been kind in his comments. The C_d curves are through experimental points, at least three sets of test results in most cases having been taken, and scatter is estimated as probably better than a band of 5% total spread. Convenience and a longer term preoccupation with errosion prompted the use of perspex for the orifices. Sharp-edged orifices were examined for interest, the others being more relevant to the duty requirement. We can offer no definite statement about edge quality. His point on comparison in section 3.1 is well taken; he is of course right. The pressure difference used in calculating the coefficients was from point 1 to point 6 in Fig. 1. We can make no useful comment on the disparity between the results for a sharp orifice, Fig. 12, and the others except to agree that more study is needed in this area.

REFERENCE

PEARCE, I.D., and LICHTAROWICZ, A.: 'Discharge Performance of Long Orifices with Cavitating Flow', 2nd Fluid Power Symposium, British Hydromechanics Research Association, January 1971

MEASUREMENT OF TRANSIENT MASS FLOW RATE OF A GAS

N. Mustafa, I. Birchall and W.A. Woods

Abstract: The paper describes the development of a device for measuring the transient flow rate of a gas in a pipe. It consists basically of an ionising source (a radioisotope), high voltage and collecting electrodes and a recording unit. The effect of pressure of the gas on the response of the flowmeter was observed and an empirical relationship between ion-current, mass flow rate and pressure was found. The apparatus was calibrated on a steady flow line. The instantaneous flow rate from a pulsating gas supply recorded by the flowmeter has been compared with the theoretical results computed by numerically solving the characteristic equations for unsteady flow.

1   INTRODUCTION

The measurement of transient mass flow rate of a gas is of even greater importance than the measurement of steady mass flow rate, particularly in the fields of gas dynamics and the study of internal combustion engines. The methods now available are all theoretical and are time consuming and tedious. The most successful of these i.e. the calculation of instantaneous mass flow rate of a gas by numerically solving the characteristic equations for one dimensional unsteady compressible flow in a pipe with appropriate boundary conditions, is expensive and complex. A computer is also indispensible for this method.

The need to have an instrument which can measure instantaneous mass flow rate of a gas in a pipe arises mainly from the research and development work being done in the above mentioned fields. For example, when designing or improving small engines, an instrument can be installed at the inlets and exhausts of the system so as to obtain the mass flow rate-time diagrams for the purpose of calculating power, efficiency etc. Apart from many other applications, this type of instrument will be far more useful in the field of super-charged internal combustion engines.

An attempt was, therefore, made to develop an instrument capable of giving instantaneous mass flow rate of a gas in a pipe. For this purpose a steady flowmeter was designed, optimised and calibrated before assessing its performance for the measurement of pulsating flow rates. The flowmeter was based upon the continuous ionisation principle.[1-2]

This version of the continuous ionisation method differs from previous work in two respects; namely, the electrodes are all flush with the surface of the pipe and the electrical field is used as a barrier between the ions and the collector in order to give a linear response at high flow rates.

The optimised steady flowmeter, after calibration, was tested for measuring gas pulses from the exhaust of a two stroke model engine. A comparison was made between the output signal of the flowmeter and the expected signal evaluated from the computed values of instantaneous mass flow rate and the steady flow calibration curve and the results explained.

Although the flowmeter, in the present form, is unable to follow the mass flow rate pulses, it can be used as a steady flowmeter.

2   DEVELOPMENT OF STEADY FLOWMETER

The flowmeter consisted basically of a radioactive source in the form of a cylindrical strip, acting as the ionizing agent, and electrodes to collect the ions. The radioisotope and the electrodes were flush with the inside surface of the pipe of 1.25 in. (3.17 cms) diameter so as to give little obstruction to the flow.

2.1 Experimental arrangement

The block diagram of the experimental arrangement is shown in Fig. 1. A potential difference of 1000 volts was applied to the electrodes and an electrometer was used to measure the ion current from the collector. A mass flow rate measuring device i.e. an orifice meter to the specification of B.S. 1042[3] was fitted in the flow line upstream of the electrode system.

2.2 Optimisation

2.2.1 Source electrodes arrangement

Figures 2 and 3 show different electrode arrangements studied and corresponding ion current-mass flow rate characteristics. The configuration (b) was the only one which gave a good linear response within the range of about 0.04 - 0.10 lbs/sec (about 18 - 45 gm/sec) and was selected for further

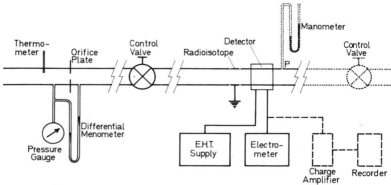

Fig. 1   Steady flow arrangement

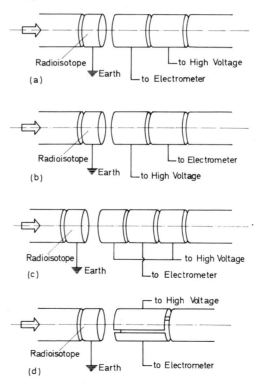

Fig. 2   Electrode configurations

224   Mustafa et al.

optimisation. The ion current remains zero until the momentum given to the positive ions is sufficient to overcome the repulsive field.

2.2.2 Ionising source

Results shown in Fig. 3 were obtained using Americium 241 (α-emitter) as the ionizing source. When this was replaced by Thallium 204 or Tritium (β-emitters) different shapes of the character-

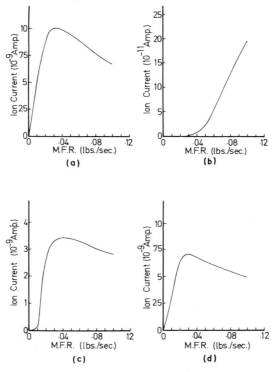

Fig. 3   Response of the electrode systems shown in Fig. 2

istic curve were obtained. Figure 4, shows the responses of the flowmeter with these radio-isotopes acting as the ionizing agents. With the β-emitters there is some ion current at zero flow. After passing through a minimum it starts to increase with increase of flow rate. Because of this odd shape of the response the use of Thallium and Tritium sources was abandoned.

2.2.2.1 Intensity distribution across the pipe due to different sources

It was thought that a better understanding of the interaction between the electric field, mass flow rate and ionisation could be obtained if the radiation intensity distributions across the pipe were measured. Photo-sensitive films were exposed (perpendicular to the axis) to different sources used in the flowmeter. The density values were converted to relative intensities with the use of a characteristic curve of the film for the corresponding radioisotope. The characteristic curves were drawn using standard Americium and Thallium sources. The intensity distribution of alpha particles from Americium 241 and of beta particles from Thallium 204 are shown in Figs. 5a and b. Figure 5c represents the density distribution for Tritium since the characteristic curve for this source could not be obtained due to the non availability of a standard Tritium source.

2.2.3 Potential on h.t. electrode

Effect of the high voltage electrode potential on the response of the flowmeter is shown in Fig. 6.

The ion current decreases at higher voltages because of a stronger repulsive field between the source and the high voltage electrode.

2.2.4 Size of the electrodes

Figures 7a and b show the results when the length of one electrode was changed while that of the other was kept the same. As any of the electrodes was decreased in size the number of ions collected also decreased. This was due to the decrease in ion collecting area as the length of collecting electrode was reduced and due to the increased potential gradient of the repulsive field as the high voltage electrode was decreased in length. The selected length was 1.5 in. (3.81 cms) for both the electrodes.

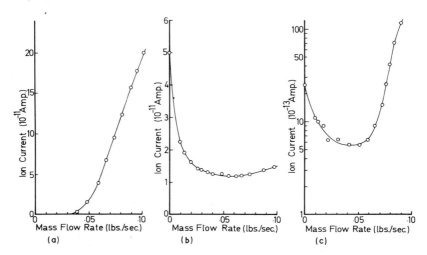

Fig. 4   Comparison of response with various ionising sources

    a   Americium 241
    b   Thallium 204
    c   Tritium 3

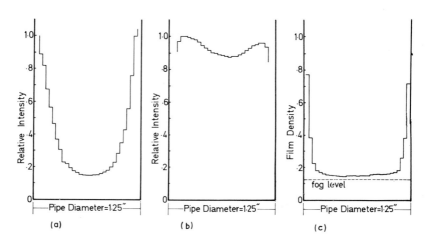

Fig. 5   Intensity/density distribution over pipe cross-section due to various ionising sources

    a   Americium 241
    b   Thallium 204
    c   Tritium 3

2.2.5 Source of high voltage electrode spacing

As the distance between the source and the electrode was increased from 0.125 in. (0.32 cm) to 1.0 min (2.54 cm) the ion current was found to increase but further increase in spacing produced a reduction in current. Figure 8 shows this effect. This behaviour might be due to the fact that the cylindrical electrode system acts as a decelerating and an accelerating electrostatic lens[4]. The electric field pattern due to the electrodes is shown in Fig. 9. Most of the ions in the experimental condition, might be focused on to the collecting electrode when the spacing was 1.0 in. (2.54 cm) and on either side of it when the spacing was less than or greater than 1.0 in. (2.54 cm). To keep the size of the flowmeter small a value of 0.5 in. (1.27 cm) was selected.

2.3 Final design

The final electrode assembly is shown in Fig. 10. The electrodes and insulating rings were threaded so as to be held together tightly and accurately aligned. Two B.N.C. connectors, for the high voltage electrode and collector were fitted on to the shielding.

Specifications are as follows:

Source of ionisation	Americium 241 (alpha emitter)
Source of high voltage electrode distance	0.5 in. (1.27 cm)
Length of high voltage electrode	1.5 in. (3.81 cm)
Length of collecting electrode	1.5 in. (3.81 cm)
Potential on high voltage electrode	+ 1000 volts
Material of electrode	copper
Insulating material	cobex
Shielding (electrical)	iron

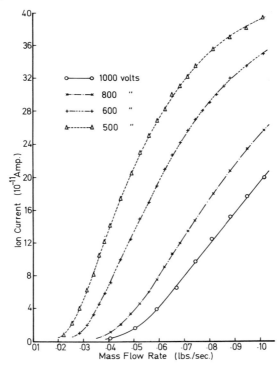

Fig. 6   Effect of voltage on the collector response

2.4 Effect of pressure of gas on the flowmeter response

Pressure dependence of the continuous ionisation flowmeter has been mentioned by Brain and Cameron[5]. This effect was studied with the present flowmeter. An additional control valve downstream of the meter was used to control the pressure at P (Fig. 11 dotted lines). Figure 11a shows the flowmeter response at different values of static pressure of the gas at P. It was observed that all the four curves in Fig. 11a nearly coincide with each other when ion current was plotted against the factor $(W - 0.05)\sqrt{P}$

where W = mass flow rate (lbs/sec)

and P = absolute static pressure (p.s.i.)

Figure 11 (b) shows the unique curve obtained in this way.

3   CALIBRATION

So that the flowmeter could be used after calibration for measuring pulsating flow rates, the electrometer was replaced by a charge amplifier (10 millisecond time constant) followed by a U.V. recorder. This is shown by broken lines in Fig. 1. The deflection on the recorder was plotted against mass flow rate. The system was calibrated for four different values of static pressure as shown in Fig. 12a.

Referring to Section 2.4, a single calibration curve (Fig. 12b) was obtained by plotting the deflection against $(W - 0.05)\sqrt{P}$.

4   MEASUREMENT OF PULSATING FLOW

The system was fitted at the exhaust of a pulse generator—a two stroke engine model, as shown in Fig. 13. The instantaneous value of pressure was also included in experimental measurements as well as in theoretical calculations.

4.1 Theoretical calculations

Instantaneous pressure and mass flow rate in the exhaust pipe of the pulse generator were

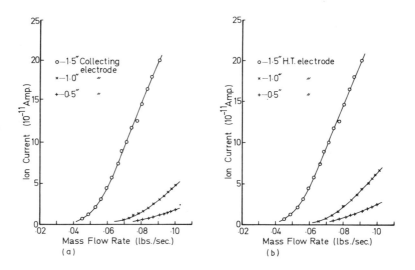

Fig. 7   Effect of electrode size on collector response

    a   Collecting electrode
    b   H.T. electrode

calculated by[6] :–

(i) numerically solving the equations of one dimensional unsteady compressible flow in the pipe based upon the method of characteristics and

(ii) formulating quasi steady flow at various physical configurations representing the pipe boundaries and matching these boundary conditions with the unsteady flow calculations.

A computer programme developed by Daneshyar[7] was used, after modification, to calculate the mass flow rate and pressure-time diagrams at the collecting electrode and point P respectively.

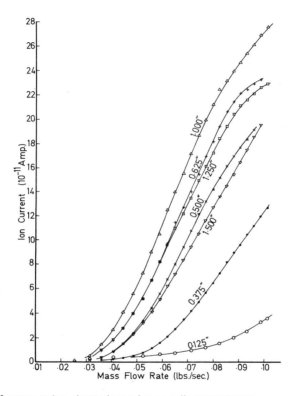

Fig. 8   Effect of source to h.t. electrode spacing on collector response

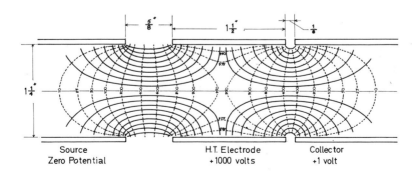

Fig. 9   Electrostatic field inside the electrode system
———— lines of forces
– – – equipotential lines

4.2 Measurements

Two pre-calibrated transducers were used to record the pressure at P and the cylinder pressure. The output from the flowmeter was also recorded on the same recording paper.

4.3 Analysis of the results

The cylinder pressure-time diagram was used to calculate:—

(a) the frequency of the pulse generator

(b) cylinder release pressure

which were then fed into the computer as data to perform the calculations.

The agreement between measured and calculated pressure-time diagrams is shown in Fig. 14. Figure 15 represents the mass flow rate pulse as computed. These values of mass flow rate and pressure were used in evaluating the factor $(W - 0.05) \sqrt{P}$. The calibration curve of the flowmeter (Fig. 12b) was then used to obtain the expected deflections on the recorder. The resulting expected trace is plotted in Fig. 16 along with the actual trace given by the flowmeter.

The shape of the observed trace was found to be totally different from that expected.

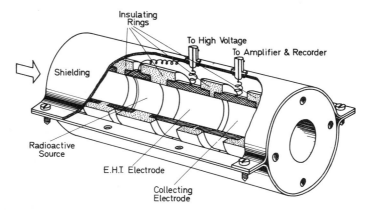

Fig. 10   In-line cylindrical electrode system

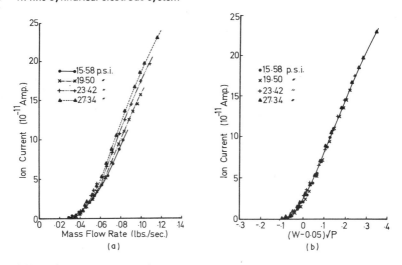

Fig. 11   Effect of gas pressure on collector response

5 DISCUSSION AND CONCLUSIONS

5.1 Explanation of the Behaviour of the Flowmeter

The unexpected response of the flowmeter in unsteady flow conditions may be explained in the following way.

When there is no gas flow the negative and positive ions drift under the influence of the electric field, towards the high voltage electrode and the source respectively. Because of the difference in the mobility of positive ions and electrons there will be a positive space charge formed near the source.

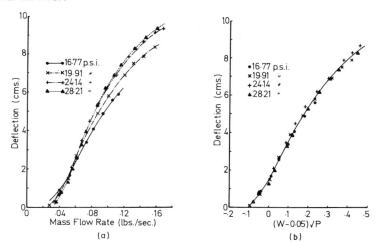

Fig. 12   Calibration of flowmeter with u.v. recorder

Fig. 13   General assembly of equipment used for pulse flow measurement

As the first portion of the pulse comes in, most of this positive space charge passes by the collector because of the large pressure wave. When the charge approaches the collector it induces a negative charge which is detected by the charge amplifier and results in a negative shoot at the start of the pulse. As this negative charge is detected a portion of it passes away through the circuit. As the positive space charge moves away from the collector it induces a positive charge on it the amount of which is greater than the previously induced negative charge because a portion of the latter has already leaked away through the circuit of the charge amplifier. The result is a net positive charge on the collecting electrode which produces the positive shoot immediately following the negative one.

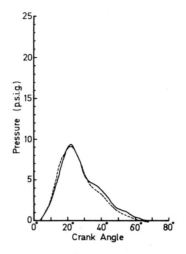

Fig. 14   Measured and calculated pressure/time diagrams
———— experimental
— — —computed

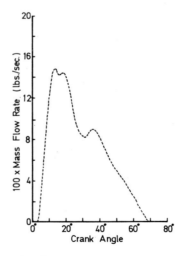

Fig. 15   Calculated mass flow rate pulse

After this a somewhat steady condition is reached at the maxima of the flow rate. This is represented by the small region where the experimental and calculated traces are closest to each other.

When the flow is decreasing the positive ion density in the vicinity of the collector is being reduced (though it is due to the decreasing flow rate) and has an induced effect which is positive.

This is why an increasing deflection is obtained when the flow is decreasing.

After the flow has stopped the charge on the collector starts decaying according to the time constant of the charge amplifier and results in the tail of the pulse.

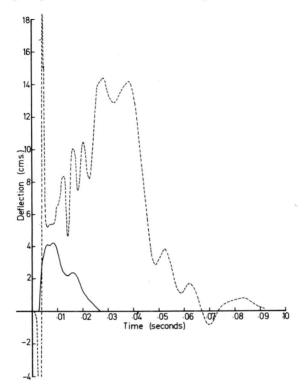

Fig. 16   Comparison of calculated and observed response of flowmeter for pulsating flow
——— expected trace
— — — observed trace

5.2 Conclusions

As a result of the steady flow experiments a steady flowmeter, based upon the continuous ionisation principle, has been achieved. The advantages of this flowmeter over the previously studied versions are:—

(i) as it gives practically no obstruction to the flow, the head losses will be just the same as due to a smooth pipe,
(ii) the size of the flowmeter is small,
(iii) a convenient empirical correlation between mass flow rate, pressure and ion current has been found.

From the unsteady flow experiments it is apparent that with the present version of the flowmeter, using the charge amplifier, the steady flow calibration curve cannot be used for measuring pulsating flow rates.

It might be possible to improve the present version of the flowmeter for measuring transient mass flow rates but the following phenomena will have to be taken into account.

(i) the formation of positive space charge,

(ii) the induction phenomenon.

The formation of a positive space charge may be reduced by using some shield (wire gauze etc.) between the source and high voltage electrode. This would terminate the electrical lines of forces and there will not be any current between source and high voltage electrode; hence no space charge.

Increasing the distance between source and high voltage electrode would also reduce the formation of space charge but the size of the flowmeter would increase.

It might be worthwhile replacing the charge amplifier with a fast d.c. amplifier to reduce the induced effects.

6   REFERENCES

1   LOVELOCK, I.E. and WASILEWSKA, E.M.: 'An ionisation anemometer'. J. Scient. Instrum., 1949, 26(11), pp. 367-370

2   CLAYTON, C.G., and WEBB, J.W.: 'The measurement of mass flow and linear velocity of a gas by continuous ionisation', Int. J. appl. Radiat. Isotope, 1964, 15, pp. 603-610

3   BSI042:1964: Pt. 1: 'Methods for the measurement of fluid flow in pipes — Orifice Plates, Nozzles and Venturi Tubes'

4   LIVINGSTON, M.S. and BLEWETT, J.P.: 'Particle Accelerators' (McGraw-Hill, 1962) pp. 102-106

5   BRAIN, T.J.S. and CAMERON, E.M.R.: 'Performance of a large diameter gas ionisation flowmeter'. National Engineering Laboratory, East Kilbride, Glasgow, 1970

6   BENSON, R.S., GARG, R.D. and WOOLLATT, D.: 'A numerical solution of unsteady flow problems', Int. J. Mech. Sci., 1964, 6, pp. 117-144

7   DANESHYAR, H.: 'Computer programmes for the analysis of gas exchange processes in multicylinder engines'. University of Liverpool, 1966

DISCUSSION

F.H. Huyten: I would like to make a general remark on the use of this kind of device for flow measurement.

The devices described belong to a group of detectors originally introduced by Lovelock for the measurement of minute quantities of impurity in gases after mixing with argon or helium.

These detectors have proved to be very sensitive but at the same time capricious. Keeping this in mind, I wonder if, when using this type of detector system for flow measurement, one would expect severe interferences from impurities in the gas. Possibly this could explain the differences between experimental and calculated results shown in Mr. Mustafa's work.

C.G. Clayton: I think that it is erroneous and misleading to compare gas composition, ionisation detectors of the type first introduced by Lovelock with the ion collector systems used in gas ionisation flow meters.

The former are designed specially to detect very low compositions of particular gaseous impurities by using specific properties of an ionised gas molecule, such as charge exchange, electron capture and mobility. The gas ionisation flow meter, on the other hand, uses high collecting fields and generally is designed to avoid interaction phenomena of this type: it is less sensitive to impurities by several orders of magnitude (approximately 10^5 to 10^{10}), than the composition detector device.

In the limit, the precision of a continuous ionisation gas flow meter is related to variations in composition of the gas flowing through the meter: temperature variations and moisture content are other important factors. But within well-defined limits there is no reason to expect unstable operation.

Authors' reply: As pointed out by Mr. Clayton, the electric field used in the continuous ionization flowmeters is always sufficiently high to collect the ions in relatively short time. Therefore, small amounts of impurities have a negligible effect on the performance of these meters.

As regards our experimental results, these were obtained using a compressed air supply which provided dust free dry air and the same supply was used for steady as well as unsteady flow experiments. The lack of correlation between experimental and calculated results cannot, therefore, be attributed to the variation in the degree of impurity of the gas used. This is further confirmed by the fact that the ionization potentials for a wide range of gases only differ at most by a factor of 2.5 *. For the gases and impurities (O_2, H_2, N_2, CO_2, NO_2, etc.) found in internal combustion engines, the difference is less and could not account for the difference between theoretical and experimental results.

J.W. Webb: From my own experience using ionization flow meters, I would say that the same mass flow-current characteristics as shown in Fig. 3(b) could be obtained using an electrode configuration as in Fig. 2(d) but with the radioactive source section at high voltage instead of at earth potential. To obtain the required flow rate both the radioactive source strength and the distance between the source and collector should be increased but not excessively. A meter of this type need only be about 12 in. long compared to 4 in. for the present one, with about ten times the source strength.

The advantage of this configuration is that the electrostatic field is far more even and all ions are collected on a single electrode system. This would reduce the space charge effect and thus the meter would be more likely to follow rapid changes in flow.

Authors' reply: We partly agree with Miss Webb.

It seems possible to obtain a curve similar to that of Fig. 3(b) by increasing the distance between the radioactive source and the electrodes and by using a stronger source but, as pointed out by Miss Webb, the length of the meter will be increased to about three times that at present. However, as has already been mentioned, we had as one of our design objectives the production of a physically small flowmeter.

As regards applying high voltages to the radioactive source section, this would definitely increase the formation of space charge as a result of ion current flowing between the source and the pipe on the upstream side.

T.J.S. Brain: One of the main advantages of the pulsing technique (described in the next paper) is that the output signal is not affected by the composition of the gas. It is worth noting that in the work described, while the current at the second collector varied with flowrate from approximately 1.10^{-7} to 6.10^{-7} Amp. the level of the signal in the detection circuit remained unaltered.

---

* VON ENGEL, A., 'Ionized Gases', Oxford, 1965, p. 59

5·2

DEVELOPMENT OF A PULSED GAS IONIZATION FLOWMETER

T.J.S. Brain

Abstract: Tests have been carried out to assess the feasibility of a novel gas ionization technique for flow velocity measurement. Air at conditions close to ambient was used as the test fluid. The device was located in a 152 mm (6 in) diameter pipeline and a Borda inlet was used to calibrate the instrument against flow velocity within the range 5-25 m/s (16-82 ft/s). The results show that even at this preliminary stage the meter is capable of accuracies better than ±2.5 per cent. The instrument has a linear calibration and, since it presents little obstruction to the flow, head losses are small.

1   INTRODUCTION

Ionization techniques can be used to measure gas flowrates and most of these involve continuous ionization of the flow by radioactive means with collection of the ions by electrode arrangements. Lovelock and Wasilewska[1] were among the earlier experimenters to use such systems. For their basic method the meter can be arranged as shown in Fig. 1. The outer cylindrical electrode is coated with a radioactive substance which emits either α or β radiations which ionize the gas stream. When an electrical potential is applied between the outer electrode and the central electrode a current flows through this circuit, positive ions being collected at the central electrode. The velocity of the ions in the space between the electrodes is determined by two components. One which is directed along the lines of force of the field is dependent on the field strength and the mobility of the ions. This component may be estimated from

$$v = \mu H \qquad\qquad ... (1)$$

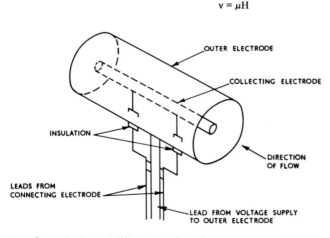

INSULATION

OUTER ELECTRODE

COLLECTING ELECTRODE

DIRECTION OF FLOW

LEADS FROM CONNECTING ELECTRODE

LEAD FROM VOLTAGE SUPPLY TO OUTER ELECTRODE

Fig. 1   Lovelock and Wasilewska's basic meter

where v is the ion velocity along the lines of force; μ is the ion mobility and H is the field strength. The other component is the velocity of the flow and as the gas flows through the meter it carries away a number of ions from the interelectrode space. As the flow velocity is increased the number of ions swept away increases, causing the collector current to decrease. This current can therefore be used to give a measure of the flow velocity.

Since the output signal decreases as the gas velocity increases this method is only suitable for low velocities. However Iordan[2] has described a system in which high mean gas velocities can be determined. In this flowmeter, outlined in Fig. 2, a voltage is applied at the electrode E_1 and a differential coupling between the collecting electrodes E_2 and E_3 is arranged and balanced at zero flow. When the gas velocity is increased the number of ions arriving at the collector E_3 increases and the number at E_2 decreases. This causes an out of balance in the currents which is proportional to the speed of the flow.

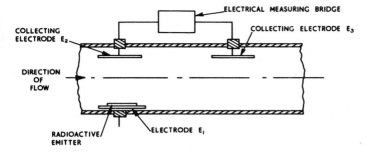

Fig. 2   Iordan's method

More recently Clayton and Webb[3] have published work on an instrument which can be used to measure either low or moderate gas velocities. The basic meter is arranged as shown in Fig. 3. As in the previously mentioned methods the gas is ionized by a radioactive source placed inside the pipe. The source is shown in Fig. 3 as a strip of foil located around the inner circumference of the pipe. The ions leave the ionizing section and are swept downstream with the flow towards the collecting section which consists of two electrodes E_1 and E_2. The electrode E_2 is a solid cylindrical rod and E_1 is the pipe wall. Provided all the ions are collected, the number arriving at the collecting section is found by measuring the collection current and, as the gas moves from the ionizing section to the collector the ions tend to recombine, the number reaching the collector will depend on the transit time. When the gas is moving slowly a large proportion of the ions will have had time to recombine before reaching the collector and a small current will result but at higher velocities less recombination will take place and a higher current will be measured. This current therefore gives an indication of the gas velocity and Clayton and Webb have shown that the meter performance may be estimated using the following equation:

$$I = \frac{AJf\sigma u^2 e}{(1 - \beta L)\left\{u^2 + \lambda Jf\sigma L(1 - \tfrac{1}{2}\beta L)\right\}} \qquad \dots (2)$$

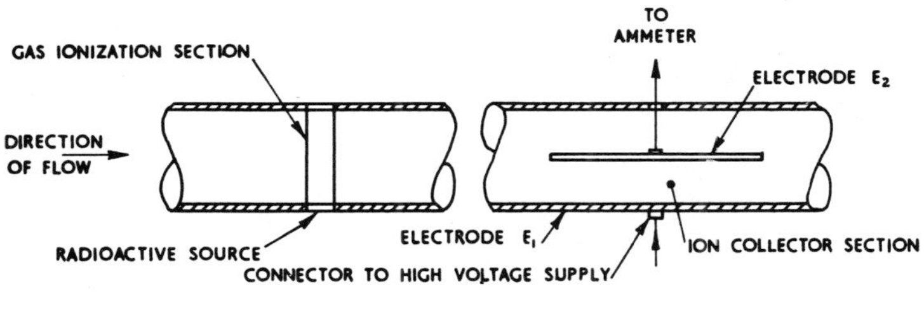

GAS IONIZATION SECTION

TO AMMETER

ELECTRODE E_2

DIRECTION OF FLOW

RADIOACTIVE SOURCE

ELECTRODE E_1

ION COLLECTOR SECTION

CONNECTOR TO HIGH VOLTAGE SUPPLY

Fig. 3   Clayton and Webb's basic meter

where I is the meter output current; A is the cross-sectional area of the meter; Jf is the ionizing potential of the source; σ is the ratio of the density of the gas at the ionizing section to the density at N.T.P.; u is the mean gas velocity at the ionizing section; $\beta = 1/p\ dp/dx$ where x is the axial distance from the source and p is the gas pressure; λ is the ion recombination coefficient and L is the distance between the source and the collector.

This technique was further studied by the author[4] using a 152 mm (6 in) diameter flow-meter. The instrument was found to have certain distinct advantages over conventional flowmeters: it presented little obstruction in the pipeline, the effects of wear were negligible and no conversion of a mechanical signal to an electrical signal was necessary.

However the main disadvantage of all the methods so far described is that the readout signal is also dependent on the state of the gas and variations in the conditions of the gas inside the pipe can cause serious errors. The present paper deals with a novel method, devised by the author, in which this disadvantage is largely overcome and hence the accuracy of the meter is not greatly affected by the state of the gas. A pulsing technique is used and since pulsing is achieved by purely electrical means the problems associated with mechanical pulsing devices[5] are not encountered.

Figure 4 illustrates the method. The gas flow is ionized by a radioactive source situated inside the pipe and to collect ions downstream of the source two sets of electrode systems are placed a specified axial distance apart. The voltage across the first set of electrodes is pulsed and a stable d.c. voltage is applied across the electrodes of the second collector. At the first collector ions are collected for the duration of each voltage pulse. This removal of ions from a section of the flow by the first collector causes a decrease in current at the second collector when this part of the flow passes through the second electrode system. The time interval between the release of a pulse at the first collector and the corresponding fall in current at the second collector gives a measure of the gas velocity.

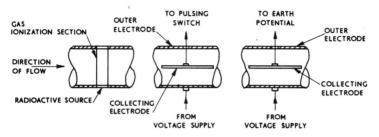

Fig. 4   Pulsed ionisation flowmeter

The tests reported in this work were undertaken to assess the feasibility of this technique and also to obtain design data for future instruments.

2   EXPERIMENTAL APPARATUS

2.1 The meter

The meter tested is shown in Fig. 5. It was essentially the device already described in Ref. 4 but altered to incorporate the additional electrodes and new electrical circuits necessary for this novel method. A radioactive foil which contained the isotope Americium 241 (activity = 24 μCi/mm)

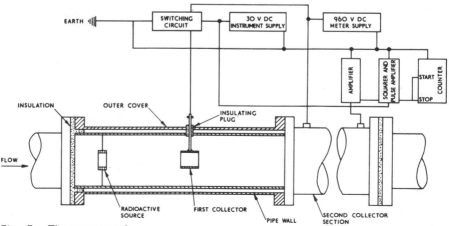

Fig. 5   The meter tested

formed the two-sided hollow cylindrical source and the distance between the source and the second collector was arranged to be 380 mm (15 in). An additional collector was inserted as shown such that the distance between the collectors in the calibration tests was 253 mm (9.95 in). Tests were also carried out where this distance was reduced to 170 mm (6.69 in).

The construction of the additional collector is shown in Fig. 6. The electrode E_2 was a 51 mm (2 in) diameter copper cylinder 51 mm (2 in) long which was joined to the supporting rod by an insulating plug. The electrode E_1 was the pipe wall, which was insulated from the rest of the pipework as shown in Fig. 5. A stable d.c. potential of 960 V was applied at E_1 and pulsing was achieved by manually switching E_2 from a potential of 960 V to earth potential. This collector arrangement, while offering little obstruction to the gas stream, gave sufficient field strength to collect ions from a section of the gas which was large enough, even at the highest velocities in the range, to give a readily detectable decrease in current at the second collector.

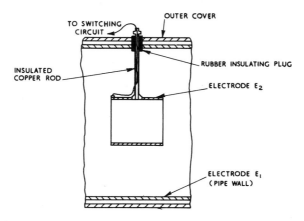

Fig. 6   Additional collector construction

A cross-section of the second collector is shown in Fig. 7. The pipe wall also formed part of the outer electrode of this system and a stable d.c. voltage of 960 V was applied at this electrode. The central electrode was permanently connected through the amplifier to earth potential.

The basic electrical circuit is also shown in Fig. 5. In the detection circuit at the second collector an a.c. coupling was incorporated so that only sudden changes in current were detected and the level of the signal remained unaffected as the flowrate was altered during experiments. The trigger level at the counter was set just below the level of this signal and outwith the range of the noise on the signal. The timer was started, and the central electrode at the first collector switched to earth potential, by manually depressing a push button which actuated a non-bouncer switch (mercury-wetted relay) connected in line with the central electrode of the first collection

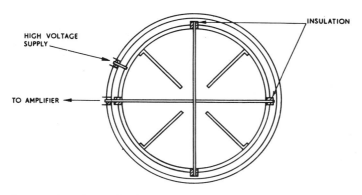

Fig.   7   Cross-section of second collector

system. The timer was stopped when the level of the signal at the second collector dropped to the trigger level.

2.2 The calibration line

Figure 8 shows the layout and relevant dimensions of the calibration line. A Borda inlet was positioned at entry to the system with a screen around it to prevent any variation in inlet conditions caused by adjacent rigs.

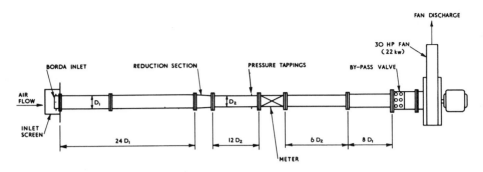

Fig. 8   The calibration line

Since the Borda inlet was designed for 120 mm (7.5 in) diameter pipework a reduction section and a length of 152 mm (6 in) diameter pipe were inserted before the meter. Another length of 152 mm (6 in) diameter pipe connected the meter outlet to the inlet of a 22 kW (30 hp) fan. The velocity of the air through the meter was altered by adjusting the bypass valve.

A piezometer ring and a thermometer were located just upstream of the meter so that the pressure and temperature of the air at entry to the meter could be determined. All pressure measurements were read using Betz manometers and temperature measurements were made using mercury and glass thermometers which could be read to ±0.1°C.

3   EXPERIMENTS

3.1 Flow measurement tests

Two sets of tests are described. For the main calibration tests the distance between the first and second collectors was 253 mm (9.95 in) but tests were also carried out where this distance was reduced to 170 mm (6.69 in).

The experimental procedure was as follows. The fan was switched on and the bypass valve fully opened. When flow was steady the push button switch was actuated and the pulse transit time noted. This was repeated until ten timings had been recorded and the average was then calculated. At the same time as these were being taken the differential and static pressures at the Borda inlet were also noted. Other readings recorded were the temperature and pressure at inlet to the meter and the line temperature at the Borda inlet. When sufficient data to obtain at least one calibration point had been noted the bypass valve was adjusted to increase the flowrate. At steady conditions readings were again taken and this procedure repeated until a maximum velocity of approximately 25 m/s (82 ft/s) was reached. After the maximum flowrate was reached the flow was gradually reduced, readings again being taken at regular intervals. In each set of tests several runs were completed and the barometric pressure and wet and dry bulb temperatures were recorded before and after each set of tests.

Densities and mass flowrates were computed as recommended in BS 1042. From the cross-sectional area and air density at the meter entry, the air velocity through the meter was calculated using the mass continuity equation.

3.2 Head loss tests

Tests, similar to those already reported in Ref. 4, were undertaken to assess the head losses across the meter. Using pressure tappings located upstream and downstream of the instrument, pressure differentials across the device were measured over the range of velocities covered in experiments.

The results are shown in Fig. 9 where they are compared with the losses which would be experienced if an orifice plate of area ratio 0.7 (the largest permitted in BS 1042) was employed and also those which would occur in a straight length of pipe.

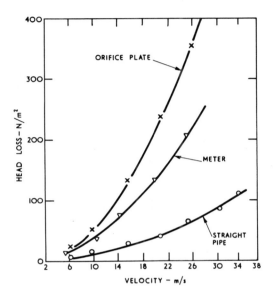

Fig. 9   Head loss tests

The losses were due mainly to the obstruction caused by the second collector. They were, however, small and they could be made even smaller by reducing the length of the second collector. Since the original purpose of this electrode system was to collect nearly all of the ions reaching it (see Ref. 4), and the velocities to be measured were in excess of those covered in the present tests, it was considerably longer than necessary for the experiments reported here. Its length could therefore be substantially reduced with consequent reduction in head losses.

4   DISCUSSION OF RESULTS

The calibration test results and the effects of shortening the distance between the collector sections may be seen from Fig. 10 and 11.

A linear calibration was obtained and it is important to note that while the current at the second collector varied with flowrate from approximately 1.10^{-7} to 6.10^{-7} amps the level of the signal in the detection circuit remained unaltered.

The results of the calibration tests were fitted to a straight line equation of the form

$$1/t = kv \qquad \qquad \text{... (3)}$$

where t(s) is the transit time; $k = 3.975 \ (m^{-1})$ is the calibration constant and v(m/s) is the mean flow velocity. The deviation plot is shown in Fig. 10b and the standard deviation was found to be ±1.1 per cent. This should of course be multiplied by 2 to give the usual 95 per cent confidence limits.

By examining the pulse transit traces on a storage oscilliscope it was noted that the accuracy of the leading edge technique employed in these experiments could be improved by sharpening the definition and steepening the gradient of the leading edge of the signal which stops the timer. This gradient could be steepened by removing a larger number of ions from the gas at the first collector and this would also allow the gain of the amplifier to be reduced with consequent reduction of the noise on the signal. A meter is being constructed in which ions will be removed across the complete cross-section at the first collector.

The effects of shortening the separation distance between the collectors can be seen from Fig. 9 and 10a. The short transit results were also fitted to a straight line equation and in this case the equation was

$$1/t = 5.680V \qquad\qquad\qquad ... (4)$$

The repeatability of the points at the shorter distance lay within a considerably wider tolerance than the long transit results the standard deviation of the points from a linear characteristic being ±2.3 per cent. The main reason for this was that while the transit times were shorter the timing errors did not reduce and hence the percentage errors in timing increased. The selection of the

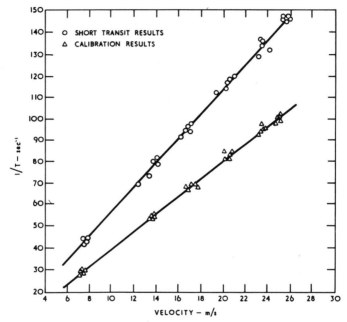

Fig. 10   Calibration and short transit results

separation distance between the two collectors is clearly important. The spacing should be distant enough to ensure that timing errors are small compared to the transit time but since the ions tend to quickly recombine after they are formed (see Ref. 3) this distance must be limited if a strong signal is to be obtained at the second collector. In addition if an adequate pulsing rate is to be achieved the separation distance cannot be excessive.

5   CONCLUSIONS

These tests have illustrated the feasibility of this technique and have shown that even at this preliminary stage accuracies of better than ±2.5 per cent can be attained. In previous continuous ionization flowmeters the main disadvantage has been that the output signal is also dependent on the state of the gas but in these experiments the level of the signal at the meter detection circuit, and hence the meter calibration, remained unaffected by variations in the number of ions reaching the second collector. In addition, since pulsing was achieved by purely electrical means the problems associated with the mechanical devices used in other pulsed ionized gas techniques were not encountered.

The tests have also shown that the device has certain distinct advantages over conventional flowmeters. The meter presents little obstruction to the flow and hence head losses are small. Calibration was linear and there was no need to convert a mechanical signal to an electrical signal.

The method is however not suitable for gases at high temperatures since the 'life' of the ionized molecules becomes shorter as temperature is increased. In addition since a radioactive

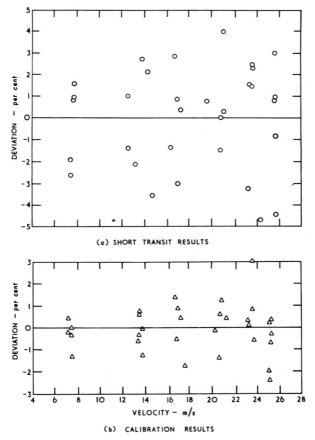

Fig. 11   Deviation plots from linear characteristics

    a   Short transit results
    b   Calibration results

source was employed certain safety precuations had to be observed although, by careful design of the ionizing section, radiation and contamination hazards could be virtually eliminated.

Different collector designs and electronic circuits could be investigated during the development phase to increase the accuracy of the flowmeter. In addition the device can also be made to operate automatically: to do this a circuit such as that outlined in Fig. 12 may be used.

A pulse generator is employed to switch the voltage at the first collector. When the effect of the removal of ions at the first collector is noted by a sudden decrease in the number of ions collected at the second electrode system, a signal from the detector circuit is led to a trigger device which then actuates the pulse generator. The time intervals between pulse generator actuations depend on the velocity of the flow.

The signal from the triggering circuit is tapped-off before the pulse generator and led to the frequency meter and recording instrument. The basic measuring signal is a series of voltage pulses which have a frequency proportional to the flow velocity. Flowrate may be integrated by incorporating a counter to sum the number of pulses released.

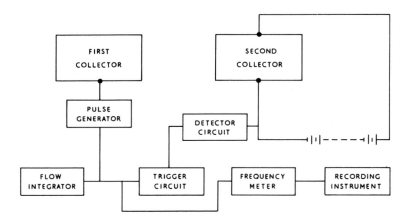

Fig. 12   Electrical circuit for meter automation

6   REFERENCES

1   LOVELOCK, I.E., and WASILEWSKA, E.M.: 'An ionization anemometer', J. scient. Instrum., 1949, **26**, (11), pp. 369-370

2   IORDAN, G.G.: 'An ionization flowmeter' (in Russian), Priborostroenie, 1955, (11)

3   CLAYTON, C.G., and WEBB, J.W.: 'The measurement of mass flow and linear velocity of a gas by continuous ionization', Int. J. appl. Radiat. Isotopes, 1964, **15**, pp. 603-610

4   BRAIN, T.J.S., and CAMERON, E.M.R.: 'The performance of large diameter gas ionization flowmeter'. NEL Report No 445, East Kilbride, Glasgow: National Engineering Laboratory, 1970

5   SHUMILOVSKII, N.N., and MEL'TTSER, L.V.: 'Automatic monitoring of gaseous flow on the "marked molecules" principle' (in Russian), Priborostroenie, 1956, (3), pp. 8-10

DISCUSSION

K.F.A. Walles: Presumably the best form of electrode would be a grid across the pipe, to minimize the varying ion path length and thus sharpen the signal. Is the technique intended to take a narrow slice out of the ion cloud, then to measure the transit time and shape of the slice downstream?

I agree that grid electrodes would have an undesirable pressure drop.

Author's reply: In the experiments described the leading edge of the pulse was used to trigger the timer. The technique mentioned by Mr. Walles could certainly be used but the pressure drop across the grid electrodes would give rise to an undesirable effect.

MEASUREMENT OF GAS FLOW BY RADIOTRACER METHODS

G.V. Evans, R. Spackman, M.A.J. Aston and C.G. Clayton

Abstract: The radioisotope constant-rate-injection method and the integrated-pulse-velocity method of measuring accurately the flow rate of gases are described. The accuracy of these methods has been established by comparisons with a primary standard of gas flow measurement based on a gas collection technique and associated with 5 cm. diam. and 10 cm. diam. pipelines having flowrates of air up to 20 m³/min. Accuracies of better than ±0.2% at the 95% confidence level have been demonstrated. The application of these techniques to the in-situ calibration of flow meters and to the performance testing of large fans is also described.

1   INTRODUCTION

Although radiotracer methods have been in occasional use for some years to measure the flowrates of gases, these methods have not achieved wide acceptance, mainly because techniques have not been fully developed and the operational advantages and accuracy of the methods have never been reported. This paper sets out to correct this situation.

A new technique which allows the constant-rate-injection method to be used for gas flow measurement has been developed and this now enables the well-known advantages of this method[1,2,3] to be used for accurate gas flow measurement.

The integrated-pulse-velocity method is well developed as a means of measuring water flow-rate[1] and the present paper demonstrates the simplicity of the method for gas flow measurement.

The accuracy of both the integrated-pulse-velocity method and the constant-rate-injection method has been examined by systematic comparisons against a primary standard of gas flow measurement[4] and it is shown that, for the integrated-pulse-velocity method an accuracy of within ±0.2% can be obtained and for the constant-rate-injection method accuracies of within ±0.5% are now possible. The development of relatively simple techniques and the demonstration of high accuracy, together with the fact that these methods can be operated equally well in many industrial sites as in a calibration laboratory, enable them now to be considered for many important problems in flow measurement.

2   DESCRIPTION OF THE INTEGRATED-PULSE-VELOCITY METHOD

2.1 Principles and theory

The volume flowrate of gas in a pipe can be determined by measuring the mean residence time of a tracer, between two positions a known distance apart. Any concentration-time function can be used for the injected tracer[5] but, in practice, a rapid injection is most convenient.

The flowrate Q is given by

$$Q = \frac{LA}{t} \qquad\qquad \text{... (1)}$$

where L is the distance between the two measurement positions, A is the mean cross-sectional area of the pipe and t is the mean residence time of the tracer.

The passage of the tracer is measured by radiation detectors coupled to integrating count rate meters and recorders, the signals being displayed as a trace in the form shown in Fig. 1. It has been shown that the distance between the half-amplitude points corresponds to the mean velocity of the tracer[1] and this distance can be found by a simple geometrical construction.

This method, which is a variation of the pulse-velocity method, has particular advantage for the measurement of gas flow. Apart from the greater precision which can be obtained in locating the characteristic point on the recorded trace, considerably less tracer is required. Also the response time of the electronic equipment used in the integrated-pulse-velocity method is practically zero and this results in negligible distortion of the recorded trace, even at the high velocities often encountered in gas flow measurement.

To achieve an accurate measurement of transit time, the detectors should be well separated and a distance at least equal to the mean spatial distribution of the tracer is preferred. This represents a

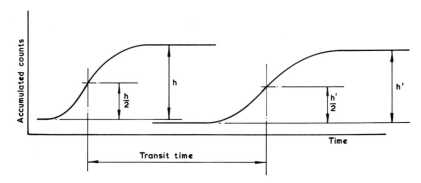

Fig. 1   Typical trace obtained in integrated-pulse velocity method

practical limit when the detectors are operated with a single electronic recording and display system.

The dispersion of tracer flowing in a straight pipe can be calculated from the expression[6]

$$T \cong \frac{6}{U} \sqrt{Xa} \qquad \qquad ...(2)$$

where T is the time taken for the tracer to pass a detection position a distance X downstream from injection, U is the mean velocity of the gas and a is the radius of the pipe.

The mean dispersion S of tracer between two positions at distances X_1 and X_2 from the injection position, is given by the equation,

$$S = 3 \sqrt{a} \ (\sqrt{X_1} + \sqrt{X_2})$$

and the ratio, n, of the distance between detectors to the mean spatial dispersion of the tracer is given by,

$$n = \frac{\sqrt{X_2} - \sqrt{X_1}}{3\sqrt{a}}$$

If the distance between the injection position and the first detector (X_1) is N pipe diameters, then the distance, in pipe diameters, between the first and second detector positions can be expressed as

$$\frac{X_2 - X_1}{2a} = \frac{(3n \sqrt{a} + \sqrt{2aN})^2}{2a} - N$$

$$= 4.5n^2 + 4.25n \sqrt{N} \qquad \qquad ...(3)$$

This relationship between the distance to the first detector, the distance between detectors and the ratio of the mean transit time to the mean dispersion time is shown graphically in Fig. 2.

2.2 Equipment and measurement procedure

2.2.1 Injection

For the measurement of flowrates at low pressures ($< 20N/cm^2$) an isotope injector, previously described elsewhere[7], is used. This injector which is shown diagramatically in Fig. 3, consists of a cylinder having a volume that can be varied from zero to 1 ml, and a spring-operated piston. On

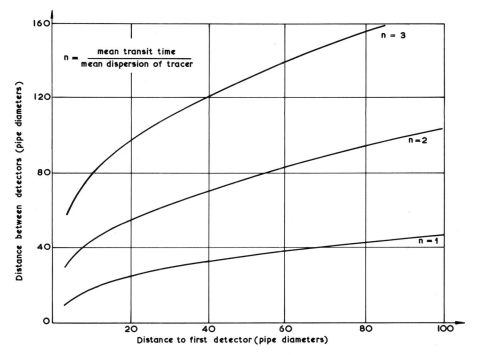

Fig. 2   Effect of measuring length of straight pipe on ratio of mean transit time to dispersion of tracer

raising the piston, the cylinder is automatically filled from a separate reservoir which contains the radioactive gas at the required concentration and which initially is at atmospheric pressure.

The tracer may be injected directly into the flow to be measured, or, if it is more convenient, it may be injected into a secondary flow before being introduced into the main flow.

This latter method is particularly useful for measuring the flow of gases at high pressures. An injection chamber is filled at atmospheric pressure with the tracer gas and this is injected into the flow by gas supplied from a cylinder and at a pressure above that of the main gas flow. The functional aspects of this injection system are shown in Fig. 4.

2.2.2. Detection

When a γ-emitting isotope is used as the tracer, the detectors can be positioned on the outside of the pipe a measured distance apart. Each of the detectors comprises a NaI(T1) scintillator, photo-multiplier and power supply, and is connected to an integrating countrate meter. This instrument has a digital input, the electronic pulses being accumulated in an integrating circuit so that there is no time constant effect.

For β-emitting tracers, a plastic scintillator is generally used as the detector and for sensitive detection it must be in close proximity to the tracer gas. A diagram of a suitable detection system is shown in Fig. 5.

It is convenient to connect the output from each integrating countrate meter to separate channels on a multichannel u/v galvanometer recorder and, so that the transit time of the tracer between the detectors can be derived independently of chart speed, to display timing signals from a crystal oscillator (with a recurrence period of 1 ms and 10 ms e.g.) on one of the recorder channels.

The effect of background countrate on the recorded trace can be eliminated by adjusting the

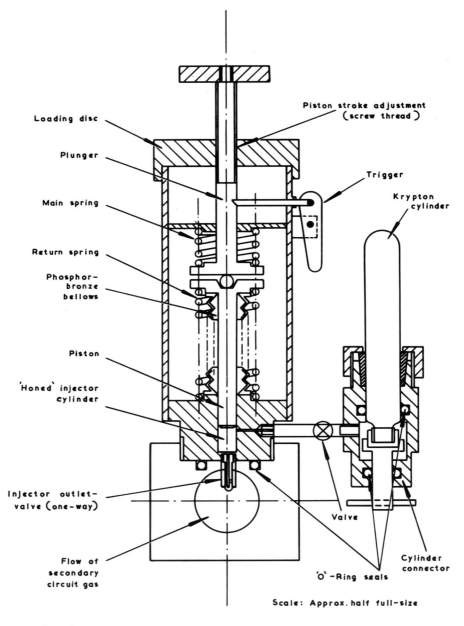

Fig. 3 Gas injector

output from the integrating ratemeter until a constant output level is achieved before radioactivity is introduced into the flow.

3   DESCRIPTION OF THE CONSTANT-RATE-INJECTION METHOD

3.1 Principles and theory

If a gas containing a concentration c_1 of tracer is introduced at a known mass flowrate m, into a pipe containing a gas with mass flowrate M, the rate of injection of the tracer is equal to the rate at which the tracer passes the sampling point. So that

$$mc_1 = \int_A \rho v c_2 \, dA \qquad \qquad \dots (4)$$

where ρ is the density of the gas in the pipe after the tracer is mixed, v is the local velocity of the gas at that point, c_2 is the local mass concentration of tracer and A is the cross-sectional area of the pipe.

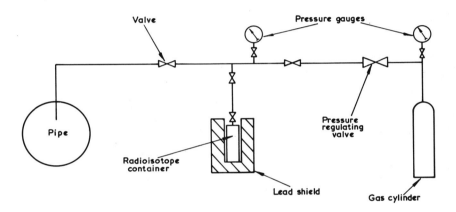

Fig.   4   High pressure gas injector

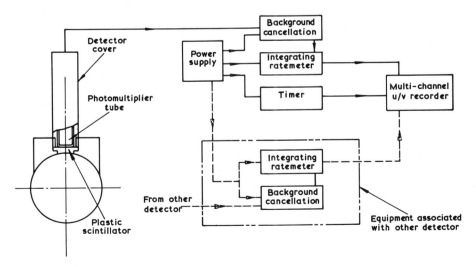

Fig.   5   Detection equipment

If the gas density is constant, equation 4 can be expressed as

$$mc_1 = \frac{M}{\overline{v}A} \int_A vc_2 \, dA$$

or

... (5)

$$M = \frac{mc_1}{\frac{1}{A} \int_A \frac{v}{\overline{v}} c_2 \, dA}$$

where v is the mean velocity of the gas flowing in the pipe. In a perfectly-mixed system c_2 is constant over the area of the pipe and equation (5) becomes

$$M \simeq m \frac{c_1}{c_2} \simeq \frac{m}{\beta}$$

... (6)

where β is the mass concentration of the injected gas in the gas flowing in the pipe.

When a radioactive tracer is used, the value of β can be determined by comparing the count-rate c_1 from a known mass of the injected gas with the countrate c_2 obtained from a known mass of gas extracted from the flow, provided that the countrates are measured by the same radiation detector containing a constant measurement volume.

Then

$$\beta = \frac{C_2}{m_s} \frac{m_i}{C_1}$$

... (7)

where m_i and m_s are the masses of the injected gas and sample gas respectively, measured by the radiation detector.

Substituting equation (7) in equation (6), the mass flowrate of gas M is given by

$$M \simeq m \frac{C_1}{C_2} \frac{m_s}{m_i}$$

... (8)

If the volume flowrate v of the injected gas is known at some reference conditions of gas pressure and temperature, the volume flowrate V through the pipe (at these reference conditions) is given by,

$$V = v \frac{C_1}{C_2} \frac{V_s}{V_i}$$

... (9)

where V_s and V_i are the volumes, at reference conditions, of the sample and injected gas measured in the radiation counter.

3.2 Equipment and measuring procedure

3.2.1 Injection

The tracer is injected from a high-pressure gas cylinder connected through a pressure-regulating valve to a 1 mm diameter critically-operated orifice (Fig. 6). The gas pressure upstream of the orifice is measured, during the injection period, with a calibrated, high-precision pressure gauge and the temperature is measured with a calibrated thermocouple. The pressure downstream of the orifice is also measured during the injection period to ensure that the orifice is operated in the critical flow regime (i.e. the ratio of the gas pressure downstream of the orifice to the gas pressure upstream of the orifice, is less than the critical pressure ratio).

When this condition is achieved, the mass flow of gas through the orifice is independent of the pressure and temperature of the gas downstream of the orifice and directly proportional to the ratio of the absolute pressure to the square root of the absolute temperature upstream of the orifice.

The injection rate through the orifice is determined for various upstream conditions by measuring the volume of gas discharged over a measured period by means of a wet-meter calibrated by the meter-proving system shown diagramatically in Fig. 7.

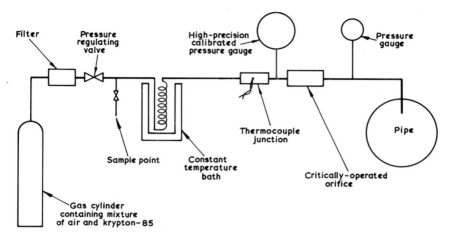

Fig. 6   Constant-rate-injection system

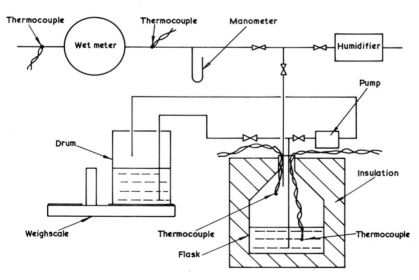

Fig. 7   Meter-proving system

3.2.2 Sampling

When the injected tracer gas is completely mixed with the flow to be measured, samples of the mixture are collected in previously-evacuated plastic bags of approximately 25 litres capacity. Usually ten samples are taken during the injection period.

To enable the samples to be collected during a period when the tracer is at a constant concentration, a continuous indication of the concentration of the tracer in the pipe is obtained by sampling the gas which is continuously discharged to atmosphere by passing it through a flowcell containing a scintillation detector, connected to a countrate meter and pen recorder. A diagram of a typical flowcell is shown in Fig. 8.

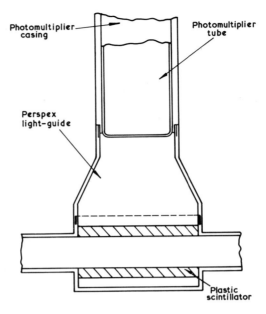

Fig. 8   Flowcell detector

3.2.3 Counting

The gas counter used when krypton-85 is the tracer gas, consists of a brass cylinder containing an axial rod of plastic scintillator connected at each end to a photomultiplier tube. Each photomultiplier is connected to a separate scaler and an automatic 'print-out' unit which records the period of the measurement and the count from each photomultiplier. (Fig. 9).

With this arrangement of a common scintillator and separate counting systems, the countrates

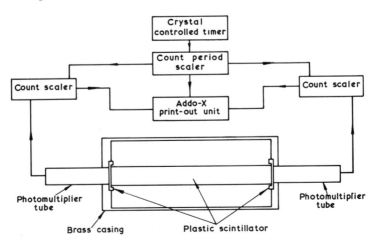

Fig. 9   Counting equipment

as measured by the two systems are not independent measurements of the tracer concentration, but they enable a check to be made on the stable operation of the concentration-measuring equipment.

The internal volume of the gas counter, which is approximately 5 litres, was determined accurately by measuring the amount of distilled water required to displace air from a flask into the gas counter to a measured pressure using the meter-proving apparatus shown in Fig. 7. This value was checked by a rigorous dimensional survey of the components used to construct the counter and by subsequent calculation of volume. Values derived for the volume of the counter by both methods agreed to within 0.01%.

The countrate from the sample gas is measured by first evacuating the gas counter and then connecting it to a sample bag. Gas from the sample bag passes into the counter and, as the bag acts as a diaphragm, atmospheric pressure is obtained in the counter after filling is complete. Atmospheric pressure is measured with a calibrated, high-precision aneroid barometer.

3.2.4 Dilution

The countrate from the injected gas cannot be measured by filling the gas counter directly, in a similar manner to that used for the samples, because the much higher countrates which would be obtained would tend to produce instability in the counting equipment. Consequently, it is more convenient to reduce the countrate to approximately (within an order of magnitude) that obtained when counting the samples.

This is achieved by collecting small samples of the injected gas in a standard container of accurately-known volume. This vessel, which is shown in Fig. 10, consists of a sample chamber sealed with a valve at each end, and a gas-flushing section connecting the spaces surrounding the outside of the two sealing valves.

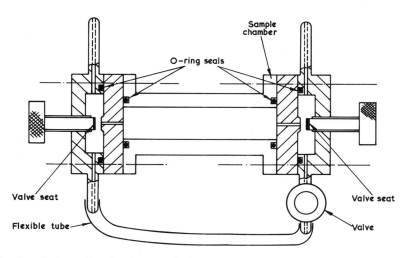

Fig. 10   Standard container for dilution of injected gas

The standard container is evacuated before filling it, to atmospheric pressure, with gas from the high-pressure cylinder on the injector. Tracer gas remaining in the gas-flushing section is then removed by evacuation and flushing with air. In order to measure the countrate from this standard volume of gas, it is transferred to the previously-evacuated gas counter by allowing air to pass through the sample chamber into the counter.

At least three separate samples of the injected gas are measured in the gas counter to check that the tracer is mixed within the injection cylinder.

4   CHOICE OF TRACER

The choice of a suitable radioactive isotope for use in gas flow measurement depends mainly on:—

(a)   the stability of the tracer in the gas flow and the absorption of tracer on to pipe walls or to the walls of any measuring equipment which is used;

(b)   the maximum permissible concentration of tracer;

(c)   the type and energy of the radiation emitted;

(d)   the half-life of the radioactivity;

(e)   the cost of the isotope.

Several isotopes which have been used for gas flow measurement are given in Table 1.

TABLE 1 Properties of Isotopes Used for Gas Flow Measurement

| Isotope | Type of Radiation | | | |
| | Beta | | Gamma | |
	Energy	Abundance	Energy	Abundance
	MeV		MeV	
Argon−41	1.20	(99.1%)	1.29	(99.1%)
Half-life 110 min	2.48	(0.9%)		
	0.35	(3%)	0.56	(45%)
Arsenic−76	1.20	(6.5%)	0.66	(6.3%)
(as $^{76}AsH_3$)	1.75	(3.5%)	1.21	(5.3%)
Half-life 26.5 hours	2.41	(31%)	1.44	(0.8%)
	2.97	(56%)	1.79	(0.3%)
			2.08	(1.0%)
	0.44	(100%)	0.55	(75%)
			0.62	(42%)
Bromine−82			0.70	(28%)
(as C_2H_5 ^{82}Br or CH_3 ^{82}Br)			0.78	(83%)
Half-life 36.0 hours			0.83	(25%)
			1.04	(29%)
			1.32	(28%)
			1.48	(17%)
Krypton−85	0.15	(0.4%)	0.51	(0.4%)
Half-life 10.6 years	0.67	(99.6%)		
Xenon-133	0.34	(100%)	0.081	(35%)
Half-life 5.27 days				
Sulphur-35	0.167	(100%)		

For the integrated-pulse-velocity method, where access to the pipe flow is difficult or restricted, γ-emitting isotopes are preferred since the passage of tracer may be observed by radiation detectors positioned outside the pipe. It has been found that Bromine 82 (γ energies of 0.55–1.48 MeV, half-life of 36 h) in the form of an organic compound such as ethyl bromide or methyl bromide is often a convenient tracer.

For the constant-rate-injection method, where samples are taken from the flow, Krypton-85 is recommended because it is an inert gas with a half-life of 10.6 years, and this enables it to be stored for long periods before use. Also, as ^{85}Kr is mainly a β-emitter (only ∼ 0.4% of the disintegrations result in γ rays) it can be shielded easily.

5   HEALTH AND SAFETY

When a radioisotope is used as a tracer, the radiological hazard associated with the ingestion and inhalation of the isotope and from direct exposure to the emitted radiations must be considered.

The maximum permissible levels of exposure to ionising radiations are given in the Recommendations of the International Committee on Radiological Protection[8]. In the United Kingdom, these recommendations are incorporated into the Ionising Radiations (Unsealed Radioactive Substances) Regulations 1968 which also include statutory requirements relating to the use of radioactive tracers in industry.

The amount of radioactivity required for the integrated-pulse-velocity method is generally considerably less ($\cong <$ 1%) than that required for the constant-rate-injection method; although the latter method can be used with concentrations of ^{85}Kr below the recommended maximum permissible concentration (calculated on the basis of continuous exposure over a period of fifty years) for unclassified workers.

Apart from hazards due to exposure to the radioactivity in the gas flow, consideration must also be given to radiation arising from the storage, handling and injection of the tracer. Consequently, the vessel in which the tracer is stored is normally kept behind a lead shield to reduce the exposure from γ-radiation. When ^{85}Kr is used, the β-particles are absorbed completely by the walls of the storage containers and injection devices, and radiation is only due to the small amount of γ-rays and resultant bremsstrahlung.

The magnitude of the radiation dose from these sources, with the activities used, can be made negligible, and are usually considerably below the recommended maximum levels.

6   DESCRIPTION OF AIRFLOW TEST LABORATORY

The test laboratory contains two airflow systems connected to a common volume-measuring system. One test system consists of a 10 cm (4 in) diameter test pipe, 30 m long, with a maximum airflow rate of 18 m^3/min supplied from a water-sealed rotary air compressor. The other test system consists of a 5 cm (2 in) diameter pipe, 16 m long with a maximum air flow rate of 4 m^3/min supplied by a separate rotary compressor. Each system contains an air receiver and pressure control valve to reduce fluctuations in flowrate through the test pipes. A diagram showing the main features of these two facilities is given in Fig. 11.

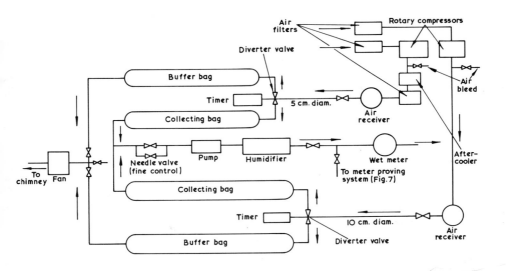

Fig. 11   Layout of air flow test systems

The airflow is determined by diverting the flow, by means of a solenoid-operated valve, into a large plastic bag (collecting-bag) for a measured period and measuring the volume of collected air, with a wet-type gas-meter of 28.3 litres (1 ft^3) capacity which is calibrated after each measurement by a meter-proving system based on weighing the amount of distilled water required to displace an indicated volume of air through the meter (Fig. 7). A full description of this method of flowrate measurement and the derivation of errors are given elsewhere[4].

When the airflow is not being transferred to the collecting-bag, it passes through the diverter-valve to another plastic bag (buffer bag). One end of the buffer bag is connected to a fan which discharges the air to atmosphere outside the laboratory. This bag is only partially inflated during operation, so as to maintain a gas pressure at the discharge of the test pipe equal to atmospheric pressure. Consequently, when the flow is diverted to the collecting bag, there is no change in back-pressure and the air flow in the test pipe is unchanged.

The period during which air flows into the collecting-bag is measured by a method based on accumulating timing signals (at 1 ms intervals) from a crystal-controlled oscillator onto a scaler. A diagram of the timing circuit is shown in Fig. 12. The display of the scaler is inhibited by the application of a bias voltage which is removed when the diverter-valve starts to open, and is re-applied when the valve starts to close.

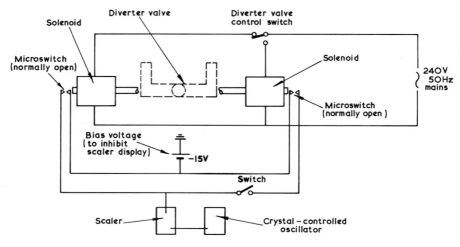

Fig. 12   Circuit diagram of system to measure collecting period

The volume of collected air is saturated with water vapour by a humidifier before measurement with the gas-meter. From measurements of air pressure, temperature and humidity in the test pipe and at the gas-meter, the measured volume can be corrected to the gas conditions in the test pipe.

The error in this method of flow measurement, which is directly traceable to standards of mass and time, has been shown to be approximately ±0.2% at the 95% confidence level[4].

7   COMPARISON OF RADIOTRACER METHODS WITH THE PRIMARY STANDARD

7.1 Integrated-pulse-velocity method

Gas flowrates were measured in both air flow test systems by using the integrated-pulse-velocity method with ^{85}Kr as the tracer gas, and the measurements were compared with the flowrates determined by the primary standard.

The tracer gas was injected into the airflow by using the injector shown in Fig. 3, and the passage of tracer was observed downstream of the injection position by scintillation detectors, as shown in Fig. 5, separated by a known distance. The volume of pipe between detectors was determined by weighing the amount of distilled water required to fill a measured length of pipe, and was also calculated from measurement of the internal diameter of the pipes at approximately 80 positions along the test length. Values of the mean cross-sectional area obtained by these methods agree to within 0.1%.

The distance between the injection position and the first detector was generally about 100 pipe diameters. However, some measurements were carried out with the first detector at only 40 pipe diameters downstream of the injection position. No difference in the measurement accuracy was observed. This result is in agreement with previous tests carried out on a water flow system[9].

where errors of less than 1% were observed for the integrated-pulse-velocity method when the first detector was at 10 diameters downstream of the injector position.

The dispersion of tracer observed during the flow measurements agreed very well with the relationship given in equation (3) and shown in Fig. 2, and confirmed its use for the selection of suitable measuring lengths.

Flowrates were measured before, during and after diversion of the flow into the collecting-bag to confirm the constancy of flow during the diversion period. The mean flowrate was compared with the flowrate determined from the collected volume and diversion period.

A total of 61 tests, involving approximately 500 measurements of velocity, were carried out on the 5 cm and 10 cm diam. pipes at flowrates varying from 3 to 300 litres/sec (0.1 to 10 ft³/sec). The mean deviation of the measurements by the integrated-pulse-velocity method from the measurements by the gas collection method was found to be 0.047%.

An analysis of the comparison of flow measurements by the tracer method with those of the primary standard is given in Table 2. This shows that approximately 70% of all the tests agreed with the standard to within ±0.2%, which is the basic uncertainty associated with the standard, and about 90% of the tests gave agreement to within only twice this error. This deviation was caused by slight fluctuations in flow during the measurement period, particularly at high gas velocities and short transit times, the precision of locating the characteristic point on the recorder trace representing the mean velocity of the gas, as well as the uncertainty in the flow standard.

TABLE 2 Comparison of Flow Measurements by the Integrated-Pulse Velocity Method With the Primary Standard

Deviation of measured flowrates derived from tracer and gas collection methods	No. of tests*
$< \pm 0.2\%$	40
$0.21 - 0.3\%$	7
$0.31 - 0.4\%$	7
$0.41 - 0.5\%$	3
$< 0.5\%$	4

*Each test involved approximately eight separate flow measurements by the integrated-pulse-velocity method

7.2 Constant-rate-injection method

The gas flow was measured with the constant-rate-injection method using ^{85}Kr as the tracer gas. A mixture of air and krypton was injected from a gas cylinder at a pressure of up to 1030 N/cm² (about 1500 lbf/in²), for a period of several minutes by means of the injection system shown in Fig. 6.

Generally 500 μCi of ^{85}Kr was used for each measurement and this was injected at four points equally spaced around the pipe wall. Samples of the flow were extracted from the pipe through four radial probes positioned 300 pipe diameters downstream of the injection position and immediately downstream of the flow diverter-valve. (A diagram of the sampling system is shown in Fig. 13).

The results obtained from these tests are compared in Table 3 with the flowrates measured by the primary standard. The random erros estimated from the measurements of countrate and the preparation of standard samples are also given and it can be seen that the deviations in the measurements from the standard flow measurements are within those expected from the basic error in the standard and the analysis of errors obtained from the results.

TABLE 3 Comparison of Flow Measurements by the Constant-Rate-Injection Method With the Primary Standard

Flowrate measurement by tracer method m³/s	Flowrate measurement by gas collection method m³/s	Deviation %	Calculated error (95% confidence level) %
0.2481	0.2475	+ 0.24	0.7
0.2503	0.2518	- 0.56	0.7
0.2485	0.2472	+ 0.52	0.7
0.2575	0.2503	- 0.46	0.7
0.2480	0.2496	- 0.65	0.7
0.2481	0.2498	- 0.69	0.7
0.2488	0.2497	- 0.36	0.7
0.2700	0.2719	- 0.69	0.7

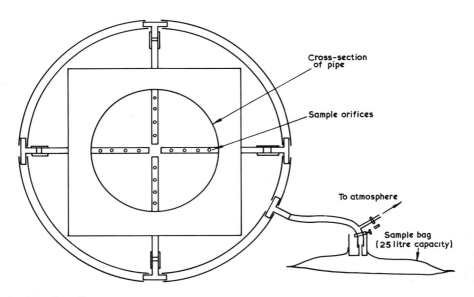

Fig. 13   Sampling arrangement

8   INDUSTRIAL APPLICATIONS

The accuracies of these methods of flow measurement can be obtained equally well on-site as in a test laboratory, provided suitable lengths of pipe are available for measurement. In particular, the integrated pulse-velocity method has been shown to be a convenient and accurate method of gas flow measurement when applied to lengths of pipe of accurately-known cross-sectional area. The practical limit on the flowrate which can be measured by this method is determined by the velocity of the gas and the length of pipe available, although the flowrates may be considerably larger than those which can be measured by a primary standard. Consequently the integrated-pulse velocity method can be used as a simple, accurate secondary standard of flow measurement for the in-situ calibration of gas flow meters.

The integrated-pulse-velocity method has been used to measure gas flow in a medium pressure gas distribution system to demonstrate the mobility of the measurement system for gas flow measurement during peak gas demand[7]. Although the accuracy of these measurements was limited by the lack of knowledge of the exact internal diameter of the pipes used, a great deal of valuable information was obtained and it was conclusively shown that the method is a convenient means of data collection for network analysis.

More recently, this technique has been used to calibrate an orifice plate installation which measures the flow of natural gas at a pressure of up to 689 N/cm^2 (1000 lbf/in^2). Bromine 82, in the form of ethyl bromide was used as the tracer so that detectors could be positioned on the outside of the pipe, access to the flow only being required for injection of the tracer gas.

The volume of the pipe between the detector positions was calculated from the length of pipe between detectors, measurements of the outside diameter of the pipe and from measurements of the pipe wall thickness by using a radioisotope γ-backscatter gauge. The estimated accuracy of the individual flow measurements, based on the errors in measuring pipe volume and transit time was less than $\pm 0.5\%$ at the 95% confidence level.

The ease of on-site calibration of meters by this technique can be indicated by the fact that only two staff were required to operate the method and the total period for installing and checking the measurement equipment, carrying out the tests and dismantling the equipment was less than 5 hours—and most of this time was taken up with adjusting the control of the flow system so as to obtain the desired flowrates.

Flowrates of over 35000 std m^3/h ($< 10^6$ std ft^3/h) were measured using approximately 3 mg ethyl bromide containing 120 μCi ^{82}Br for each injection.

The integrated-pulse-velocity method has also been used successfully at two power stations to measure the flow of superheated steam supplied to power generation plant. The mass flow of steam was determined from measurements of volume flow, steam pressure and temperature.

The constant-rate-injection method, which has the advantage of being independent of the internal dimensions and size of the flow system, has been used to measure air flow through an air-cooled heat-exchanger[10]. A mixture of ^{85}Kr and air was injected into the inlet of the cooling fan and samples were taken over the cross-section of the flow after the heat-exchanger tube bundles. Although the tracer was not completely mixed, flowrates determined from the weighted (according to local velocities) and unweighted mean concentrations (see equations (5) and (6)) differed by less than 0.5%. Values for the air flow through the heat-exchanger were also obtained from measurements of the heat lost from the process fluid and the temperature rise of the air passing through the heat-exchanger. Differences in flowrate of not more than 1.6% were observed for tests carried out at several different operating conditions of the heat-exchanger.

9   CONCLUSIONS

The integrated-pulse-velocity method and the constant-rate-injection method have been compared with a primary standard of gas flow measurement.

The integrated-pulse-velocity method has been shown to have an accuracy within $\pm 0.2\%$: the constant-rate-injection method has been shown to have an accuracy within $\pm 0.7\%$ although there is reason to believe that this can be improved.

The operational convenience of both these methods enable them to be used on-site as a secondary standard for the calibration of flow meters. Provided that adequate mixing can be achieved in the test section, the accuracy of both methods is solely dependent on the associated measuring equipment. Under these conditions the same accuracy in measurement can be obtained on-site as in a flow calibration laboratory.

The integrated-pulse-velocity method, in particular, is not limited by mass flowrate provided sufficient length of test pipe is available. At velocities at which compressibility effects are not significant, the accuracy of the method is independent of mass flow and consequently enables the method to be used at flowrates above that at which it is economic to develop a primary standard of flowrate measurement.

10   REFERENCES

1   CLAYTON, C.G., SPACKMAN, R., and BALL, A.M.: 'The accuracy and precision of flow measurement by radioactive isotopes'. Symp. on Radioisotope Tracers in Industry and Geophysics, I.A.E.A., Prague, 1966

2   CLAYTON, C.G., and EVANS, G.V.: 'The constant-rate-injection and velocity methods of flow measurement for testing hydraulic machines'. Harwell Report No. AERE – R 5872, 1968

3   CLAYTON, C.G., and EVANS, G.V.: 'Experience in the use of the constant-rate-injection method during tests of the performance of hydraulic machines' (see p. 276)

4   EVANS, G.V., SPACKMAN, R., and GARDINER, J.C.D.: 'A primary standard of gas flow measurement'. Harwell Report No. AERE – R 6011, 1969

5   BISCHOFF, K.B., and LEVENSPIEL, O.: 'Fluid dispersion – generalization and comparison of mathematical models – I. Generalization of models'. Chem. Eng. Sci., 17, pp. 245-255, 1962

6   EVANS, G.V.: 'Note on the dispersion of tracer in turbulent pipe flow', Chem. Eng. Sci., 24, pp. 1736-1738, 1969

7   CLAYTON, C.G., EVANS, G.V., SPACKMAN, R., and WEBB, J.W.: 'A mobile system for measuring flow in a gas distributor network'. Paper presented to Institution of Gas Engineers, Bristol, Jan. 1969

8   Recommendations of the International Commission on Radiological Protection (I.C.R.P.). Report of Committee II (Pergamon, 1959

9   CLAYTON, C.G., EVANS, G.V. and WEBB, J.W.: (unpublished work, 1969)

10   COWAN, G.H., EVANS, G.V., SPACKMAN, R., and STINCHCOMBE, R.A.: 'Measurement of airflow through an air-cooled heat-exchanger' (to be published)

DISCUSSION

J.C. Schuster: In studies on mixing of fluid and tracer, was account taken, especially at short mixing distances, of correspondence in the circumferential position of the injection and sampling points? Was there an effect of pressure fluctuations at the injector tip for short mixing distances?

Authors' reply: The work reported in this paper concerned the development of tracer methods for the accurate measurement of gas flow and consequently it was preferable to examine these methods under as ideal conditions as possible. This entailed using distances between injection and the positions of sampling or detection of tracer which were considerably longer than those required for adequate mixing so as to minimise uncertainties in the method. No tests were carried out with the specific purpose of examining injection configurations and mixing distances, but some measurements were made using the integrated-pulse-velocity method with distances between the injection and first detector positions of only 40 diameters. These measurements were in complete agreement with measurements made using detectors at 150 diameters and with the gas collection method.

Pressure fluctuations at the discharge of the constant-rate-injection system would not affect the mass injection rate provided that the pressure ratio across the orifice in the injector remained above the critical pressure ratio. If the pressure fluctuations are at constant mass flow then the measurement by the constant rate injection method would be unaffected. In the velocity method variations in volume flowrate would be observed as variations in the measured transit times of tracer between detector positions. The longer the distance between detectors the greater would be the effective time response of the method and so the measurements would be less sensitive to high speed fluctuations in volume flowrate.

F.C. Kinghorn: Accuracies, using the integrated-pulse-velocity method, of ±0.2% at the 95% confidence level are reported. This does not seem to tally with the data in Table II of the paper, which shows a discrepancy of greater than 0.2% between this method and the primary standard in approximately 30% of the tests.

Authors' reply: The flowrates given in Table II were obtained from comparison tests between the integrated-pulse-velocity method of flow measurement and the primary standard. Deviations between the flowrates as measured by these methods are caused by systematic and random errors obtained in both methods of measurement and, as the measurements were not made simultaneously, to differences in the flowrates existing during the two measurement periods.

As stated in the paper, the mean deviation of the measurements was approximately 0.05% and represents a negligible systematic error in the integrated-pulse-velocity method. An estimate of the random error in the measurements was derived by subtracting the variances in measurement due to the primary gas standard and the flow stability from the observed overall variance in the differences of flowrate obtained by the two methods. Assuming a normal distribution of errors, the random error, to the 95% confidence level, in the velocity method was obtained as twice the value of the square root of the remaining variance.

K.E. White: A third method, not dealt with in the paper, is the integration method proper. This has the advantage that only one detection apparatus, which can be remote from the system under study, is required and it is not necessary to determine the cross-sectional area of the pipe over a considerable distance, as in the integrated-pulse-velocity method. However, a second access to the flow may be required for a continuous sample to integrate the tracer cloud if calibration of external detection geometry is difficult. Is this why this third method is not used or are there other reasons?

Authors' reply: The total quantity or integration methods of flowrate measurement, in which a known amount of tracer is injected into the flow, have been used extensively to measure the flows of rivers, mainly because of the large degree of dispersion obtained in rivers and the need to reduce the amounts of tracer required. However, the main disadvantages of these methods, when applied to the measurement of flow through pipes, are the difficulties of integrating accurately the concentration-time relationship obtained at a sampling point and the need to obtain samples simultaneously from several positions (at least five positions for a reliable analysis of the random error of sampling) across the pipe cross-section. In addition, the velocities generally encountered in gas flow are greater than those obtained in liquid flow so that the dispersion time of the tracer 'pulse' is correspondingly reduced. Because of this, significant errors in measurement can arise when the tracer is rapidly injected into the flow. When the integration of the concentration-time distribution is obtained from a detector positioned outside the pipe (i.e. 'total count' method) errors are introduced due to the 'dead-time' of the radiation detector. These errors can be reduced by injecting the tracer over a longer period but this increases the amount of tracer required and, apart from using a simpler injection system, has no advantage over the constant-rate-injection method.

Nevertheless we intend to examine the use of these techniques for gas flow measurement as they may be particularly useful for the measurement of flowrates in systems which exhibit a large degree of longitudinal dispersion (e.g. flows through reactor vessels). The work reported in this paper concerns only our work on the constant-rate-injection and integrated-pulse-velocity methods which were considered to be more applicable to the accurate measurement of gas flows through pipes.

C.R. Brewer: Would the authors please state what the advantages and disadvantages of this method are as compared with using a non-radioactive tracer and a mass spectrometer?

Authors' reply: Any tracer gas which does not react with the gas flowing in the pipe or with contacting surfaces can be used in the constant-rate-injection method provided that the concentrations obtained are below maximum permissible levels and can be measured accurately.

There are obvious advantages in using an inert, non-toxic and non-radioactive gas (e.g. Argon, Neon, Krypton) but the main limitations in their use is the accuracy with which low concentrations can be measured and, in the use of Argon, from concentrations of gas present in the atmosphere. A mass spectrometer can be used to measure low concentrations of these gases but the error in the measurement of concentration is generally quoted as about ±2%. Thus, its use is unsuitable for accurate flow measurement.

Radioactive tracers can be measured accurately at low concentrations and a continuous record of the concentrations of tracer at the sampling point can be easily obtained over the sample period. The main disadvantages in using a radioactive tracer is the finite decay period of the isotope used, the public relations aspects of using a radioactive tracer in some production process or gas supply and, in the United Kingdom, the registration of the premises with a ministry department which must be satisfied that negligible radiological hazard will be presented. The first point may be a severe problem when Argon-41 (half-life 110 minutes) is used but it not important when using Krypton-85 (half-life 10.6 years).

The above comments also apply to the use of the integrated-pulse-velocity method but, in addition, the tracer concentrations must be measured continuously. The use of a radioactive tracer gas has the advantage that detection of the tracer can be made from scintillation detectors positioned on the outside of the pipe, thus eliminating the need for sampling the gas. This is an important consideration when measuring the flow of gas at high pressures and where the available distances between detectors allow only short transit times to be measured.

Finally the cost and suitability of equipment has to be considered when selecting the tracer to be used. Mass spectrometers are more expensive than radiation measurement equipment and because of their size cannot be easily moved to remote locations for in-situ measurements.

A.T.J. Hayward: It must require an extremely large number of measurements of diameter to obtain the volume of pipe between detectors to an accuracy of 0.2 per cent. On how many planes do you make measurements, and how many diameters do you measure on each plane?

Authors' reply: The accuracy of determining the pipe volume depends on the uniformity of size of the pipe. For uniform pipes, the volumes have been determined accurately by measuring at five cross-sections of the pipe and at four radial positions at each section. For pipes of non-uniform diameter and thickness, measurements must be made at a greater number of positions to achieve an accurate value for the internal pipe volume.

Paint or sediment on the inside of the pipe can present a source of error in this method of determining pipe volume and the internal state of the pipe must be known for reliable results to be obtained.

P.R. Sanderson: Is not the accuracy of the integrated-pulse-velocity technique when applied to flow measurement in the field dependent on the accuracy with which the pipe cross section can be measured? Is not this limiting accuracy of the order of ±3% rather than the 0.2% accuracy of the velocity measurement?

Authors' reply: Yes! The accuracy of the method when applied to the in-situ calibration of flow meters is determined mainly by the accuracy of determining the internal volume of pipe between detector positions. If the diameter of commercial grade wrought iron pipes and the tolerance of this diameter is taken from BS then the uncertainty on the volume between detectors can be as much as ±3%. However it is usually possible to calculate the internal pipe diameter by measuring the outside diameter and pipe wall thickness with external instrumentation and this can be done accurately (±0.2%) using γ-ray backscatter or ultrasonic gauges.

An alternative method, where practical, is to take out a section of pipe and to determine the volume between two sections directly by filling the section with a measured amount of water. Again it may be possible to install a pipe of accurately known cross sectional area in series with a meter in a similar way to that required using a ball-type meter prover. This is obviously more inconvenient and its practice must be influenced by the accuracy of measurement required.

T. Agar: Referring to the statements in page 257, I would like to ask what precautions and tests were carried out to ensure that the following conditions were met: (1) Constant density (2) Perfect mixing. How was the accuracy of the ratio m_i/m_s determined?

Authors' reply: The comparison tests were made in the air flow calibration facilities at Harwell, where reasonably steady flowrates could be achieved at constant temperature and pressure. Consequently the air density at the sampling section was sensibly constant during the test periods.

Mixing was achieved by injecting the tracer into the pipe through four equally spaced orifices at the pipe wall and samples were taken from multipoint sample probes at a distance of 300 pipe diameters downstream of the injection point. In addition the test pipe contained a butterfly valve and diverter valve causing additional turbulence and mixing. Samples were taken at various positions of the pipe to check that mixing was complete.

The accuracy of the ratio m_i/m_s determined from the errors in the measurements of the volumes of the gas counger and standard container and the measurements of atmospheric pressure during filling the vessels with samples and injector gas. From these component errors it is estimated that the accuracy of determining the ratio of m_i/m_s is better than $\pm 0.05\%$.

D.L. Smith: Does the accuracy of flow measurement of the velocity method depend on the flow distribution in the pipe? This would affect the application of the technique as an on-line meter prover.

Authors' reply: No, provided the tracer is sufficiently mixed with the flow at the first detector. As stated in the presentation of the paper, the method should be described as a "residence-time distribution method" rather than a "velocity method". This is because the mean residence-time of the tracer in a length of pipe between two detector positions is measured and provided the tracer represents the flow into this pipe section, the residence time at a given flowrate is independent of the distribution of velocities within the pipe.

For the successful application of tracer methods of flow measurement the tracer must be mixed with the flow to be measured. This generally restricts their use to turbulent flow conditions although with suitable injection equipment or mixing device within the flow system they may be used for the measurement of laminar flow.

J.I. Yarwood: We have been using the bag method since 1964 with a bag of about 50 m^3 capacity and a maximum rate of filling approaching 1 m^3/s. The standard deviation measuring the reproducibility is 0.081% based on 24 results with a nearly full bag and rises to 0.12% based on 42 results if we include tests in which the bag is only filled to 1/6th or 1/3rd of its capacity. In these tests the volume metered into the bag is compared with the volume metered out again. This reproducibility is comparable with the authors' claim of an accuracy better than 0.20% at the 95% confidence level. I gather that the authors believe that faults arising from their use of a wet meter in measuring the volume collected in the bag are responsible for the considerable proportion of the results in Table II which they reject in arriving at this accuracy. We have always used a dry rotary displacement meter. Of course, at sometime, this meter has to be calibrated, using the bag, against a wet meter, or other device which can be directly related to a primary standard volume and this involves complexities similar to those which are thought to cause the unacceptable results. But the calibration depends on the mean of a set of observations, so the reproducibility of individual values does not matter so much and a large enough set can be taken. Perhaps a rotary displacement meter such as we use might help to demonstrate the high accuracy claimed for the tracer method.

5·4

FLOWRATE DETERMINATION BY NEUTRON ACTIVATION ANALYSIS

C.R. Boswell and T.B. Pierce

Abstract: Neutron Activation Analysis has been used to measure the velocity of liquids, solids and slurries by counting the induced radioactivity formed in the moving material as a result of neutron irradiation. Two methods have been used, one measuring the ratio of the activity of the stream at two positions downstream of the place of irradiation and the other measuring the time taken for activity formed by a neutron pulse to travel between source and detector.

1   INTRODUCTION

Techniques of neutron activation analysis have been extensively developed to provide laboratory methods for the elemental analysis of discrete samples. The method usually employed requires irradiation of a sample in a flux of neutrons, and subsequently, the measurement of the radiation emitted during decay of the radioisotopes formed as a result of nuclear reaction of the neutrons with the elemental constituents of the sample. Neutrons are penetrating radiation, as is γ-radiation which is frequently produced as a result of radioisotope decay, so that the neutron source and γ-ray detector can be positioned outside the sample container thus permitting examination of material in containers or pipes which are not provided with access ports for more conventional analytical sensors. Further, since the energy of γ-radiation emitted during nuclear decay is a function of γ-transitions occuring within the decaying nucleus, established techniques of γ-ray spectroscopy frequently enable radiations associated with different decaying species to be distinguished separately thus enabling the behaviour of more than one component in a dynamic system to be followed. As part of a more general programme of work aimed at developing neutron techniques for the on-line analysis of moving streams of solids, slurries and liquids, some preliminary work has been carried out to assess the possibility of obtaining information about the flow of liquids and solids in closed vessels by neutron activation analysis. Methods exploit either the time-dependancy of radioactive decay which permits the transit time taken for some induced radioactivity to pass between two detectors to be calculated from the ratio of the activity measured by the two detectors, or the ability of certain portable neutron generators to produce short, intense pulses of neutrons which thus give rise to a pulse of induced radioactivity in the sample that can then be measured at some known distance downstream of the irradiation position. γ-ray spectroscopy can be used to discriminate between the γ-radiation to be measured and others induced in the samples.

2   PRINCIPLES OF THE METHODS

2.1 Basic techniques

Two separate methods were used to provide information about flowrate, both effectively providing the time taken for the irradiated sample to move between two positions. In the "activity ratio" method, the two positions were defined by two γ-ray detectors placed downstream of the irradiation position and separated by a known distance; the neutron source was operated continuously to generate radioactivity continuously in the sample stream. In the "pulsed source" method, radioactivity was produced in the moving sample in the form of a pulse by pulsing the neutron source, and a single γ-ray detector was used to monitor the variations in activity of the sample stream.

2.1.1 Activity ratio method

The disintegration rate A_1 of a source initially of a strength A_0 after a time t_1 is given by the equation

$$A_1 = A_0\, e^{-0.693 t_1/t_{\frac{1}{2}}}$$

where $t_{1/2}$ is the half-life of the radionuclide being measured. Thus the ratio of activities A_1 and A_2 recorded by two separate detectors counting the induced activity of a sample after decay times t_1 and t_2 respectively

will be
$$\frac{A_1}{A_2} = e \left\{ 0.693 \ (t_2-t_1)/t_{1/2} \right\}$$

so that
$$\Delta t = \frac{t_{1/2}}{0.693} \ \log A_1/A_2$$

where $\quad \Delta t = t_2 - t_1$

Clearly with a flowing sample and two detectors placed at different distances downstream of the irradiation position, a knowledge of Δt and of the detector spacing, permits sample velocity to be calculated.

2.1.2 Pulsed source method

The pulsed source method as applied in this work is based on the formation of induced radioactivity in a very limited volume of moving sample, and subsequent measurement of the time to elapse between formation of the radioactivity and appearance of the radioactivity at a detector placed a distance s downstream. Velocity of the sample stream is calculated from a knowledge of t and s.

2.2 Neutron sources

Application of flowrate measurement by neutron activation techniques is dependent upon suitable, portable sources being available which can produce adequate radioactivity in the sample stream examined to permit a satisfactory measurement to be carried out. In general, two types of neutron source, radioisotope and accelerator based systems are potentially useful for flowrate measurements. Standard two-component radioisotope sources, which produce neutrons by interaction of the decay radiation of a radioisotope with a second element do not provide high neutron outputs but the imminent availability of the spontaneous fission emitter, californium-252 which gives an output of 2.3×10^{12} n/sec/gm will enable much more intense isotope neutron sources to be produced. Radioisotope neutron sources are reliable, compact and have zero-power requirements but the neutron output cannot be regulated which is a disadvantage if discontinuous operation is envisaged. An alternative is a specialised version of accelerator neutron source which produces neutrons by nuclear reaction of accelerated charged particles with targets of suitable materials. In particular the reaction

$$^3H + {}^2H \rightarrow {}^4He + {}^1n + 17.6MeV$$

is particularly valuable since (a) high neutron yields can be obtained at relatively low accelerating voltages of the order of 120KV and (b) the high positive energy value of the reaction results in the production of energetic neutrons with an energy of ~14MeV which can consequently induce highly endoergic reactions in sample nuclei. Neutron outputs of accelerator sources can be controlled and switched off if necessary, but with ancilliary electronic equipment the systems are clearly more bulky than those based on radioisotope sources.

2.3 Nuclear reactions for flow rate measurement

An essential requirement for successful flowrate measurement by neutron activation analysis is that suitable radioisotopes are produced in the flowing material by neutron interaction, and possible reactions will be dependent on neutron energy. Reactions most commonly induced by high-energy neutrons, for example those of 14MeV, are (n,a), (n,2n), (n,p) and (n,n') while low-energy neutrons, or those high-energy neutrons which have been reduced to low energy by moderation, may induce radiative capture (n,γ) reactions. Examples of a number of reactions that might be used for flow-rate measurements by one or other of the methods described in sections 2.1.1 or 2.1.2 are given in Table 1 together with the half-lives and characteristic γ-ray energies of the product nuclides. The reactions given in Table 1 are by no means exhaustive and serve only to illustrate the type of reaction that can be produced by neutron irradiation. Since the short irradiation times experienced in flowing systems favour the production of short-lived nuclides only reactions leading to the production of radionuclides with half-lives of a few minutes or less are included.

TABLE 1   Examples of neutron-induced reactions leading to the formation of short-lived radio nuclides

Element	Reaction	Product $E\gamma$(MeV)	$t_{1/2}$
Oxygen	$^{16}O(n,p)^{16}N$	6.1	7.3s
Titanium	$^{46}Ti(n,p)^{46m}Sc$	0.14	19s
Fluorine	$^{19}F(n,p)^{19}O$	0.2	27s
Sodium	$^{23}Na(n,p)^{23}Ne$	0.44, 1.64	38s
Silicon	$^{28}Si(n,p)^{28}Al$	1.78	2.3m
Phosphorus	$^{31}P(n,a)^{28}Al$	1.78	2.3m
Silver	$^{107}Ag(n,\gamma)^{108}Ag$	1.77	2.3m
Chromium	$^{52}Cr(n,p)^{52}V$	1.43	3.76m
Vanadium	$^{51}V(n,p)^{51}Ti$	0.32	5.8m
Nitrogen	$^{14}N(2,2n)^{13}N$	0.51	9.96m

3   EXPERIMENTAL INSTRUMENTATION AND EQUIPMENT

3.1 Neutron source

Neutrons were produced with sealed-tube neutron generators producing neutrons by the reaction $^3H (d,n)^4He$ reaction described in section 2.2; the sealed-tube system obviated the need for continuous pumping and hence offered a convenient source of neutrons. Generators were available giving neutron outputs of from 10^8 to 10^{11} n/sec total and output could be either continuous or pulsed. The smallest neutron generator was capable only of pulsed operation giving 10^8 neutrons in a 15μsec pulse.

3.2 Counting and electronic equipment

γ-ray detection was by means of 3" x 3" thallium activated sodium iodide scintillators. Output from the detectors was fed, after amplification, to electronic equipment either for pulse-height analysis or for multiscaler examination of the time dependancy of the induced radioactivity. Out put pulses from neutron generators during pulsed operation were available to control counting equipment and to initiate operation of timing circuits.

3.3 Materials handling facilities

Examination of moving material by neutron activation analysis was carried out under controlled conditions using materials handling facilities designed to assist in the development of techniques of on-line analysis. Two systems were available, one for examining liquids or slurries, and the second for examining solids.

3.3.1 Liquid and slurry systems

The liquid and slurry system consisted essentially of a constant-speed pump which could be connected at input and outlet to vessels of different size; the vessels could also be connected, if required, to allow continuous recirculation of the sample for extended experimental tests in cases where half-life permitted. Pumping could be regulated over a range of speeds and an injection system was also available which permitted injection of additional solutions (e.g. dye or radio-active isotopes) into the main sample stream.

3.3.2 Solid system

The solid handling system was made up of a central storage hopper, capable of holding 10-15 tons of material, feeding a screw conveyor which in turn lifted material into a second hopper which fed an 18″ belt conveyor; material from the belt conveyor was subsequently returned to the storage hopper by a second screw conveyor. The rotating vane feeder to the belt conveyor and the hydro-static drive to the belt conveyor itself were variable thus permitting the total throughput of material round the system and the velocity of movement on the belt conveyor to be carefully controlled.

3.4 Flux monitoring

Measurement of the neutron output of the neutron generators was necessary to follow any changes that might occur during measurements and to permit absolute calibration of neutron output when this was necessary. During continuous operation of the neutron generators, neutron output was followed with a B F_3 detector, or a europium activated lithium iodide scintillator, surrounded with a suitable quantity of neutron moderator. A neutron sensitive scintillator was also used to monitor the output of the generator during pulsed operation, but in this case gave an output which was a function of the total number of neutrons per pulse. Absolute calibrations were carried out by irradiation of foils of pure materials or by the use of pre-calibrated neutron detectors.

3.5 Data handling

In general, the simplicity of the neutron activation method enabled flow rates to be calculated directly, either from an activity ratio or from an elapsed time. However, where more detailed examination of experimental data was desirable to support practical studies some additional data processing was necessary. Experimental information was accumulated either in a hard-wired multi-channel pulse-height analyser or directly in a PDP-8 computer interfaced and programmed to operate as a pulse-height analyser. Data from the fast store of the PDP-8 could be read out on to magnetic tape and held for storage while information from the hard-wired analysers could be punched on to paper tape and subsequently read into the PDP-8 by means of a fast tape reader. Simple data manipulation was carried out in the PDP-8 itself but where more complex data man-ipulation was necessary, data could be read from magnetic tape directly into an IBM 360/75 computer. The 360/75 was also used for computer simulation studies to assess the effect that changes in certain parameters such as neutron output, flowrate and elemental concentration occurring during a measurement would have on the calculated results.

4   RESULTS AND DISCUSSIONS

4.1 Activity ratio method

Successful application of the activity ratio method to the determination of flow-rate measurement requires that the combination of nuclide half-life and detector spacing permits the activity ratio to be calculated with adequate precision. Clearly at low velocity, measurements based on counting a short-lived nuclide require that the two detectors be placed close together in order to obtain adequate statistical precision in counts from the detector placed further down stream from the irradiation position. Conversely, if the radionuclide produced has a relatively long half-life, sub-stantial separation of detectors may be necessary to obtain satisfactory A_1/A_2 ratios, particularly at high flow-rates. Any discussion of detector spacing necessarily assumes that γ-ray detectors can be brought sufficiently near the moving sample stream for measurement of the induced radio-activity at suitable positions which may of course be complicated by the design of the container of the flowing system. Neutron activation analysis cannot, of course, be used to examine the flow rate of those materials which do not produce measurable induced radioactivity as a result of neutron irradiation, but fortunately the reaction $^{16}O(n,p)^{16}N$ is potentially useful for the determination of mean velocity over a range of possible values and the widespread need to examine flowing materials containing oxygen makes this reaction of particular interest. Consequently applications of the activity ratio method will be illustrated by considering its application to flowing, aqueous systems.

4.1.1 Method used for the determination of mean velocity by the $^{16}O(n,p)^{16}N$ reaction

A block diagram of the experimental system used for the determination of mean velocity by the $^{16}O(n,p)^{16}N$ reaction is shown in Figure 1. Neutron generator and the two detectors were placed on the outside of the pipe containing the flowing, aqueous stream with a neutron detector, positioned beside the sealed tube housing. The operation of the γ-ray detectors was checked with standard sources and single-channel analysers set to accept pulses attributable to γ-rays from nitrogen-16. A typical high-energy γ-ray spectrum obtained from the detectors is shown in Figure 2 in which the 6.1MeV photo-peak from nitrogen-16 can be clearly distinguished, together with the associated first and second isotope peaks: Single Channel Analysers accepted counts due to photo and escape peaks. The neutron detector was included in the instrumental system to monitor output of the neutron generator for although not necessary for obtaining the activity ratios it did provide a useful check on satisfactory functioning of the neutron system. Integrated counts accumulated in scalers associated with neutron and γ-ray detectors were automatically printed out after pre-set integration times.

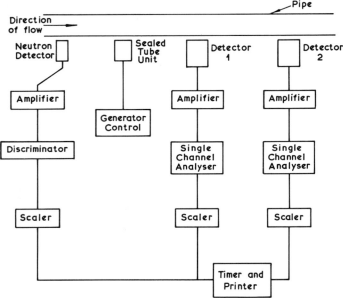

Fig. 1   Block diagram of experimental system used for determination of mean velocity of aqueous systems in pipes

4.1.2 Results obtained for the determination of mean velocity by the $^{16}O(n,p)^{16}N$ reaction

Mean velocity was measured by counting the nitrogen-16 activity induced in aqueous solutions and slurries at a variety of flowrates and in pipes of different diameters. A typical set of results is presented in Figure 3 in which values calculated from neutron activation measurements are compared with known velocities. The precision of the measurements depended on the total count accumulated which in turn was a function of such parameters as sample size, oxygen content, generator output, detector size, detector positioning, background count and counting time, but in general, precisions of better than 1% could be obtained easily and relatively rapidly and were adequate to support the analytical work for which the method was developed; there appears to be no reason why substantially better precision could not be obtained for suitable systems if required.

4.2 Pulsed source method

4.2.1 Method used for examination of liquids and slurries

The pulsed source method was somewhat simpler than the activity ratio method in that only a single detector was required for successful application of the method, but in order to make the activity wave as sharp as possible to improve the resolution of the method a short neutron pulse was required. A generator providing 10^8 neutrons in a $15\mu s$ pulse was used to provide the short neutron pulse to irradiate samples and a block diagram of instrumentation is given in Figure 4. Control pulses from the neutron generator were used to synchronise operation of the multiscaler which accepted pulses from the γ-ray detector after amplification and sorting by a single-channel pulse-height analyser; the single channel analyser was pre-set to accept only those pulses which corresponded to signals from the required γ-radiation. A flux monitor was included in the instrumental system only if the quantity of induced activity was to be correlated with amount of sample irradiated, and was omitted when only the transit time of the activity pulse between generator and detector was to be measured. The method of operation consisted of pulsing the neutron generator

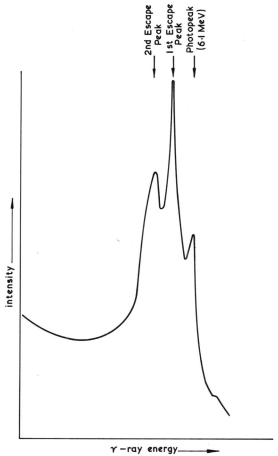

Fig. 2   Typical spectrum obtained from irradiated aqueous samples

and at the same time initiating operation of the multiscaler system with the generator control pulses. γ-counts from regions of the spectrum selected by the settings of the single channel analyser were fed sequentially into successive registers of the multiscaler so that the variation of count-rate of the sample with time could be followed.

4.2.2 Method used for the examination of solids

The equipment used for the examination of solids was similar to that employed for the examination of liquids and slurries and shown in block form in Figure 4, except that flow was not measured through a pipe but through different components of the solids handling system described in Section 3.3.2. Flow under controlled conditions was examined on a belt conveyor, where speed could be carefully controlled, but flow has also been examined through the screw conveyors and near to the walls of storage hoppers.

4.2.3 Measurement of flow in liquids and slurries based on the reaction $^{16}O(n,p)^{16}N$

The production of nitrogen-16 from oxygen in the sample stream offers a means of flow-rate determination for materials of suitable composition by the pulsed-source method just as it did for the activity ratio method described in Section 4.1.2. The variation of the intensity of high-energy

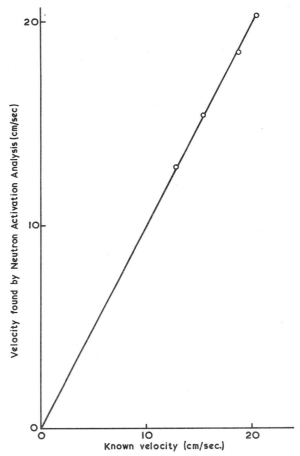

Fig. 3   Typical results for sample velocity obtained by neutron activation analysis for flowing aqueous streams

γ-radiation with time after the neutron pulse was measured and the time of arrival of the activity maximum at the detector was identified. A typical calibration curve is given in Figure 5. Again precisions of better than 1% were obtainable.

4.2.4 Measurement of flow in sand using the reaction $^{28}Si(n,p)^{28}Al$

Neutron Activation Analysis has been used to determine flow rates of sand through the solid handling system described in Section 3.3.2. Reactions were available for production of radioactivity from both oxygen and silicon in the sample, namely $^{16}O(n,p)^{16}N$ and $^{28}Si(n,p)^{28}Al$. Since flow rates were relatively low, measurement of 2.3 minute aluminium-28 was preferred to 7.3 second nitrogen-13 since this permitted greater source-detector separation. The method was applied under controlled conditions to material moving on a belt conveyor since both the quantity of sand fed to the belt and belt speed could be accurately controlled. An example of the multiscaler data accumulated from experiments on the belt conveyor is given in Figure 6; the activity peak can be clearly identified to give a measure of transit time between source and detector. A calibration curve obtained from conveyor experiments showing the correlation obtained between actual belt speed and that found by neutron activation analysis is given in Figure 7.

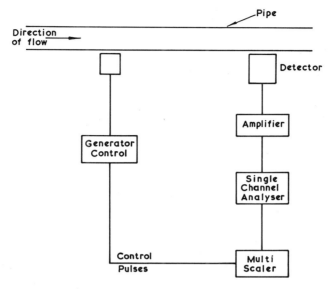

Fig. 4   Block diagram of experimental system used for pulsed-source method for examination of liquids and slurries

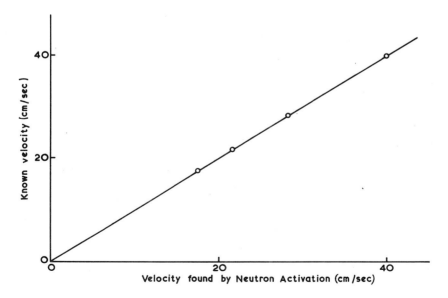

Fig. 5   Calibration curve for aqueous flow measured by the pulsed source method

5   SCOPE OF THE METHOD

Equipment for the measurement of flow-rate by neutron activation analysis as used for the investigations reported in this paper was primarily designed for other work and **was** expensive. However, equipment specially designed for flow-rate measurement, although very substantially cheaper, would still be an expensive method of measuring flow and the cost must be assessed against the special advantages of the method. The technique can be used for measuring and checking elemental flow through various components of plant handling liquids, solids and slurries; the lack of any need for special access ports enables flow to be examined at the specific point of interest without involving any modification of the containment system or causing any interference with sample flow.

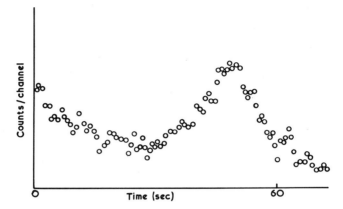

Fig.   6   Multi-scaler trace of activity from pulsed neutron irradiation of sand

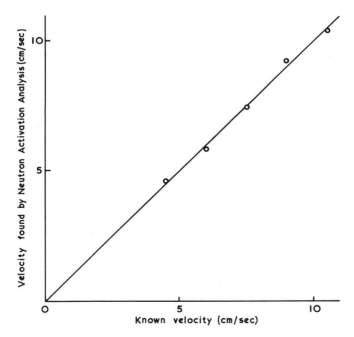

Fig.   7   Results obtained for the measurement of velocity of sand on a belt conveyor

TESTING AND COMPARISON OF THREE METHODS OF MEASURING FLOW THROUGH PIPES: RADIOACTIVITY CLOUD, ULTRASONIC, THERMODYNAMIC

L. Roche, H. Andre, S. Gamby and B. Noiret

Abstract: Three methods of measuring the flow rate through pipes were tested in the same hydro-electric plant and the results compared. Radically differing techniques form the basis of these three methods: radioactivity, ultrasonics and thermodynamics. The thermodynamic method measures the efficiency of hydraulic machines but the discharge may be calculated if the power of the machine is known.

The first comparisons show close agreement between the results obtained by using a radio-active cloud and by thermodynamic calculations. The ultrasonic method required calibration with some other method and it would be premature to draw conclusions at this stage, but the repeatability and linearity as a function of discharge appear to be within 1% of the discharge value.

1   INTRODUCTION

The first two of the three methods considered measure the speed of water and thus the discharge if the cross section of the pipe is known. The thermodynamic method is used to calculate hydraulic machine efficiency directly and the discharge may then be calculated if the power consumption or power output of the machine is known.

2   MEASUREMENT OF DISCHARGE USING A RADIOACTIVE TRACER*

The principle is that of the Allen salt velocity method. A tracer is injected into the flow and its passage through two sections circumscribing a measured volume is recorded as a function of time. The discharge is equal to the ratio of the volume to the transit-time. It is essential that the tracer concentration be homogenous through the first measurement section.

The special feature of radioactive tracers is the facility of using scintillation probes which can be fitted to the exterior of the pipe as this presents a great advantage in practical operation.

The tracer most generally used by Electricité de France is barium-137 which has a half life of 150 s. This radionuclide is obtained from a generator-injector (apparatus designed by Electricité de France, Direction des Etudes et Recherches, CHATOU) which contains a caesium-137 source with a 30 year half-life. The caesium is held in a porous resin and when the resin is washed by a liquid of suitable composition, Ba-137 atoms, generated from the caesium-137, may be separated from the caesium and then driven off. The barium solution so extracted is then injected practically instantaneously into the pipe.

For special requirements indium-113 with a 104 min half-life may be used as the tracer. This is obtained from the same type of generator charged with a tin-113 source which has a 118 day half-life.

The radioactive tracer method may be adopted in all cases where the outer wall of a pipe is accessible. It is especially suitable for measuring the discharge through multiple pipes connected to a single header.

* This method is fully described in a paper of M. SILBERSTEIN, reproduced in LA HOUILLE BLANCHE, no. 5, 1969

3 MEASUREMENT OF DISCHARGE BY ULTRASONICS

The physical phenomenon which forms the basis of this method is the difference between the propagation velocity of hypersonic waves through a fluid, depending upon whether the fluid is at rest or whether it is in motion.

The measurement process adopted* during the experiment required the use of four barium titanate transducers A, B, C and D placed in a plane formed by two generatrices of the pipe diametrically opposed, so forming a square ABCD in which the angle between diagonals AC and BD is at 45° to the flow line. A steep front wave is generated from transmitter A to receiver C following the direction of flow; a further wave is generated from B to D against the direction of flow. When a wave reaches a receiver it triggers off a further wave from its associated transmitter. Thus the difference between the two repetition frequencies is developed and this difference is proportional to the speed of the fluid. The lengths AC and BD being equal, this process has the advantage of ignoring the speed of sound factor in the differential frequency — fluid speed relationship, thus eliminating the influence of temperature.

We have not tried to use this device as an absolute meter. We first tested its reliability and repeatability under site conditions, calibration being carried out by another method.

4 INDIRECT MEASUREMENT OF DISCHARGE BY A THERMODYNAMIC METHOD

For fifteen years, the DIVISION TECHNIQUE GENERALE of Electricité de France has perfected and developed a thermodynamic method of measuring the efficiency of hydraulic machines. The difficulty of measuring water temperature to a greater accuracy than 10^{-3} °C limits the application of this method to heads exceeding 100 metres.

Practically, the method consists of determining two quantities whose ratio is equal to the machine efficiency; the available mass energy between the inlet and outlet "e_n" and the mass energy of water transformed into mechanical energy "e_u". If the mechanical power on the shaft of machine P is measured, then the flow may be calculated as the ratio of P to e_u (in the general case, where the machine is coupled to an alternator or to an electric motor, P is deduced by measuring electric power).

5 COMPARISON OF THE THREE METHODS

The three devices described above were put into operation in 1970 and 1971 on the SASSENAGE hydroelectric plant which has a 600 metre head; the turbine discharge is 2 m^3/s and the penstock diameter is 0.90 m.

A few preliminary, incomplete, conclusions may be drawn from these tests concerning the accuracy, advantages and disadvantages of each of the three methods.

There is close agreement between the results obtained by the radioactive cloud and by the thermodynamic methods, the mean difference being less than 1%. The procedure of measurement has to be improved in order to reduce the scatter which has been ±1.4% and some tests will be done to study the influence of distance between injection and the first probe.

The ultrasonic device must be calibrated by another method so that its repeatability and linearity as a function of discharge can only be characterized. Apart from some operating anomalies, which have not yet been satisfactorily explained, it seems that these two characteristics of the device are within ±1%.

In the present state of development it may be noted that each meets differing requirements:

The ultrasonic device, in its present design, is permanently attached to the pipe. The pipe must be drilled and the transducers fitted and positioned accurately, and we believe that the device has to be calibrated. Under these conditions, it gives a permanent reading of discharge in the pipe in the form of an electric current, with a good accuracy.

* Apparatus made by the firm VOITH, Germany

A similar device has been developed in Japan and this does not require drilling the pipe; consequently such a unit may be considered as an item of portable test equipment.

The radioactive cloud method may be used in any plant with the same equipment, operation requiring only drilling of the pipe for injection. Determination of discharge requires simple processing of a recorder reading. A direct reading of discharge may be considered, by using a chronometer triggered when a threshold is reached at the scintillation detector output, but the same degree of accuracy could not be expected from such measurements.

The thermodynamic method, apart from the fact that it may only be used if the flow passes through a hydraulic machine and if the head is sufficient, has a complex operation, especially with regard to the calculations which require the measurement of numerous parameters. However, when requirements justify its use, and when the necessary conditions are satisfactory, this method allows flow rates to be calculated extremely accurately.

DISCUSSION

J.E. Carrington: To what extent did the authors attempt to predict the performance of the ultrasonic flowmeter, and how did the performance compare with this prediction?

Authors' reply: This ultrasonic flowmeter has been used by the authors only on the SASSENAGE penstock mentioned in the paper.

No attempt was made to obtain an absolute value of the discharge. The device was calibrated by another method (the thermodynamic method) and our interests were restricted to observations of reliability, repeatability and linearity.

It is perhaps too early to judge reliability. For some months it has been possible to observe a good repeatability within ±1%, and a test of comparison with the thermodynamic method from 30% to 100% of the full discharge has shown a very good linearity of the output over the full flow range: differences between the two methods are less than 0.8% over the whole range.

This gives us confidence to use this device as a permanent meter for measuring the flow through pipes.

K.E. White: For regular measurements, over a long period at one site, the ^{137}Cs/Ba generator must be very convenient. How many resin washes are possible before some Cs-137 starts to leak out and have you tried steady state injection of Ba-137m by continuous washing? What resin and washing liquid do you use?

Authors' reply: In this generator, Cs-137 is chemically fixed on a resin containing "ferrocyanate de cuivre et potassium" and maintained between two screens of fritted glass or metal. To prevent Cs-137 leaving the generator, another column of resin, chemically identical, but without Cs, is placed just under the first one. Checks are made periodically on the solution drawn off. It is not yet possible to say how long work can be carried out with the same resin, but after several hundreds of injections, the concentration of Cs in the solution drawn off is much lower than the minimum admissible concentration in drinking water.

We sometimes use continuous washing when making flow measurements on cooling water circuits connected to the same header tank: it is thus possible to get injections with the same activity every 10 minutes instead of every 30 minutes. The generator is not designed for steady state injection of Ba-137m.

The washing liquid is a solution of calcium nitrate in water at a concentration of 4%.

5·6

EXPERIENCE IN THE USE OF THE RADIOISOTOPE CONSTANT-RATE INJECTION METHOD IN TESTING THE PERFORMANCE OF HYDRAULIC MACHINES

C.G. Clayton and G.V. Evans

Abstract: Operational experience of using the radioisotope constant-rate-injection method of flow measurement for testing pumps and turbines at sixteen different installations is described. Flow rates of up to 35 m³/sec have been measured. Particular emphasis is given to operational procedure, to the equipment used, the methods of data collection and to data handling techniques. It is shown that the consequences of using radioactive material, either during construction or during normal site operation do not impose important restriction to the successful application of the method. The influence of the hydraulic system on the accuracy of flow measurement is considered and examples of various types of installation are used to demonstrate how procedures can be adopted to remove potential areas of uncertainty and so ensure that high accuracy can be achieved. The results presented in this paper show that, even at the highest flow rates measured, and within the normal operating requirements, accuracies of approximately ±0.5% (95% confidence level) can be expected.

1   INTRODUCTION

The accurate measurement of large flow rates of water, such as exist through pumps and through turbines in hydroelectric power stations, is important both during acceptance tests and during performance tests to establish the operating efficiency of a power station. However, such measurements are difficult, mainly because of the large flow rates involved, the inaccessibility of the system or the need to conserve hydraulic power, and a method which satisfies all the requirements has not yet been developed.

In practice, because of these difficulties, the acceptance of large hydraulic machines is often based on model tests, even though it is well known that significant differences can occur between the performance of models and the on-site performance of large machines.

In general, many of the methods which have been available in the past to measure large flow rates to a high accuracy have proved to be unattractive, for one reason or another. Apart from operational considerations, this is mainly because the measurements on which these methods depend cannot be traced back to a primary standard, and an objective assessment of errors is not possible. Quoted accuracies are suspect and incomplete, and no differentiation is made between systematic errors and random errors which affect the precision of the method.

The radioisotope constant-rate-injection method avoids many of these disadvantages. All measurements are made outside the flow system and are independent of the flow rate. Errors can be derived according to accepted statistical techniques: to every measurement which is quoted an associated error can be given to some defined confidence level (generally 95%). The final quoted error in the measurement of flow rate is then compiled from errors in the individual measurements. An important requirement in this method is that the tracer which is introduced into the hydraulic system must be mixed with the water in the system, at some region downstream from injection. The distance required to achieve this condition is known as the mixing distance. The significance of these requirements and of other factors affecting mixing are examined in this paper and it is shown that, in general, they do not impose a serious constraint on the practical application of the method.

The main operational characteristics of the constant-rate-injection method are that only small injection volumes are required and the method can be applied with virtually no loss in pressure. Injection and sampling probes can easily be designed into existing or new conduits at very small cost. Measurements canthen be made at any time without interrupting the flow. In general, flow measurement over a range of 10^8 can be made using the same equipment.

The principal disadvantages of this method arise from the discontinuous nature of the measurement and the relatively short half-life of the tracer which it may be necessary to use. This prohibits casual measurement unless a store of activity is available. However, neither of these objections are important unless the site is particularly remote.

Radiological protection is not a problem: the concentration of the tracer in these hydraulic systems is many times less than the allowed concentration in drinking water and techniques have been developed which do not involve handling the radioactive material at injection.

The radioisotope constant-rate-injection method has now been used a significant number of times to measure flow rates during on-site acceptance and performance tests of hydraulic machines, and this paper describes some of the operational requirements and the results which have been achieved. An analysis of the errors of 283 flow measurements indicates that the method can be used with confidence to measure large flow rates with high accuracy and it is suggested that direct on-site measurements of the flow of water through large machines can now be undertaken with confidence.

2   PRINCIPLE OF THE METHOD

Measurement of flow by the constant-rate-injection method[1] is based on a comparison between the concentration C_1 of a tracer, continuously introduced at a known rate q, with the concentration of samples removed from some position beyond the mixing distance.

The rate at which activity is injected is equal to the rate at which it passes the sampling point.

Thus,

$$qC_1 + QC_0 = (Q + q)C_2 \qquad \qquad ... (1)$$

where C_0 is the initial concentration in the stream which is flowing at the rate Q.

Hence,

$$Q = q(C_1 - C_2) / (C_2 - C_0) \qquad \qquad ... (2)$$

Generally, $C_1 \gg C_2 \gg C_0$ so that equation (2) is reduced to:

$$Q = q(C_1/C_2) \qquad \qquad ... (3)$$

The flow rate can thus be determined absolutely by comparing the concentration of the injected solution with the concentration of samples removed from the conduit. Provided the flow is turbulent, the method is completely independent of the velocity of the fluid, the dimensions of the conduit, or variations in these dimensions should they occur. Since only the concentration of the tracer is required at the sampling point, samples can be taken from a branch or standpipe which is connected to the main pipe at a point at which lateral mixing is complete. This is useful if the main pipe itself is inaccessible. The velocity in the branch pipe is immaterial and the flow may be laminar.

The injected solution is introduced for a period which is sufficiently long to establish a region of uniform concentration at the sampling point and to enable the required number of samples to be withdrawn. The samples from the pipe are collected in any suitable container and the concentration C_2 is compared with a known dilution of samples of the injected solution (C_1) in the same high-grade counting system. The concentration of the diluted sample of the injected solution, used to obtain C_1, is made about equal to the concentration C_2 (to equate counting rates, approximately).

3   INFLUENCE OF THE HYDRAULIC SYSTEM ON THE ACCURACY OF FLOW MEASUREMENT

3.1 The measuring section

When a tracer is used to measure the flow of fluid in a conduit, there must be sufficient distance between the region in which the tracer is injected and the region from which samples are removed for the total variation in concentration of the tracer over the cross-section to be less than some predetermined value. For the highest accuracy in flow measurement, this variation in concentration must be small and this implies a long mixing distance.

3.1.1 Multi-orifice injectors

In practice, however, the mixing distance due solely to turbulent dispersion is strongly influenced by the method of injection. As can be seen from Fig. 1, the worst case occurs when a single injector on the pipe wall is used but some improvement can be expected by using an injector on the main axis of the conduit. Further improvement can be achieved by using a multi-orifice injector and, for pipes of circular cross-section, the mixing distance is reduced if most of the orifices are equally spaced and at a radius of 0.63 of the pipe radius.

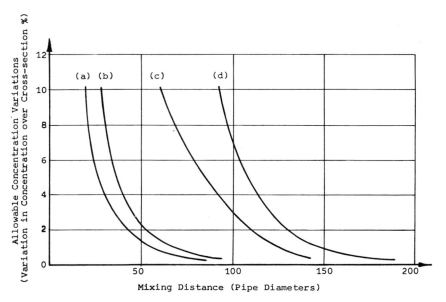

Fig. 1 The effect of allowable concentration variations on mixing distance for four injection systems at R_e = 7.7 x 10^4

    a   4 orifices at 0.63 radius
    b   4 orifices at the pipe wall
    c   Single orifice on the pipe axis
    d   Single orifice at the pipe wall

3.1.2 High-velocity jet injectors

If the tracer is injected as a jet against the flow, with a velocity which well exceeds the mean velocity of the fluid in the conduit, impact mixing is induced at the termination of the jet. The reduction in mixing distance depends on the number of jets used, their inclination to the direction of flow and to the momentum in each jet. Accurate information on the effect of these parameters on the mixing efficiency of high velocity jets is only just being obtained, but a reduction in mixing distance to significantly less than 30% of the value for a single central injector appears possible with a simple configuration of jets.

3.1.3 Obstructions in the flow system

Mechanical obstructions in the conduit, such as bends and valves, introduce additional turbulence and tend to reduce the mixing distance. Quantitative information on this type of mixing promoter is not available but, in general, measuring sections which include these devices are preferred.

3.1.4 Effect of condensers

Although relatively little mixing can occur in the tubes of condensers, it has been found that the high degree of turbulence in the flow at the "end-boxes" of the condenser contributes appreciably to the mixing of tracer in the flow. Large, modern condensers generally contain several flow-paths and associated end-boxes and whenever possible, are included in the measuring section for measurements of flow through cooling-water systems.

3.1.5 Effect of pumps and turbines

If the tracer is injected upstream of a pump, a considerable reduction in mixing distance can be achieved and information on mixed-flow pumps[2] indicates that this type of pump has an "effective mixing distance" of about 100 diameters.

No systematic information on the mixing efficiency of turbines is yet available, but from recent measurements of flow through pump-turbines[3], mixing in the turbine direction was seen to be less than when the same machine was operated as a pump. This observation is consistent with general hydrodynamical considerations. In practical applications therefore, the tracer is injected upstream of a hydraulic machine whenever possible.

3.1.6 Multi-orifice sampling

An apparent reduction in mixing distance can be obtained if samples are recovered simultaneously from a number of positions across the conduit and mixed prior to measurement. For example, at $Re = 10^5$, six sampling points having the same discharge rate and equally spaced across the conduit at 50 diameters downstream from injection are equivalent to a single sampling point at the wall at about 100 diameters from injection.

3.2 Effect of dead-spaces in the flow system

Large volumes (such as mixing chambers in multi-pump installations) in the flow system which do not achieve the same steady concentration of tracer as the main conduit constitute a potential source of error but close examination of the problem can indicate to what extent this is important.

Generally, the flow measurement results themselves serve as a critical diagnostic test of the effect of dead-volumes in the flow system. An example from measurements on a pumped storage scheme[3] will illustrate how this problem can be accommodated.

In this system, which is shown diagrammatically in Fig. 2, water flowed through one arm only of the bifurcation during the tests; the other arm being closed by a valve immediately upstream of the machine. The water in the closed arm thus constituted a dead-volume.

The worst situation would have arisen if the total volume of water in the dead zone of the closed arm had exchanged with the main flow during the measurement period. The size of the gross effect can be obtained by comparing the total dead-volume with the volume of water which passes the bifurcation during the injection period. For an injection period of 20 min. and a dead-volume of 800 m^3 this corresponds to a total change of 3% in the concentration of the mixed fluid at the lowest flow rate of 2.5×10^4 kg/sec in these tests. This change would easily be seen as a systematic change in the count rates associated with the 20 samples extracted during each sampling period.

As the exchange-rate is likely to be exponential, the greatest change in count rate would arise in the first few samples and these can be compared critically to see if the effect of the dead-space introduces a change in the count rate from the samples outside the variations expected from statistical considerations alone.

The influence of the bifurcation was also examined experimentally by model tests using dyes to observe the degree of exchange of fluid between the two arms of the bifurcation. It was found that, over a wide range of flow conditions, the volume of fluid which exchanged was limited to less than one quarter of the dead-space and the total exchange occurred over a short period following the arrival of the tracer.

From both considerations, it was concluded that the bifurcation would only affect the tracer at the sampling position for a short period (one or two minutes) and that this would be apparent from a careful numerical examination of the count rates of the samples.

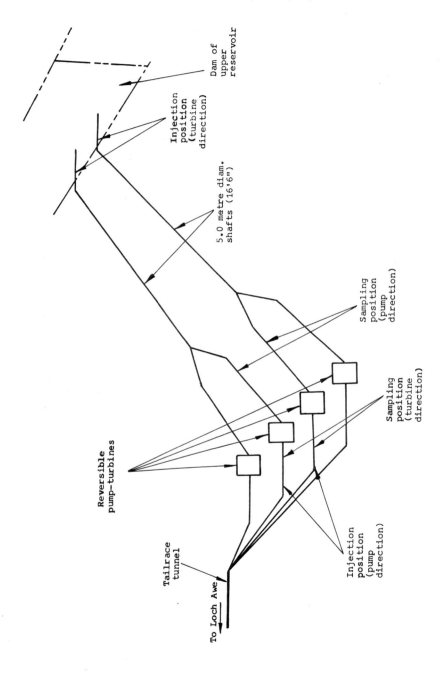

Fig. 2 Diagram of the hydraulic system at Cruachan

3.3 Recirculation

In power stations equipped with cooling-towers, recirculation of the cooling-water during a flow measurement series can result in a build-up in the concentration of the tracer in the injection section. Although an increase in background concentration may limit the number of measurements which can be made, two factors generally limit the importance of this effect.

The volume of water in the cooling-tower pond is generally a large fraction of the total volume in the cooling-system and, as a cooling-tower and pond constitute a very good mixing system the background count rate caused by the injection of tracer increases at a low rate depending on the total activity injected, the flow rate and the volume of the cooling-water system. The rate of build-up of radioactivity is also restricted by the choice of radio-tracers with short half-lives (^{24}Na, $T_{1/2}$ = 15 hr: ^{82}Br, $T_{1/2}$ = 36 hr). In practice, in a 2,000 MW station when using ^{24}Na, an interval of 2 days in the middle of a series of 24 measurements was found adequate.

Re-circulation also arises in pump-storage schemes but this problem can be avoided by ensuring that measurements in the turbine phase precede measurements in the pump phase, and either by introducing an interval between the two series of measurements or by arranging that water which has been discharged into the lower reservoir during the turbine tests is moved from the entry to the tail-race tunnel.

If re-circulation is a serious problem then it is possible to use a tracer with a very short half-life from a parent/daughter system, but as yet there is no reported experience on the use of this type of arrangement for accurate flow measurement using the constant-rate-injection method.

3.4 Compatibility of the tracer

The tracer should be soluble and stable when introduced into the hydraulic system. By using ^{24}Na and ^{82}Br as tracers in the form of ^{24}Na HCO_3 and K ^{82}Br, the solubility is adequate for all the preliminary operations and for use in the flow system.

The problem of stability is often regarded as stringent with radioisotopes owing to the high sensitivity of radioactive methods in terms of the weight of material which is used. Consideration has to be given therefore to the possible occurrence of wall absorption and precipitation and non-radioactive material may have to be introduced as a "carrier" into the flow simultaneously with the radio tracer. However, in all the tests carried out so far in conduits lined with steel or concrete and hewn from natural rock there has been no evidence of adsorption of the tracer.

In some power stations cooling-water systems, chlorine is introduced as a fungicide and, if ^{82}Br is used as tracer, free bromine may be liberated giving erratic flow measurement results. However, this problem can be avoided by arranging that chlorine is not introduced during flow measurement tests.

3.5 Branches in the measuring section

Provided the tracer is fully mixed, water can be removed from the measuring section without affecting the tracer concentration and hence without affecting the accuracy of the measurement. However, if water is removed before mixing is complete, the tracer concentration can be affected in an unknown way and a serious error in the flow measurement could result.

If water is introduced into the measuring section at a point which is sufficiently far from the sampling region for it to be mixed, then the derived value will refer to the gross flow rate. If mixing is not complete, the variations in concentration of the tracer at the sampling region will be reflected in significant errors in the recorded flow rate.

4   ORIGIN OF ERRORS

The overall error in the measurement of flow rate is made up of random and systematic errors which may arise during injection, sampling, diluting the injected solution (to compare concentrations), measuring the relative radioactivity of the various solutions (counting) and from instability and incomplete mixing of the tracer in the fluid.

It is a unique characteristic of the constant-rate-injection method that every measurement which is required to derive the flow rate can be checked, or repeated several times so that its associated error to a predetermined confidence limit (generally 95%) can be obtained. The methods for carrying out this procedure in the various stages of the analysis of the data are described in detail elsewhere[1].

In all cases where systematic errors might occur, repeated calibrations are carried out so that the probability of an error occurring and passing unnoticed is believed to be extremely small. For example, three independent counting systems are used and each is calibrated using the same standard source.

One of the counting systems is used principally to measure dilutions of the injected solution although one of these samples is also measured on each of the other two systems. These counting systems are used to assay the samples of water from the conduit, each series of approximately 20 samples, corresponding to one flow measurement, being counted on each system. This procedure provides a check against systematic errors occurring in the counting systems by providing two independent estimates of the same flow rate. The final quoted flow rate is the mean of these two measurements, the associated error being compounded statistically from the errors of each of the two measurements.

The possibility of a systematic error occurring at injection is examined by calibrating the injection pump immediately before and after each series of measurements. The frequency of the electrical supply to the injection pump is measured using a crystal-controlled oscillator so that the mean injection rate is accurately known.

As mentioned above, the possibility of adsorption of the tracer on the walls of the conduit is examined critically and if necessary sufficient quantity of a soluble non-radioactive compound of the tracer is introduced into the water. But so far this has not proved to be necessary in our experience. Instability of the tracer has only been observed when free chlorine was introduced into the water as a fungicide and ^{82}Br was used as tracer. This was rectified simply by discontinuing the release of chlorine during the flow measurement period.

If mixing is incomplete, because the measurement section is too short, the resultant variations in concentration of the tracer over the cross-section of the conduit at the measuring region appear as an increase in the overall error of the measurement. The probability of systematic variations in concentration occurring over the cross-section which are stable in time is thought to be small if the tracer is injected upstream of the machine. If the tracer is injected downstream of the machine, the probability of a non-random concentration distribution occurring can be estimated from hydraulic considerations and from the mixing distance corresponding to the injection system and Reynolds number of the flow[4,5]. A multi-orifice sampling system can be used to examine symmetry in the tracer distribution.

5   FACILITIES REQUIRED ON SITE

In general, the requirements are few. It is preferable for the injection equipment and for the associated crystal-controlled oscillator, used to measure variations in injection rate, to be protected against bad weather, and a small hut is adequate. It is also necessary for the samples from the conduit to be collected in a covered area so that the possibility of dilution by raindrops, for example, is avoided. Each sample, which is about 2.5 litre, is immediately covered to prevent evaporation.

To indicate the arrival of the tracer and build-up to uniform concentration, a radiation detector is either mounted in the conduit or in a flow cell through which water from the conduit is passed. This detector is coupled to a count rate meter and paper-chart recorder: a time mark is automatically added to the chart as each sample is collected. Suitable accommodation for this equipment and the operator is also required.

When measuring large flow rates, the samples removed from the injected solution are diluted by a factor of about 10^7 so that a comparison with the samples collected from the conduit can be made in the same counting system at approximately the same count rate. Extreme care is required during the dilution operation and this procedure should be carried out in isolation in a small well-lit room. The dilution factors are determined by weighing using accurate balances and these must be mounted in the same room on vibration-free surfaces. A water supply is useful but not essential, but a drain is required; the only other service necessary is an electricity supply.

The three counting systems are best mounted together in one room and this should be large enough to accommodate one or two desks with hand calculators.

Apart from the provision of agreed tapping points for injection and sampling, all other equipment is transported to site.

6   ORGANISATION OF TESTS

The normal procedure during acceptance tests is for equipment and staff to arrive on site one day before the tests are due to start. The equipment is installed and calibrated and preliminary measurements are carried out during this period.

It has been found by experience that the most important single organisational factor contributing to success is effective oral communication between staff. Although this is an elementary point, it is one which can be overlooked and is therefore worth emphasizing. Telephone communication is arranged between staff at the injection and sampling points. Staff at the sampling point are in telephone communication with the Chief of Test. It is equally important to have good communication with other teams on site responsible for measuring the hydraulic head and the power delivered or supplied by the machine. Trials of the degree of co-ordination and of operating procedure for the flow, head and power measurements should be made before the first test and are vital to a successful operation.

Injection periods generally vary between about 10 and 20 minutes independent of flow rate and the time required to complete measurements on the series of samples from a single test, is usually about 2 hours. This limits the number of flow measurements which can be made in one day to about six. If the accuracy can be relaxed, then the number of flow measurements which can be completed in one day may be increased substantially.

It should be noted that the random error is derived from repeated measurements, so that the greater the number of measurements which can be made, the smaller the random error. In practice, a limit is imposed by the number of measurements which can be carried out in a test series, and in general, a compromise is made between the acceptable accuracy and the lowest rate at which tests can be carried out economically. The sacrifice in accuracy by increasing the number of tests can be seen from the fact that when the number was increased in one series from four to five tests in one day the error in the measurement of flow rate increased by less than 0.1%.

Preliminary values for the flow rates can be derived on the same day as the tests are carried out and results are generally within about 0.5% of the final value which can be made available a few days later.

The radioactive solution is either delivered by road in a container of the type shown in Fig. 3 or as tablets which are dissolved in an aspirator and transferred to an injector such as is shown diagrammatically in Fig. 4. Additional consignments of tracer are normally delivered every day or every other day depending on the isotope, the number of tests to be made in one day and the distance from site. From a base in the United Kingdom, operation is possible anywhere in Europe. The isotope would be flown to the nearest airport (in the form of tablets) and delivered to the site by road: the use of ^{82}Br allows this procedure. Operation outside Europe is more difficult but not impossible, even if the isotope is supplied from Harwell. Although not concerned directly with tests of a hydraulic machine, flow rate measurements have been carried out in Malaysia during measurements of the performance of a hard-rock tunnel[6].

As the operational efficiency has improved, the number of staff required to carry out this method has been reduced to four, but it is not expected that this number can be reduced further if the highest accuracy during full-scale acceptance tests of a large installation are required. Some assistance from site staff is necessary apart from providing the facilities required on-site: in particular to help with connections to the injection and sampling points and to arrange for the supply of sample collecting bottles at the sampling point and for their removal, when filled, to the measurement laboratory.

The adopted procedures must comply with any statutory regulations applying to the site of the measurements.

In the United Kingdom, these regulations require that the premises at which the measurements are to be carried out must be registered, one month before the tests, for the storage and use of radioactive material and for the disposal of radioactive waste. Areas used for the storage and injection of the radioisotope and for the dilution of the injected solution must be identified and access restricted to personnel classified for handling radioisotopes. In practice, because of the consequences on the accuracy of measurement, of slight cross-contamination between these areas and the measurement laboratory, access is restricted to only particular members of the flow measurement team.

The concentration of tracer in the pipe after injection of the trace is usually about 1-2% of the

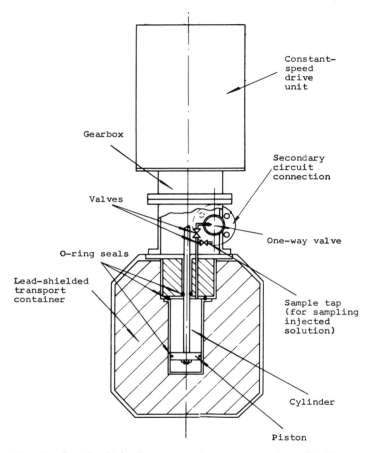

Fig. 3   Combined injection pump and transport container with drive unit attached

recommended maximum allowable concentration of tracer in drinking water, based on the continuous exposure to radiation for a period of fifty years, and consequently represents a negligible radiological hazard over the duration of the tests.

7   ANALYSIS OF RESULTS

Altogether flow measurements have been carried out at 16 separate installations. At several of these installations more than one machine was tested and at three, various combinations of several machines were examined.

The results presented in this paper refer to measurements made since 1965 when the present operating technique and analysis of errors was generally established as a result of experience previously gained. A total of 283 separate measurements of flow ranging from 660 kg/sec to

36,500 kg/sec have been made. Of these measurements, 251 were made under conditions when the mixing distance was regarded as adequate and this assumption was subsequently verified by inspection of count rates from individual samples. These showed that variations in count rate were attributable to counting statistics alone.

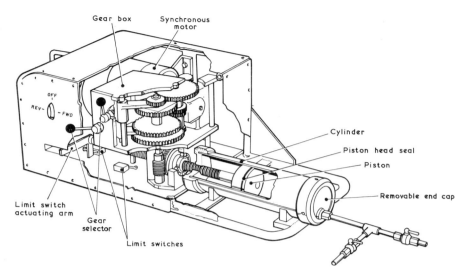

Fig. 4   Constant-rate-injection pump

The distribution of errors is shown in Fig. 5. The shaded portion of the histogram refers to measurements made with the constant-rate-injection method and it can be seen that the mean random error in 137 measurement is about ±0.6%. The errors exceeding 1% were obtained during tests at Eggborough Power Station and are attributable to an insufficient number of samples of the injected solution being available (2 only) so that the application of small sampling statistics resulted in a large error.

The unshaded portion of the histogram refers to a series of measurements at Tilbury Docks using the calibrated-scaler method[1], which is a variation of the constant-rate-injection method suitable for making a large number of flow measurements over a short period of time.

Of the total measurements made, 32 were carried out at Calder 'A' and 'B' stations when the flow system consisted only of a "3-pass" condenser, the tracer being introduced through a multi-orifice injector into the inlet manifold of the condenser and samples being obtained from the outlet. With this system, incomplete mixing and significant variation in the count rates from the samples were expected. This in fact occurred and the distribution of errors can be seen in Fig. 6. An analysis of the errors, given in Table 1, indicates that a very large contribution to the total error is made up from uncertainty resulting from variations in the sample concentration.

From theoretical considerations, only a very weak relationship between the random error and the flow rate is expected, a higher accuracy being associated with the lower flow rates. This effect is in opposition to the variation of error with flow rate in nearly all other methods of flow measurement in which the error tends to increase with decrease in flow rate. Results obtained for the constant-rate-injection method do not contradict this prediction as can be seen in Fig. 7 in which no systematic variation in the magnitude of deserved random errors with flow rate can be obtained.

At one station, comparative tests were carried out with current-meters and the results are given in Table 2. It is evident that fair agreement is obtained at the higher flow rates, but the accuracy of the current-meter installation falls rapidly at the lower flow rates, as expected.

8   CONCLUSIONS

(1)   The constant rate injection method is now established as a convenient and accurate

method of measuring large flow rates of water during tests of hydraulic machines.

(2)   Results obtained during the past five years are highly consistent and indicate that when adequate mixing can be achieved, the most probable error is about 0.6% (95% confidence level) and errors above 1.0% are unlikely.

(3)   One of the main advantages of this method is that an objective assessment of errors can be carried out. In this respect the method is unique for the measurement of large flow rates.

(4)   The consistently high accuracy in flow measurement which can be achieved suggests that increased emphasis can now be placed with confidence on in-situ measurements as a method of accepting and measuring the performance of hydraulic machines.

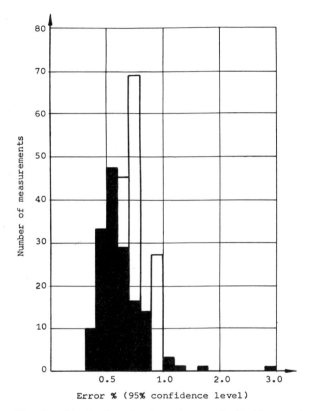

Fig. 5   Distribution of estimated errors obtained in tests since 1965

Total number of flow measurements: 251 (excluding Calder A and B)

▬ constant-rate-injection method

▢ constant-rate-injection calibrated scaler method

9   ACKNOWLEDGEMENTS

The work reported here results from the combined efforts of a number of people and, in particular, M.A.J. Aston, I.S. Boyce, W.E. Clark and R. Spackman. The authors acknowledge with gratitude that without their efforts the material for this paper would not have been available.

10   REFERENCES

1   CLAYTON, C.G., and EVANS, G.V.: 'The constant-rate-injection and velocity methods

for flow measurement for testing hydraulic machines'. Harwell Report No. AERE-R 5872, 1968

2   CLAYTON, C.G.: 'The use of a pump to reduce mixing length in the dilution method of flow measurement'. Harwell Report AERE-R 4623, 1964

3   ASTON, M.A.J., BOYCE, I.S., CLARK, W.E., CLAYTON, C.G., EVANS, G.V., and SPACKMAN, R.: 'Measurement of flow through the reversible machines at Cruachan Pumped-Storage Scheme by using the radioisotope constant-rate-injection method'. Harwell Report No. AERE-R 5816

4   EVANS, G.V.: 'A study of diffusion in turbulent pipe flow', Trans. ASME Jnl. Basic Eng., **89,** Series D No.3, 1967

5   CLAYTON, C.G., BALL, A.M., and SPACKMAN, R.: 'Dispersion and mixing during turbulent flow of water in a circular pipe'. Harwell Report No. AERE-R 5569, 1968

6   WRIGHT, D.E.: 'Full-scale tests to determine the hydraulic performance of a lined-invert rock tunnel'. CIRIA TN 16, March 1971

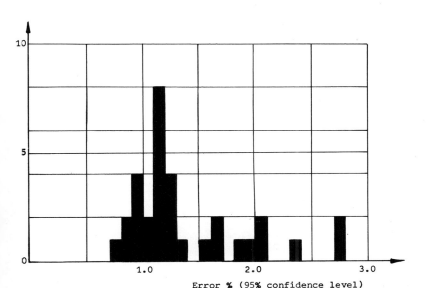

Error % (95% confidence level)

Fig. 6   Distribution of estimated errors obtained using condensers as mixing devices

Flow measurements at Calder A and B (total number: 32)

DISCUSSION

C.A.E. Clay: I found this paper a very interesting summary of the results of a considerable number of flow measurements carried out using the isotope dilution technique over the past few years. At one time the A.E.R.E. were not prepared to offer this method as a commercial service, so the C.E.G.B. established a small team of engineers to use this method (along with others) for the measurement of large water flows. Considerable assistance was received from Mr. Clayton and his staff in setting up the necessary equipment and techniques. In fact, the combined injection pump shown in Figure 3 of the paper, was developed by the A.E.R.E. for the C.E.G.B. because of our concern over the safe handling of radioactive tablets at site. I would like to record my appreciation of the help received. Since then A.E.R.E. policy has changed and I understand that a commercial service is now available from them.

High accuracies are shown for this method but as no absolute method of flow measurement is normally available for these high flow rates it is difficult to fully substantiate the accuracies claimed. The paper by Fisher and Spink on the ultrasonic method also claims very high accuracies, and a comparison between these two methods would be of considerable interest.

TABLE 1 Comparison of total error obtained, with error due to sample concentration variation during tests at Calder A and B

Flow rate Kg/sec.	Total error (95% conf. level)	Error due to variation in sample concentration (95% conf. level)
kg/sec	%	%
660	0.89	0.79
670	1.22	1.15
680	1.16	1.08
680	1.13	0.99
690	1.09	1.01
690	1.11	0.96
710	2.74	2.68
710	0.97	0.85
720	2.73	2.67
730	1.62	1.50
730	1.25	1.16
730	1.17	1.07
740	1.02	0.82
750	1.17	1.07
760	1.32	1.17
770	1.18	1.01
860	1.00	0.91
870	1.29	1.22
880	0.97	0.87
880	0.72	0.58
890	0.86	0.75
890	1.21	1.13
890	1.59	1.53
900	1.20	1.15
910	2.10	2.06
910	1.12	1.06
910	1.64	1.60
920	0.96	0.89
940	2.02	1.42
940	1.99	1.38
940	2.35	1.85
950	1.84	1.14

Has an allowance been made in the quoted accuracy values for possible systematic effects of incomplete mixing?

The technique for checking the effect of dead spaces on the calculated flow rate by examination of the results from the first few samples would seem to depend for its success on the time for fully establishing the plateau being appreciably shorter than the time for dead space effects. Perhaps the authors would comment on this factor.

For the Calder Power Station results shown in Table 1 and Fig. 6, were multiple sampling devices used at the condenser discharge to reduce the effect of inadequate mixing?

Authors' reply: We should like to record our thanks to Mr. Clay for his most generous remarks.

It would be interesting to compare the constant rate injection and ultra sonic methods, but a difference in results would inevitably lead to the question of which one was right. A comparison of both methods with a calibrated weigh tank would be more valuable.

Systematic errors from incomplete mixing could not arise in the systems referred to: the use of long lengths of pipe, pumps, condensers valves and bends in the test sections eliminates this possibility. The magnitude of random errors due to incomplete mixing can be derived from

TABLE 2 Comparison of flow measurements by the constant-rate injection method and current
meters

Flowrate as measured by constant-rate-injection method	Flowrate as measured by current meters*	Difference
Kg/s	Kg/s	%
14410	14240	1.18
14040	14000	0.29
13510	13430	0.59
13310	13280	0.23
12050	11920	1.08
11760	11670	0.76
9210	8990	2.4
9030	8720	3.4

* An array of 37 calibrated current meters were used in an octagonal culvert of approximately
2.5 m maximum width.

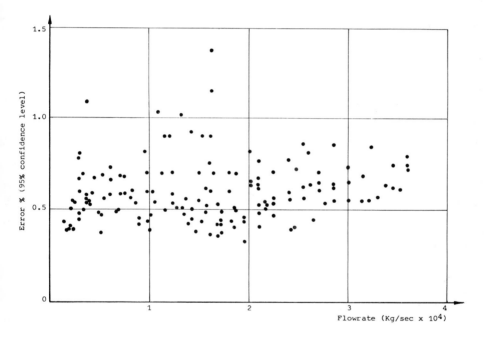

Fig. 7   Measurements of flowrate with associated error

variations in the contrates of the samples removed from the flow.

The dead space could only be ignored because its effect survived for a time which was short
compared with the time for which a constant concentration was maintained. This problem was
chosen to illustrate how it was dealt with and how the technique can be self-diagnostic for effects
of this kind.

Multiple sampling devices were used during the tests at Calder Power Station. Each sample was
collected and measured separately so that an analysis of errors due to incomplete mixing could be
made.

A.M. Crossley:

1   Does the identical theory apply to the use of chemical tracers such as lithium chloride

2   Is a peristallic pump considered to be a "constant-rate injector"?

3   Should the strength of the injected solution be as strong as possible or only sufficiently strong so as to enable the dilution to be determined downstream?

4   Is there any reason to suppose that a long injection period enables better mixing to take place compared with a short period?

5   In which sense are errors obtained? For example, a thermocouple cannot produce more millivolts than its known output, provided the cold lunction is correct, i.e. a thermocouple cannot read high, In which sense does the method tend to error, that is, do the measurements of flow rate tend to be high or low?

Authors' reply: The underlying theory of the method is the same as when using chemical tracers, but the analysis of errors is different. In this respect there is a significant advantage in using radio-isotope techniques because of the digital nature of the result.

A peristalltic pump could be a constant rate injection device if the mean delivery rate were constant. The periodicity in delivery is hardly likely to be a problem and generally would be removed by dispersion and by using an adequate sampling period. We ourselves have no experience of the ultimate precision of peristalltic pumps and their value in this method of flow measurement.

The activity of the injected solution is controlled by the activity after dilution in the fluid and this should be sufficient to give the acceptable statistical error in samples removed for assay.

Systematic errors during the operational stage could arise from leakage in the injection pump (minimised by calibration before, during and after tests), leakage into the conduit (the flow measured is then no longer referred to the injection region but to some intermediate position), leakage out of the conduit (only important if it occurs before mixing is complete) and adsorption of tracer to the walls of the conduit (flow measurements are only undertaken if the fluid contains an adequate concentration of carrier). Each of these possibilities tends to give an apparent increase in true flow rate. The possibility of systematic errors arising during sample measurement and manipulation is small but if they occur they lead to a high or low result. In the limit there could be a random distribution of systematic errors.

C.J. Causon: I am surprised at the statement that the required mixing length is more or less independent of Reynolds' number.

Mixing can only be produced by an input of energy creating turbulence. (We assume diffusion is of no account.) The necessary energy for mixing derives ultimately from friction effects. The magnitude of the available friction induced turbulence depends upon the pipe surface and upon the liquid velocity. Hence, it would seem that mixing lengths ought to depend on velocity i.e. on Reynolds' number and on the friction coefficient. It is worth pointing out that modern pipe coatings of spun enamel for example, give very low friction coefficients and there is evidence for relatively poor mixing as a consequence. Care must be exercised in translating tests on pipes with very smooth surfaces.

Authors' reply: The mixing distance has been shown to be only weakly dependent on Reynolds' number[5] but it is more strongly dependent on pipe friction factor[4]. Provided differences in Reynolds' number and friction factor are taken into account, there is no reason to believe that derived values for mixing distance are not independent of pipe diameter.

K.E. White: In the method, referred to and in BS 3680, it is necessary to take and weigh samples of radioactive solution. To reduce radiation dose and the risk of contamination we have developed a method where it is not necessary to find the density of the injected solution or to weigh a sample directly. The quantity qC_1 is determined by taking a sample of tracer solution via the peristaltic injection pump for a measured period of time t. The sample, v_1 is passed directly into a previously prepared dilution volume, V_1 of several litres and this provides direct self-shielding. Samples are taken in this way before and after the injection, and second dilutions can safely be prepared in the usual way. The quantity v_1 need not be known with great precision since the uncombined factor cancels in the equation: —

$$qC_1 = \frac{v_1}{t} \frac{(V_1+v_1)}{v_1} \frac{(V_2+v_2)}{v_2} R_1 = \frac{(V_1+v_1{}^*)}{t} \frac{(V_2+v_2)}{v_2} R_1$$

Where R_1 is the countrate given by the second dilution (v_2 into V_2). Concentrations can be represented by countrates only, there being no necessity to establish absolute radioactivity levels. Although v_1 is very small compared to V_1 it can be retained in the equation, $v_1{}^*$ to give a more precise value to the numerator — the value being a close estimate derived from non-active tests on the pumps. All volumes refer to liquid of effectively the same density and can be determined, as you have done, as weights of arbitrary volumes which is most convenient for field work.

5·7

FLOW VELOCITY AND MASS FLOW MEASUREMENT USING NATURAL TURBULENCE SIGNALS

M.S. Beck, J. Coulthard, P.J. Hewitt and D. Sykes

Abstract: Flow velocity can be derived directly from the transit time of natural turbulence signals between two positions, spaced along the direction of flow. Suitable transducers are used to measure the turbulence signals and the transit time is determined by cross correlating the transducer outputs with either a special purpose correlator developed at the University of Bradford or an on-line computer. Specific advantages of this method of flow measurement are that the time delay can be measured absolutely by the correlator, transducers can be used which do not obstruct the flow and gaseous suspensions and slurries can be metered.

1   VOLUMETRIC FLOW MEASUREMENT OF GASES AND GASEOUS SUSPENSIONS

1.1 Introduction

One old-established method of measuring flow is that of tagging by timing the passage of some measurable property down a pipe. The direct way of putting this principle into practice would be to inject some marker, such as a small, light body or a puff of smoke, which will be carried along by the flow and its progress timed over a measured distance. Unfortunately the apparent simplicity of this method is out-weighed by a number of shortcomings. Firstly, the rate of flow will usually vary from point to point across a pipe section; secondly, there will usually be diffusion effects present; thirdly, the marker must be easily detected. In practice, therefore, use of such tagging methods is restricted to the measurement of low flows in the laboratory, some applications in which high accuracy is not required and also cases in which the flow can be made visible, such as, for example, in surveying ventilation currents in a building[1]. In general, the tagging method in its simple form cannot easily be applied to pipe flow measurements directly.

A more powerful alternative is the principle of using a naturally occurring variation in some fluid property, for example turbulence, as a continuous marker. Using the correlation techniques to process the signals obtained from suitable transducers it is possible to time the passage of the measured quantity with considerable accuracy. As an example of this approach, consider suspension flow measurements: in a pipe conveying a suspension of solids in a gas the solid distribution will take the form of "clouds" due to the turbulence of the conveying air. We have used capacitance transducers[2], see Fig. 1, to detect the clouds at an upstream point A and downstream point B, the transit time of the clouds between these two points can be measured by a cross correlator. Provided the particle size is small the clouds travel at essentially the same speed as the gas and hence: —

$$\text{fluid velocity} = \frac{\text{distance between A and B}}{\text{transit time of clouds from A to B}}$$

$$\text{volume flow of liquid} = \frac{\text{volume of pipe between A and B}}{\text{transit time of clouds from A to B}}$$

Typical wave forms at the transducer outputs are shown in Fig. 2, because these are random wave forms the time delay τ between them cannot be measured directly and a cross correlator must therefore be used. A block diagram of one such correlator is shown in Fig. 1, and a typical cross correlation function is shown in Fig. 3. The time delay of the fluid property is found by comparing a time-delayed version of the output from the upstream transducer A with the output from the downstream transducer. The time delay introduced, to obtain the best match between the two outputs, is a measure of the transit time of the fluid flow. The cross correlator performs this process by multiplying together the output n(t) of the downstream transducer with a time-delayed version of the output from the upstream transducer m(t - β) as shown in Fig. 1, where an adjustable time delay β is inserted in the output from the upstream transducer. The product m(t - β). n(t) is then integrated over a period of time T to give its mean value which is called the cross correlation function. When the cross correlation time delay β is different to the fluid transit time delay τ, the mean value of the product is small. However, when the fluid time delay and the cross correlation time delay are equal, the signals $m(t - \beta)$ and $n(t)$ will have their closest possible similarity and the mean value of their product will be large. Hence the time delay of the maximum value of the cross correlation function, shown in Fig. 3, uniquely defines the transit time of the fluid from position A to position B. A cross correlator specifically designed for flow measurement work outputs the value of this time delay to a simple digital or analogue circuit which calculates the flow of fluid using the equation given earlier.

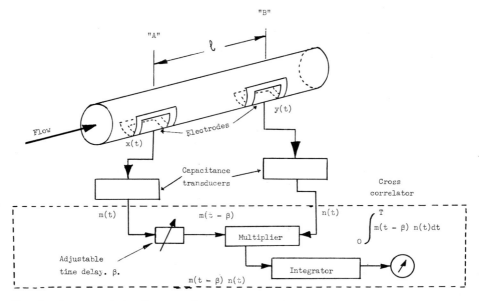

Fig. 1   Cross correlation flowmeter for gaseous suspensions

1.2 Practical applications

In many industrial situations reliable measurement of gas flow is often not possible because of solid matter suspended in the fluid. The proportion of solid material present in a gas flow system depends on the particular case considered but can vary over a quite wide range; examples include: the exit gases from steel refining processes and cement kilns, through to the extreme case in which the solid is a process material being transported by means of a pneumatic conveyor.

Fortunately, by making use of the cross correlation approach, it is possible to turn the presence of solid material in a flow to good advantage. Experience has shown that the flow rate in any fluid system can be measured by a cross correlation flowmeter, provided that a measurable variation is present in the flow stream, or that some suitable variation can be introduced into the flow.

In a number of applications the variation in solids concentration has been measured by using two electrodes installed in the wall of the pipe as shown in Fig. 1, these being connected to simple and robust capacitance transducers[3]. Examples of applications making use of the capacitance method include pneumatic conveying systems transporting materials ranging from finely powdered flour, coal and cement to granules such as wheat and barley. The capacitance type of transducer is particularly suitable for use with systems having high solid/air loading ratios, such as those normally encountered in industrial conveyors.

The capacitance method has been used for solids loadings between 0.005 and 1.5[2]. For lower solids loadings and with clean gases, an ultrasonic cross correlation method has been developed[4,5] and some results using this method will be given in this paper.

The ultrasonic transducers measure the transit time of the density variations that result from the turbulence of the fluid. These density variations can be detected by measuring the associated variations in the transmission of acoustic energy across the diameter of the pipe. Preliminary results of this method, obtained from tests on clean air flow, have been satisfactory and a major investigation into its use is currently in progress.

1.3 Accuracy considerations

A feature of the ultrasonic cross correlation method which interrogates the flow over a whole pipe diameter is that it appears to give an absolute measurement of flow velocity. In measuring volumetric flow, therefore, calibration of a cross correlation flowmeter is dependent primarily on the

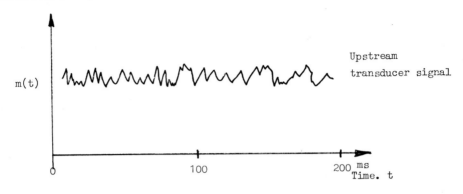

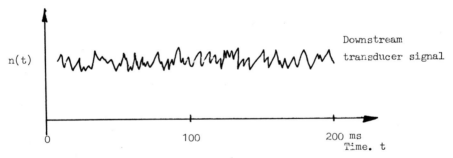

Fig. 2   Transducer output signals

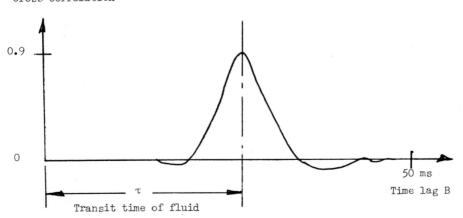

Fig. 3   Typical cross correlation

volume of the pipe. Calibration of the transducers is not required, since only the time delay between them is used for flow measurement and this delay can be measured absolutely by the cross correlator.

In practice, the accuracy of the cross correlation method depends on:

(a)   the relationship between the mean velocity of flow and the velocity of the measured disturbance

(b)   the statistical accuracy associated with the signal processing.

The relationship between the mean flow velocity and the velocity of the measured disturbance is due to a combination of velocity profile, diffusion and other effects. The relative importance of these various effects will depend upon the type of transducer used and the details of their installation. For example, tests have shown that with liquids for flow velocities above Reynolds number 8000 a correlation flowmeter can be calibrated solely on the basis of pipe volume and measured transit time[6].

The standard error, $\epsilon(\tau)$, in time delay measurement is given by the following equation (proof in Appendix 6.1).

$$\epsilon(\tau) = \frac{0.249}{B^{1.25} T^{0.25}} \left[1 + \left(\frac{1}{\rho^*} \right)^2 \right]^{0.25}$$

where B is the bandwidth of the system

             T is the correlation integration period

             ρ^* is the normalised cross-covariance of the transducer output signals at the correlation peak.

Numerically, $\epsilon(\tau)$ is an estimate of the standard deviation of the measured value of transit time delay τ, based on known system parameters. These parameters can, in fact, be readily determined from the measured corrleation curves[7].

The accuracy is a function of the following parameters

(a)   The normalised cross-covariance, ρ^*, can be increased by reducing the axial distance between the measuring points. (The cross-covariance can be regarded as a measure of the information lost between the two channels being correlated, the theoretical maximum for a pure time delay is 1.0). Reducing the distance between the measuring points will improve the degree of correlation. However, reducing the measuring length will also reduce the measured time delay and hence increase the problem of resolution. The choice of measuring distance will, therefore, be a compromise of several factors[7]. Transducer spacings between 1 pipe diameter and 5 pipe diameters seem to be reasonable in practice.

(b)   Since the improvement in accuracy is a function of the quarter power of T, it will be seen that a large increase in data collection time would be necessary in order to give a significant increase in accuracy. In practice, the integration time is limited by the response time which can be allowed for the flow measurement.

(c)   Increasing the system bandwidth, B has the most marked effect on accuracy. The bandwidth can be increased by careful choice of transducers and appropriate system design, to ensure that the small high frequency turbulences are detected.

1.4 Results – ultrasonic cross correlation

The arrangement of the transducers used is as shown in Fig. 4. At each of the measuring points a matched pair of ultrasonic transducers is used (magnetostrictive transducers for the results reported here), together with their associated electronic amplifiers and signal processors. Each measuring point has a transmitting transducer, T, which is driven by an oscillator at a suitable frequency for the flow media. The acoustic beam which is propagated across the diameter of the conveyor is received by the associated receiving transducer R. The received signal is amplified and is input to the signal processing unit, together with a reference signal from the transmitter oscillator. The signal output from each processing unit is then suitably amplified for feeding the cross correlator, a fuller description of the ultrasonic flowmeter can be found in ref.[8].

Tests on the ultrasonic flowmeter have been carried out using the flow proving system shown diagrammatically in Fig. 5. The measuring section comprised a 1 m long by 0.14 m inside dia. perspex pipe, having the two pairs of transducers flush mounted with their centres 0.25 m apart. The actual air flow was measured by means of a water manometer mounted across a Dall tube, corrections were made for compressibility of the air in the flow line.

The results which are shown in Fig. 6 show that the maximum error was ±4% of the actual reading.

2   SOLIDS MASS FLOW MEASUREMENT IN PNEUMATIC CONVEYORS

2.1 Mass flowmeter for pneumatic conveyors

Pneumatic conveyors are widely used in modern processes as a means of transport for granular and powdered solids in various stages of manufacture. However, although pneumatic conveyors enable solids to be handled almost as easily as liquids, there has hitherto been no entirely satisfactory means of measuring the quantity of material flowing.

Accepted methods of solids flow measurement usually involve the diversion of the material from the conveyor into a weighing system. Such methods have several disadvantages: indication of flow is not continuous; weighing systems tend to be complicated, with attendant problems of high maintenance costs and mechanical wear of moving parts; most methods obstruct the system with consequent risk of blockages.

The new flowmeter described in this paper overcomes the above problems. In a pneumatic conveyor, the velocity of the solid flow is substantially constant over a wide range of solids loading. Therefore, to determine the mass flow rate, it is only necessary to measure the mass of solid present in the fixed length of the conveyor, i.e. the density of the cloud of solid being blown along. As mentioned in Section 1.1, the solids in the conveyor will take the form of clouds due to the effects of turbulence; the average density of these clouds can be measured by using the arrangement shown in Fig. 7 which gives an output reading proportional to the mass flow of solids passing the transducer. The flowmeter uses a capacitive transducer which is connected to an insulated

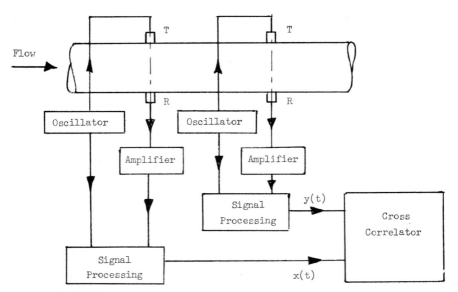

Fig.   4   Transducer arrangement

electrode forming part of the wall of the conveyor, the earthed body of the conveyor acts as the other electrode. Flowmeters to fit pipes from 2 cm to 40 cm in diameter have been produced. The theoretical basis of the system has been given in previous papers[9]. Briefly the principle is that small variations in capacitance, caused by the transit of particles past the electrode section, are converted to a.c. voltage by a capacitive transducer. The output from the transducer is then rectified and smoothed so that short-term flow fluctuations are averaged out. The smoothing time is adjustable so as to cover a wide range of flow conditions. The output signal is a standard d.c. current in a range such as 0-10 mA suitable for driving a number of remote-mounted indicators or recorders.

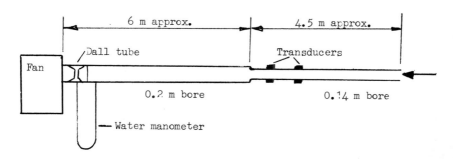

Fig. 5   Air flow test rig

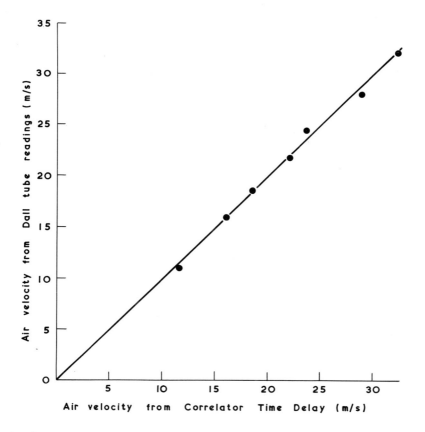

Fig. 6   Calibration curve for air flow

A particular feature of the new capacitive transducer is that it is only sensitive to the variations in capacitance due to the transit of particles and is not affected by changes in standing capacitance due to slow-changing effects such as powder sticking to the electrode, hence no zero adjustment is required in use.

2.2 Results

Investigations in industrial pilot plant, generally as shown in Fig. 8, have been carried out using a variety of chemicals and foodstuffs. The plant used provided wide flexibility in powder flow rates and air flow.

A pipe of 0.15 m diameter was used and solid feed rates of 50 to 4000 kg/hr, air velocities of 20 to 55 m/s and powder loadings of 0 to 1.5 were typical. Mass flow rates were checked by weighing and air velocities determined by using a pitot-static tube.

Calibration curves of the meter reading against measured flow rate are shown in Fig. 9. Tests on solid samples showed that the moisture content of the materials were reasonably constant during the tests and therefore that the permittivity of each solid was also constant. If large changes in permittivity are likely to occur, an arrangement using a permittivity measuring electrode could be used for compensation; such a system has been described in an earlier paper[10].

The flowmeter was also tested on a positive pressure system, conveying cement subject to wide variations in feed-rate. Powder loadings as high as 10 kg solids/kg of air were used, the meter being required to monitor the flow rate continuously with the object of observing trends relative to optimum conditions. The electrodes have not shown any significant wear in use. A reasonably long smoothing time constant was used so as to average flow pulsations and hence obtain an accurate indication of process trends. This on-line installation in an industrial plant has operated satisfactorily with no maintenance problems.

The trials just described have all involved the use of highly turbulent mixtures with air flows greater than 10 m/s. In a further application, the meter has been tested in an inclined pipe down which powder flowed by gravity feed only. The electrode was placed at the under-side of the pipe which was substantially full of powder. Capacitance variations, caused probably by natural packing density variations in the sliding bed, gave a signal which enabled mass flow to be measured. The equipment operated successfully to indicate process trends.

In cases where an on/off indication is required (flow failure alarm), the meter drive circuit, Fig. 7, is replaced by an alarm circuit[11]. The device senses only the capacitance changes due to flow of material and gives an alarm in the event of flow failure, even if the pipe remains full of static powder.

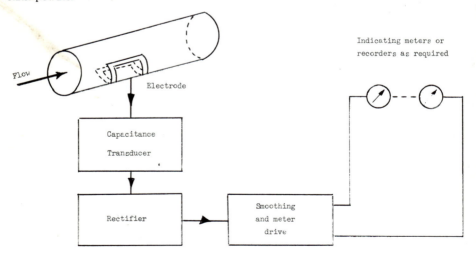

Fig. 7   Solids flowmeter for pneumatic conveyor

3   CONCLUSIONS

The objective of our program has been to apply our flowmeter systems to industrial applications, those so far studied include:

(a)   Flow of gaseous suspensions[9].

(b)   Flow of non-conducting liquid slurries using capacitance transducers[14].

(c)   Flow of single-phase liquids and gases using ultrasonic transducers[8].

(d)   The use of tribo-electric effects for flow measurement[15].

(e)   Heat flow in district heating schemes[16].

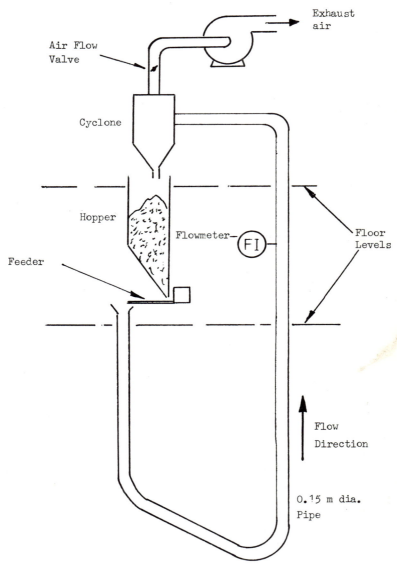

Fig. 8   Pilot plant conveying system

Volumetric flowmeters using the cross correlation technique for measuring single component and two component fluid flows are capable of a high degree of accuracy. In the case of gas/solid systems, transducers developed at the University of Bradford are commercially available.

A mass flowmeter for solids in pneumatic conveyors has been described which is becoming well established for a wide variety of industrial applications and the instrument is now on the market. A solids mass flowmeter for hydraulic conveyors, based upon similar principles, is at an advanced stage of development.

A low-cost cross correlator, which has been designed to give output flow readings directly, will shortly be available[17,18].

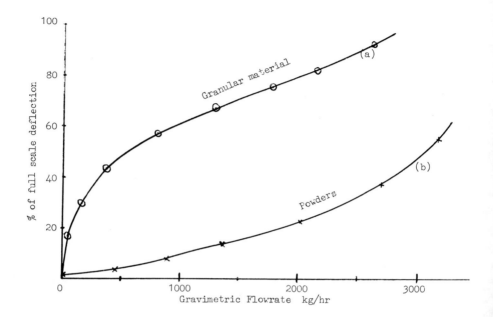

Fig. 9   Flowmeter calibration curves

4   REFERENCES

1   OWER and PANKHURST: 'Measurement of Air Flow' (Pergamon, 1966)

2   BECK, M.S., HOBSON, J.H., and MENDIES, P.J.: 'Measurement of Mass and Volume Flows of Slurries and Gaseous Suspensions', Proceedings Inst. M and C, Conference on Process Instrumentation in the Metals Industry, May 1971

3   'A Solution to Difficult Flow Measurement Problems', NRDC Bulletin — Inventions for Industry, No. 37, April 1971

4   COULTHARD, J.L.: 'Ultrasonic Cross Correlation Flowmeters for Liquids', U.K. Patent 35263/70

5   COULTHARD, J.L.: 'Improvements in or relating to the Measurement of Fluid Flow Rates', U.K. Patent 2593/71

6   ABEYSEKERA, S.A.: 'Fluid Flow Measurement using Correlation techniques', Ph.D. Thesis, School of Control Engineering, University of Bradford, Feb. 1971

7   BECK, M.S.: 'Powder and Fluid Measurement Using Correlation Techniques', Ph.D. Thesis, School of Control Engineering, University of Bradford, 1969

8   CALVERT, G., and COULTHARD, J.L.: 'The use of Ultrasonics in Cross Correlation Flow Measurement', I.Chem.E. and Inst. M and C, Symposium of the Flow Measurement

of Difficult Fluids, University of Bradford, 15/16 April 1971

9   BECK, M.S., GOUGH, J.R., HOBSON, J.H., JORDON, J., and MENDIES, P.J.: 'Solid/ fluid Two Phase Flow Measurement using Correlation Techniques'. CHEMCO 70, Australian I. Chem. E., Aug. 1970

10 BECK, M.S.: 'Powder or Grain Level can be Measured accurately with Capacitance probes and a Digital or Analogue Computer', Control, July and Aug. 1967

11 BECK, M.S., and WAINWRIGHT, N.: 'Flow-failure detector for powdered and granular materials', Control, Jan. 1969

12 ROWE, H.E.: 'Signals and Noise in Communication Systems' (Van Nostrand, 1965)

13 BENDAT, J.S., and PIERSOL, A.G.: 'Measurement and Analysis of Random Data' (Wiley, 1966)

14 BECK, M.S., GOUGH, J.R., and MENDIES, P.J.: 'Flow Measurement in a Hydraulic Conveyor, Hydrotransport 1', Sept. 1970, British Hydrodynamics Research Assoc., Cranfield

15 BECK, M.S., CALVERT, G., and HEWITT, P.J.: 'Electrodynamic methods of measuring flow and moisture in pneumatic conveyors and non-conducting liquid streams', Patent pending

16 BECK, M.S., MUSGRAVE, G., and PLASKOWSKI, A.: 'Centralised monitoring and control of district heating schemes', Patent pending

17 BECK, M.S., MUSGRAVE, G., and WORMALD, C.N.: 'Improvements in or relating to cross correlation', U.K. Patent 3193/71

18 HAYES, A.M.: 'Low-cost Cross Correlator for Flow Measurement', Internal Report, School of Control Engineering, University of Bradford

5   ACKNOWLEDGEMENTS

We are grateful to Fielden Electronic Ltd. for their continuous technical cooperation, to the National Research Development Corporation, Henry Simon Ltd. and I.C.I. Ltd. for their help with the practical applications and to the Science Research Council for financial support. The use of George Kent Ltd. facilities, together with their other support is gratefully acknowledged. We also thank Mr. J.H. Hobson for allowing the use of his solids mass flow results. The N.R.D.C. hold British, American and other patents on many of the systems described.

6   APPENDIX

6.1 Derivation of the standard error in time delay measurement

The normalised cross correlation $Q'_{mn}(\beta)$ of the transducer outputs can be approximated to the impulse response of the system model, giving[12] –

$$Q'_{mn}(\beta) = \frac{\text{Sin}\left\{2\pi B(\beta - \tau)\right\}}{2\pi B(\beta - \tau)} \qquad \text{... (1)}$$

where B =   the system bandwidth   (Hz)
        τ =   the mean transit time   (sec)

For small deviations from the cross correlation peak the numerator of equation (1) can be expanded, giving

$$Q'_{mn}(\beta) = 1 + \frac{\left\{2\pi B(\beta - \tau)\right\}^2}{3!} \qquad \text{... (2)}$$

Now the normalised standard error $\epsilon\left\{Q^*(\beta)\right\}$ in estimating the peak of the normalised cross correlation function $Q^{\epsilon}_{mn}(\beta)$ is (by definition) given by

$$\epsilon^2 \left\{Q^*(\beta)\right\} = \underset{\beta \to \tau}{\mathrm{Var}} \left\{Q'_{mn}(\beta)\right\}$$

$$= \underset{\beta \to \tau}{\mathrm{E}} \left[\left\{1 - Q'_{mn}(\beta)\right\}^2\right] \qquad \ldots (3)$$

Substituting from equation (2) gives

$$\epsilon^2 \left\{Q^*(\beta)\right\} = \left(\frac{2\pi^2 B^2}{3}\right)^2 \underset{\beta \to \tau}{\mathrm{E}} \left\{(\beta - \tau)^4\right\} \qquad \ldots (4)$$

and noting that, for a normal distribution

$$\mathrm{E}\left\{(\beta - \tau)^4\right\} = 3.0 \left[\underset{\beta \to \tau}{\mathrm{E}} \left\{(\beta - \tau)^2\right\}\right]^2$$

$$= 3.0\, \epsilon^4(\tau) \qquad \ldots (5)$$

where $\epsilon(\tau)$ is the standard error in time delay measurement, we can rewrite equation (4)

$$\epsilon^2(\tau) = \frac{0.0877}{B^2}\, \epsilon\left\{Q^*(\beta)\right\} \qquad \ldots (6)$$

where the normalised standard error in estimating the cross correlation function $\epsilon\left\{Q^*(\beta)\right\}$ is given by (13)

$$\epsilon\left\{Q^*(\beta)\right\} = \frac{1}{(2BT)^{0.5}} \left\{1 + \left(\frac{1}{\rho^*}\right)^2\right\}^{0.5} \qquad \ldots (7)$$

where T =   the correlation integration period (secs)
and ρ^* =   the normalised cross covariance of the transducer output signals at the correlation peak.

Substituting (7) into (6) gives

$$\epsilon(\tau) = \frac{0.249}{B^{1.25} T^{0.25}} \left\{1 + \left(\frac{1}{\rho^*}\right)^2\right\}^{0.25} \qquad \ldots (8)$$

DISCUSSION

K. Komiya: It is necessary to develop a simple correlator for practical applications of this method. What sort of correlator did the authors use in these experiments?

Authors' reply: For practical application of the cross-correlation method to flow measurement, a simple correlator is a necessity. Such a device is currently being developed at Bradford University for use with an industrial flowmeter. It is expected that the correlator will soon be commercially available at a low cost. The instrument used for the experiments described in the paper was a Hewlett Packard 3721A correlator.

J.E. Roughton: A problem which often arises in industry is the measurement of flow distribution in pipe networks when considerations of cost, access etc., preclude the fitting of individual flow-meters in each pipe. A low accuracy is often acceptable and the primary requirements are that the measurements can be made from outside the pipe wall and that the equipment is portable and easy to use. It would appear that the ultrasonic cross-correlation technique might be suitable for such a purpose and I should like to have the authors views on this. Other points of interest are whether the measurements could be made from one side of the pipe only, using a reflection technique, and whether for liquid flows, consistent calibrations can be obtained at Reynolds' numbers below 8000.

Authors' reply: Mr. Roughton's comments are of interest, his remark on acceptable accuracy agrees with our experience.

The ultrasonic technique has been successfully used for measurements made from outside the pipe wall and this aspect of the method is currently being developed. Measurements could be made from one side of the pipe if so desired.

Consistent calibrations have been obtained for both laminar and turbulent flow regimes: however, measurements in the transition region (Reynolds' numbers of 1500 - 5000 say) would not be advised.

F.C. Kinghorn: I am not sure if the mass flowmeter described works in the way I think it does. Does the device effectively measure the mass of solid material present in the region of the pipe beside the detector, and obtain a relative mass flowrate on the assumption that the particle velocity is constant?

A calibration graph is given for the instrument, but there is no indication of absolute accuracy. Would the authors indicate what the accuracy is?

Authors' reply: The mass flowmeter does in fact work in the way described. The instrument would normally be calibrated on site and is suitable for the many applications where ±5% accuracy is acceptable.

G.E.B. Loxley: What is the best accuracy that can be obtained using the cross correlation method to measure water flow?

Authors' reply: We cannot give a definite answer: the best accuracy is not known. However, we have obtained a figure of ±2% in laboratory tests and think this can be improved upon.

R. Theenhaus: You need a certain time to find the maximum of the correlation-integral. What is the time-resolution of this instrument?

What velocity do you really measure? Is it the mean velocity of the velocity of one component; for example, in the case of a mixture of liquid and solids?

Authors' reply: The integration time required will depend on a number of factors, in the case of gas flow a time of 5 to 15 seconds would be typical.

The velocity measured by the correlation method depends on the particular transducer system used. In the case of the capacitive transducer, since the field extends only a limited distance into the flow, the velocity measured will be related to the solids velocity local to the pipe wall. With regard to the ultrasonic system, flow tests using clean air indicate that the mean velocity of flow will be measured.

G.V. Evans: Your paper describes a tracer technique where the tracer is inherent in the flow. Nevertheless for successful operation, as with other tracer methods, the tracer must be representative of the flow being examined (i.e. it must be mixed). Theoretically, the phenomenon being used in the cross-correlation method should originate in the flow at a distance upstream of the measuring section at least equal to the mixing distance. In practice, this condition may be relaxed depending on the volume of fluid sensitive to the detector.

For example, one would not expect to obtain a reliable (i.e. accuracy better than a few percent) measurement of mean fluid velocity from correlating point fluctuations in temperature at two positions immediately downstream of a variable heat source. This method cannot then be called "absolute".

When this method has been applied to flowrate measurement, has tracer dynamics been considered in relation to dispersion properties of the flow and what practical limitation is there in the achievement of high accuracy?

Authors' reply: We agree that for successful operation of the correlation method the observed property must be representative of the flow being measured. We also agree that this method cannot be called "absolute"; but it is absolute in the sense that the measurement of velocity involves the use of the absolute quantities length and time. However, we accept that the problem of "what velocity is actually measured" is one that requires careful handling and is one of the questions that our research is attempting to answer.

Tracer dynamics as such have not been considered in the earlier work that has been carried out. However, calculations indicate that the error due to diffusion is rather less significant than errors due to other causes. This aspect of the work is now being considered more fully and will be dealt with in a future paper.

One limitation to the achievement of high accuracy is that resulting from the fact that the correlation method is essentially a statistical process. In any practical situation there are limits to the time available for data collection and processing, some of these aspects are discussed in the paper. Another limitation could arise from the nature of the fluid being measured and the circumstances under which the measurements are made. Work is still being carried out, for example, on the effect that flow disturbances have upon accuracy.

F.A. Inkley: I am interested in the accurate volumetric flow measurement of clean, homogeneous liquids; in particular, crude oil and hydrocarbon products. In the conclusions of this paper it is inferred that flowmeter systems have been used for measuring the flow of single-phase liquids using ultrasonic transducers. However, in Reference (8) it is specifically stated that "Clean homogeneous fluids are thus unsuitable for measurements in this manner" (that is, ultrasonic using *natural* turbulence). Would the authors please comment.

I would like to ask whether the results of air velocity from correlation time delay shown in Fig. 6 were derived using a *measured* value for the separation of the two transducers.

Authors' reply: When the paper was first being prepared we thought that only gases or two component liquid flows could be measured using ultrasonic cross correlation systems. More recently we have realised that the turbulent eddies in any fluid flow will give suitable signals for cross correlation and ultrasonic transducers have been successfully used for single-phase liquids and gases in pipes from 1 in. to 24 in. diameter.

The air velocity results were derived using a measured value for the separation of the two transducers.

FLOW CHARACTERISTICS OF TURBINE FLOWMETERS

V.R. Withers, F.A. Inkley and D.A. Chesters

Abstract: Extensive experiments have been carried out on five three inch nominal bore commercial turbine meters using a gravimetric flow rig with flows up to 36 ℓ/s (470 Imp.gal/min). Water, kerosine, gas oil and spindle oil have been used as test liquids and characteristic curves of "meter coefficient" versus flow rate have been obtained at temperatures between ambient and 60°C, covering a total viscosity range of 0.5 to 27 cSt. This paper presents a full discussion of these experimental results, and shows that the various meters exhibit differences in their sensitivity to flow rate, viscosity and temperature. Some preliminary attempts towards analysing the effect of these variables are also discussed.

1 INTRODUCTION

Over the last few years, the applications of liquid flow meters have increased considerably. Reasons for this are the steady trend towards automated control of plant and operations, and the preference of many government authorities for the use of meters in the place of tank dipping for custody transfer measurements. In these contexts, there are many possible applications for turbine flow meters. Further, in the Petroleum Industry, where positive displacement meters are widely used and accepted, there are many instances where turbine meters could be used to advantage.

The main advantages of turbine meters over positive displacement meters are cost (for the larger sizes), physical size and weight, ease of application in electronic control systems, simplicity in maintenance, and, perhaps, mechanical reliability. However there is uncertainty in their accuracy over a wide range of fluid conditions, and there exists only limited literature describing detailed performance data of turbine meters.

For the above reasons, it was decided that a thorough experimental investigation should be made into the performance of a number of different commercially available turbine meters. The aims of this work were not only to investigate the detailed effect of liquid temperature and viscosity on performance, but also to see if any scientific generalisations could be made from the results obtained. Three inch, nominal bore meters were selected as the largest size which could be easily handled experimentally, and these were tested with four separate test liquids, covering a wide range of flow rates and temperature.

This paper contains a description of the nature of these tests and the flow rig used, a discussion of the results obtained and details of the interpretation of these results so far.

2 EXPERIMENTAL PROCEDURE

A flow rig was designed and constructed at the BP Research Centre, initially for comparing three inch turbine meters over a wide range of fluid conditions. As such, it is the accuracy of comparisons between the performance of meters under different conditions of liquid, temperature and flow rate, which is important. This implies that the rig should have very good short term repeatability, but that the absolute calibration accuracy, providing that it is constant, is comparatively unimportant.

A gravimetric type of rig was decided upon, operating with a simple "stop and start" procedure, in which the meter is at rest at the beginning and end of each test run. A schematic diagram of this rig is shown in Fig. 1. During the course of a single test, a quantity of liquid is pumped, at a steady preset rate, through the meter section and into the tank on the weighing scales, where it is weighed. The meter performance is then obtained by comparing this measured mass of liquid collected with the volume indicated by the turbine meter for the particular conditions of flow rate, liquid and temperature.

The pump used in the rig was centrifugal type, with a maximum flow rate of 36ℓ/s (470 Imp. gal/min). This is just short of the normal rated maximum of 38ℓ/s (500 Imp. gal/min) for three inch bore (76 mm) turbine meters. In the meter section, no flow straighteners were employed, but there were straight pipe lengths of 30 pipe diameters upstream and 12 pipe diameters downstream of the meter under test. The weighing scales could weigh up to 1200 kg, which is equivalent to about 1.6 m³ of kerosine or 1.2 m³ of water.

In a gravimetric flow rig the measured volume is calculated from the measured mass of liquid and the density of the liquid, at the temperature at which the liquid passes through the meter. For this reason temperature was measured accurately (to ±0.1°C). The relationship between density and temperature was obtained with standard hydrometers on a sample of the test liquid.

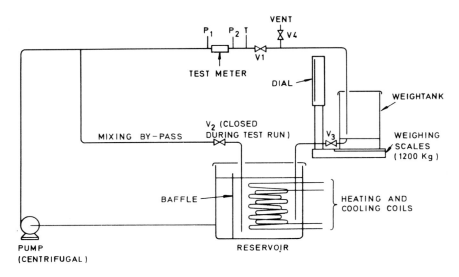

Fig. 1   Gravimetric flow measurement rig: schematic diagram
Flow range 0-36 l/s (0-470 Imp. gal/min)
Temperature range ambient – 60°C

Pipework, 4 in and 3 in stainless steel
P_1, P_2 = pressure measurements
T = temperature measurement

V1 = control valve (ball)
V3 = gate valve (quick acting)
V4 = solenoid valve

The volume indication of a turbine meter is in terms of pulses produced by an electromagnetic detector, which senses the movement of the rotor blades past it. The simplest way of expressing the meter performance is in terms of a meter coefficient, which is the number of pulses per unit of volume measured under specified conditions (i.e. liquid, temperature, flowrate). The meter coefficient is given by:

$$X = \frac{N\rho}{M}$$

X = Meter coefficient (pulses/m³)

N = Number of pulses counted

ρ = Liquid density (in air) at the temperature of measurement (kg/m³)

M = Measured mass (kg in air) of liquid collected for the N pulses counted. This is corrected for the effect of the filling and drain pipes being partially immersed in liquid.

Because the scales are calibrated in terms of 'mass in air', for ease of calculation, the liquid density is expressed in terms of 'density in air'

Generally, it is easier to appreciate changes in performance in terms of a percentage difference (Δ) from some fixed reference point.

$$\Delta = 100 \ \frac{X - X_0}{X_0}$$

X_0 is the fixed reference point, which may have some physical meaning or which may be arbitrary. For purposes of comparison the value of this reference point is not important, providing that it is within the range of the variation of the meter coefficient, and that this range is small. In the text, results will be referred to in these percentage terms.

For the experimental work proposed, it was necessary to obtain for each meter tested a complete set of meter coefficients for a wide range of measured conditions of flow rate, temperature and viscosity. This range of test conditions was achieved as follows:

2.1 Meters tested

Five different three inch meters, from four different manufacturers, designated A, B, C, D and E were tested. They were all very similar both in design and in their claimed performance. The rotors showed small variations in shape, while the bearings were of the simple journal type in all cases except one, where ball bearings were fitted.

2.2 Viscosity variation

Viscosity was varied by changing the test liquid, and by changing the liquid temperature. Tests were carried out in the viscosity range of approximately 0.5 to 27.0 cSt. To this end four test liquids were used in the flow rig, kerosine, gas oil, spindle oil (a light lubricating oil) and water. The properties of these liquids are shown in Table 1.

TABLE 1 Physical properties of test liquids

Liquid	Flash Point $^\circ$C	Approximate density (in air) Kg/m^3			Kinematic viscosity cSt			Vapour pressure psia
		20°C	40°C	60°C	20°C	40°C	60°C	50°C
Water	-	996	991	982	1.0	0.65	0.47	1.8
Kerosine	72	780	766	752	2.5	1.7	1.3	2.8
Gas Oil	72	828	814	800	4.1	2.6	1.8	0.4
Spindle Oil	105	857	844	830	23	11	6.2	0.1

N.B.: Densities were measured to an accuracy of ± 0.02%.
      The above are approximate values.

2.3 Temperature variation

Steam heating and water cooling coils were provided so that the liquid temperature could be varied between 15°C (ambient temperature) and 60°C. For safety reasons, the kerosine and gas oil were not heated above 50°C. From this range five test temperatures were selected:

(15 to 25°C), 30°C, 40°C, 50°C and 60°C (Water and spindle oil only at 60°C)

At 30°C and above, the temperature could be controlled to within ±1°C but, at ambient air temperatures, the heating effect of the pump was greater than the cooling effect of the cooling coils, so that there was a steady increase in temperature during the course of a series of tests. This leads to an increase in the scatter of the results for the low temperatures.

2.4 Flow rate variation

Flow rate could be set, and measured accurately, down to 10% of the maximum flow rate of the rig. The flow rate was calculated from a stop-watch timing and the total volume of liquid collected. For convenience in analysing the data, twelve nominal flow rates were selected to form a complete set of tests for each combination of meter, liquid and temperature. Expressed in terms of percentages of the maximum flow rate of the rig (36 ℓ/s) these flow rates were:

100, 90, 70, 50, 30, 10, 15, 20, 40, 60, 80, 100%.

Tests were always performed in the above sequence.

For each meter, the complete combination of tests, outlined above, was covered. This implies a total of 216 individual determinations of meter coefficient for each meter. In addition to this, a series of repeatability tests were performed, using a single test condition (kerosine at 40°C). These involved making about ten repeat measurements under as near identical conditions of flow rate and temperature as possible. These were repeated for three separate flow rates namely 90, 65 and 30% of the maximum of 36 ℓ/s.

3 PRECISION

The precisions for the five meters and the three flow rates mentioned above were not significantly different from each other. In all cases they gave a figure for short term precision of ±0.05% or better (95% confidence limits). This figure embraces short term variability of both the meters and the test rig. The latter was calculated from the individual sources of error and found to be about ±0.02%, which is small by comparison. This implies that the short term precisions of all of the meters tested are within ±0.05% (95% confidence limits), which is the value claimed by the manufacturers.

The absolute accuracy of a single value of meter coefficient was estimated from the individual sources of error and found to be ±0.05%.

Another source of error could occur during the acceleration and deceleration of the rotor at the beginning and end of each run. Calculations showed that such errors would be insignificant. This was confirmed experimentally by performing test runs which were interrupted by a number of additional stops and starts.

4 RESULTS

The experimental results are plotted in the form of characteristic meter curves. These relate the measured meter coefficient, expressed in terms of a percentage difference from an arbitrary reference, to flowrate, for fixed conditions of liquid and temperature. For each meter and each liquid a family of characteristic curves is obtained for the different temperatures of measurement.

In order to facilitate comparisons, particular curves from one family are collected and compared with the corresponding curves from the other families in the same figure. Likewise, similar data are collected and compared in Tables. Thus it is possible to see clearly the effect of a particular variable, or combination of variables, on meter performance.

The three independent measurement variables affecting the performance of each meter are flow rate, liquid and temperature. The effects of these variables are very much inter-related. This is particularly true for liquid and temperature because they both effect viscosity, an important parameter in flow measurement.

The results will now be considered in turn, under the headings of the three measurement variables.

4.1 Effect of flow rate

Figures 2 and 3 respectively show the families of characteristic performance curves representing the least and greatest variability of meter coefficient with flow rate obtained with all five meters and four liquids.

In an ideal meter, the meter coefficient does not change with flow rate. Figures 2 and 3, and later figures, illustrate the departure from ideality of actual meters. The deviation from the ideal response is usually referred to as the linearity of the meter, and is defined as the maximum percentage change in meter coefficient, within a specified flow range, under specified liquid conditions. The flow range is usually expressed in terms of the ratio of the maximum to minimum flow rate ('turn-down ratio'), where the maximum is the rated maximum for the meter, which is 38 ℓ/s or 500 Imp.gal/min for three inch bore turbine meters. Table 2 shows linearities for the five turbine meters for both 10:1 and 3:1 flow ranges under four fairly extreme liquid conditions. It can be seen that there is considerable variation between meters, and that only occasionally was the linearity over a 10:1 flow range within 0.5%, which is frequently claimed by manufacturers. However, over a 3:1 flow range, most meters showed linearities within 0.5%.

A study of Table 2, in conjunction with Figs. 2 and 3, shows that the shapes of the performance curves vary considerably, not only from meter to meter, but also for changes of liquid and perhaps temperature for any one meter. Several of the meters do not exhibit the 'accepted' standard performance curve, in which the meter coefficient is a maximum at about 10% of maximum flow rate, falls gradually to a steady value at the maximum flow rate, and drops away very steeply at the lower flow rates. These shapes will be demonstrated in the graphs referred to in the following sections.

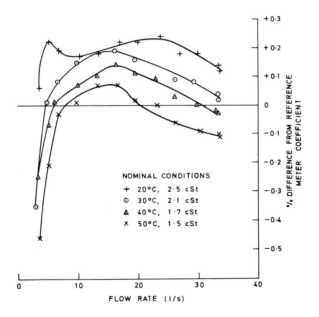

Fig. 2   Meter A: kerosene results

4.2 Effect of physical properties of liquids

Viscosity is the physical property which differs most between the test liquids. However, density and other fluid properties, which do not differ so much, may also be significant. In fluid dynamical problems it is the kinematic viscosity (the ratio of absolute viscosity to density), which is usually referred to. Table 1 shows these properties for all of the test liquids.

The effect of the different test liquids is shown in Figs. 4 and 5. These figures contain characteristics for the four test liquids at a temperature of 30°C. Figure 4 shows Meter B, whose meter coefficient was least changed by the different liquids, while Fig. 5 shows Meter D, which was most affected by viscosity. From these graphs it can be seen that at maximum flow rate a twenty fold variation in viscosity caused a 0.3% variation in meter coefficient for Meter B and a 2.6% variation (in the opposite direction) for Meter D. The other meters lay within these two extremes.

Numerically, these variations are shown in Table 3. Here the percentage change in meter coefficient for each meter at flow rates of 80%, and 25% of maximum and at a temperature of 40°C are shown. The comparisons are between the extremes of the hydrocarbon liquids (spindle oil and kerosine) and between a hydrocarbon and water (demonstrated by kerosine and water). Considerable differences between the various meters can once again be seen, which, for any particular meter application, could be very important.

TABLE 2 Linearity of the meters

Linearities are expressed as the percentage change in meter coefficient, within the stated flow ranges, i.e. 10:1 and 3:1 of the rated maximum flow of 38ℓ/s.

Meter	Water at 60°C 0.47 cSt		Kerosine at 50°C 1.5 cSt		Gas oil at 30°C 3.2 cSt		Spindle oil at 30°C 15 cSt	
	Flow range		Flow range		Flow range		Flow range	
	10:1	3:1	10:1	3:1	10:1	3:1	10:1	3:1
A	~ 0.8	0.15	0.45	0.20	0.40	0.15	1.6	0.80
B	0.40	0.35	0.60	0.30	0.40	0.25	1.2	0.45
C	0.80	0.40	0.50	0.30	1.1	0.35	1.6	0.60
D	2.9	0.30	2.6	0.20	1.3	0.70	3.3	1.6
E	0.80	0.80	0.60	0.50	1.4	0.35	2.2	1.5

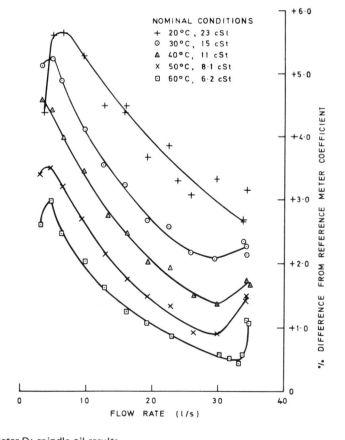

NOMINAL CONDITIONS
+ 20°C , 23 cSt
⊙ 30°C , 15 cSt
△ 40°C , 11 cSt
✕ 50°C , 8·1 cSt
▢ 60°C , 6·2 cSt

% DIFFERENCE FROM REFERENCE METER COEFFICIENT

FLOW RATE (l/s)

Fig. 3   Meter D: spindle oil results

TABLE 3 Relative sensitivity to the different liquids

The relative sensitivities are expressed as the percentage difference between the meter coefficients for the stated pairs of liquids, at a temperature of 40°C.

Meter	Spindle oil-kerosine		Kerosine-water	
	Flow rate		Flow rate	
	30ℓ/s	10ℓ/s	30ℓ/s	10ℓ/s
	~80% of max	~25% of max.	~80% of max.	~25% of max.
A	+0.40	+1.2	+0.25	+0.2
B	−0.40	+0.15	+0.15	+0.5
C	+0.45	+0.70	+1.0	+0.85
D	+1.7	+3.5	−0.20	−0.10
E	+0.45	+1.6	+0.60	+0.10

4.3 Effect of temperature

Temperature may alter the performance of a turbine meter in two ways; firstly, on account of changes in the dimensions of the working parts and their clearances, due to thermal expansion; and, secondly, due to change in liquid viscosity with temperature.

Thermal expansion causes an increase in both the area of the meter bore and the radius of the meter rotor, and both of these decrease the meter coefficient. The API Turbine Meter Standard [1]

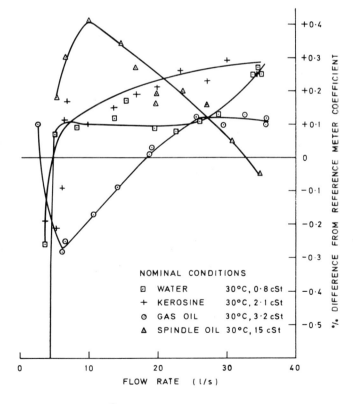

NOMINAL CONDITIONS

⊡	WATER	30°C, 0·8 cSt
+	KEROSINE	30°C, 2·1 cSt
⊙	GAS OIL	30°C, 3·2 cSt
△	SPINDLE OIL	30°C, 15 cSt

% DIFFERENCE FROM REFERENCE METER COEFFICIENT

FLOW RATE (l/s)

Fig. 4 Meter B: all liquids at 30°C

shows how these effects can be related to the thermal expansion of steel and the change in temperature. Figure 6 shows the magnitude of the cnange in meter coefficient plotted against temperature. Typical values are about -0.004%/°C.

In addition, thermal expansion may cause changes in clearances between the rotor tips and the meter body, and between parts of the bearings, both of which could alter the meter coefficient. However, these effects are both very much influenced by the fluid viscosity, which is likely to change considerably more with temperature than will these clearances. As such, the effect of these clearances is best considered in terms of fluid viscosity only, any effects due to thermal expansion are relatively small and will not be discussed further in this paper.

The characteristics for the different meters, at extremes of temperature, are shown for kerosine and spindle oil in Figs. 7 and 8 respectively. In addition, the effect of temperature and liquid on meter linearity is demonstrated in these two graphs. Numerically, these variations with temperature are shown in Table 4. Here the percentage increase in meter coefficient for a 10°C decrease in temperature is presented for each meter and each liquid at two typical flow rates (80% and 25% of nominal maximum).

Once again there is considerable variation amongst the different meters. In most cases the variation is much greater than the value of -0.04%/10°C, which might be expected from thermal expansion considerations (the value of -0.04%/10°C is only an estimate, because of uncertainty in the types of steel used for the meter bodies and rotors). This suggests that in most cases the effect of changing viscosity could account for a large proportion of the variation. Indeed, the meters which exhibited greater sensitivity to the different liquids, also show the greater dependence on liquid temperature. However, the effect on blade tip and bearing clearances must not be precluded.

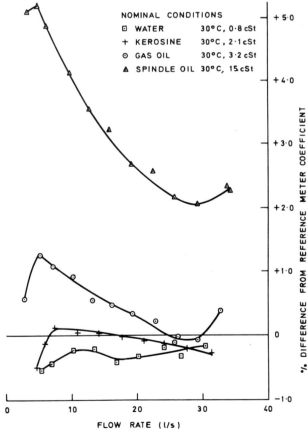

NOMINAL CONDITIONS
☐ WATER   30°C, 0·8 cSt
+ KEROSINE   30°C, 2·1 cSt
⊙ GAS OIL   30°C, 3·2 cSt
▲ SPINDLE OIL 30°C, 15 cSt

% DIFFERENCE FROM REFERENCE METER COEFFICIENT

FLOW RATE (l/s)

Fig. 5   Meter D: all liquids at 30°C

TABLE 4 Sensitivity to change in liquid temperature

The sensitivity to liquid temperature is expressed as the percentage difference in meter coefficient per 10°C *decrease* in temperature, over the stated range and at the stated conditions.

Meter	Water 20-60°C 0.47-1.0 cSt		Kerosine 20-50°C 1.5-2.5 cSt		Gas oil 20-50°C 2.2-4.1 cSt		Spindle oil 20-60°C 6.2-23 cSt	
	Flow rate		Flow rate		Flow rate		Flow rate	
	30ℓ/s 80% of max.	10ℓ/s 25% of max.	30ℓ/s 80% of max.	10ℓ/s 25% of max.	30ℓ/s 80% of max.	10ℓ/s 25% of max.	30ℓ/s 80% of max.	10ℓ/s 25% of max.
A	0.081	0.091	0.097	0.060	0.10	0.16	0.30	0.16
B	0.051	0.042	0.043	0.057	0.083	0.050	0.15	0.14
C	0.15	0.13	0.14	0.12	0.12	0.20	0.38	0.33
D	0.037	0.16	0.12	0.29	0.20	0.43	0.64	0.81
E	0.050	0.062	0.061	0.12	0.083	0.25	0.36	0.50

5   ANALYSIS OF RESULTS

It is probable that most of the dependence of meter coefficient on liquid and temperature follows from the change in liquid viscosity. From the raw results it is not possible without further analysis to distinguish between the direct effects of liquid and temperature, or their combined effect via liquid viscosity or any other fluid property. This analysis will be carried out by the authors and may form the basis of a future publication. However, a common factor for all meters tested was that thermal expansion effects could not fully account for the observed changes in meter coefficient with temperature.

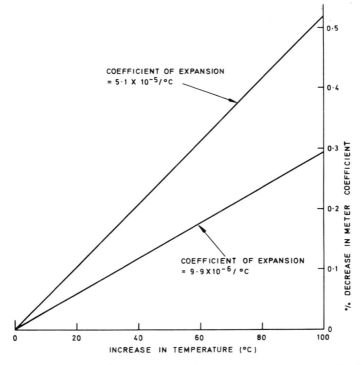

Fig. 6   Change in meter coefficient due to thermal expansion of meter rotor and body

It is often claimed that a turbine meter follows a universal characteristic curve, in terms of Reynolds' number. This suggests a more simple approach to the understanding of the separate effects of viscosity, flow rate and temperature on meter performance.

Reynolds' number is a dimensionless group, containing both a velocity (and therefore flow rate) and a viscosity term, which is frequently used to characterise and explain conditions in fluid flow.

It is defined as:

$$\text{Reynolds' number} = \frac{u\ell}{\nu}$$

where   u = a fluid speed (m/s)

    ℓ = a linear dimension (m)

and ν = kinematic viscosity (m^2/s = 10^6 cSt)

In calculating any dimensionless group, such as Reynolds number, it is important that the units of the three terms are all from the same rational absolute system. Therefore in the SI system, kinematic viscosity must be expressed in m^2/s and not in centistokes.

This idea of a universal Reynods' number characteristic was applied to the experimental data. For turbine meters, the speed was taken to be the mean liquid speed past the rotors, while the dimension was the width of the rotor blades from leading to trailing edge. The selection of the

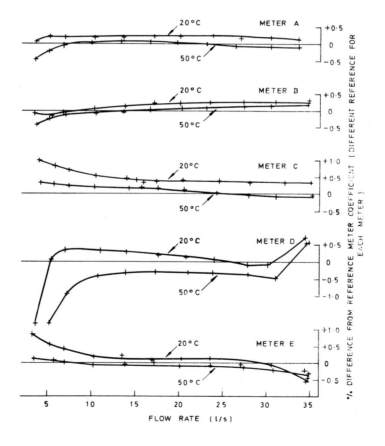

Fig. 7 All meters: kerosene at extreme temperatures

dimension is a little arbitrary for turbine meter problems, because the exact nature of the viscous effects are far from clear. Following the use of Reynolds' number in pipe flow, the meter bore might have been equally suitable. However, the choice of dimension is not important, because, with respect to flow rate and viscosity, it is constant over the conditions covered.

Figures 9 and 10 show typical graphs in which meter coefficient is plotted against Reynoids' number. Figure **9** contains the data for meter D with spindle oil, previously plotted against flow rate in Fig. 3. For meter D, the length of the rotor blades from leading to trailing edge was 22 mm. The effect of combining flow rate with viscosity clearly causes the curves for the different temperatures to merge together. The vertical separation of the curves over a range of $40°C$ and a proportion of the conditions covered has now reduced from about 2.7% to about 1%. This suggests that a part of the viscosity dependence of meter coefficient has been accounted for. The residual separation might then be associated with the effects of temperature on meter dimensions, though this separation is equivalent to $-0.25\%/10°C$, which is still larger than would be expected for simple thermal expansion effects $(-0.04\%/10°C)$.

Figure 10 contains the data for meter B, which was previously plotted against flow rate in Fig. 4, showing all liquids at $30°C$. For meter B, the length of the rotor blades from leading to trailing edge was 10 mm. The data suggests that the curves for the four liquids merge together, though with the very much smaller variation in meter coefficient this is not as pronounced as in the previous figure. Further, there is no temperature variation because all curves were at $30°C$.

Similar graphs were plotted for all of the meters tested, and they all showed similar tendencies. A broad conclusion is that there is some indication of a universal Reynolds' number characteristic,

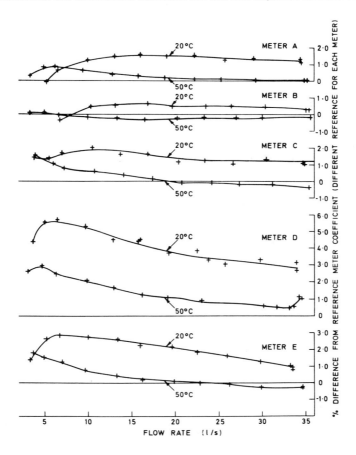

Fig. 8   All meters: spindle oil at extreme temperatures

though there is a departure from this at low flow rate. When the results have been investigated more thoroughly it is hoped that more definite conclusions will follow.

A number of authors have published attempts at analysing the behaviour of turbine meters, some based upon a full theoretical treatment and others based upon more empirical approaches. References 2 to 5 are some typical examples. All of these rely on many assumptions, and frequently

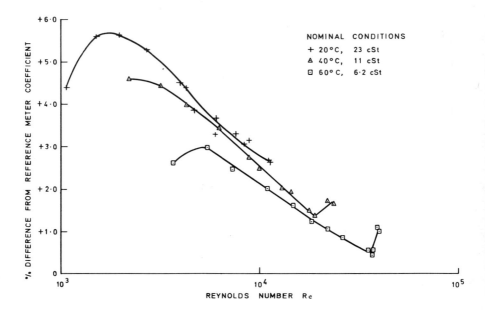

Fig. 9   Meter D: spindle oil results

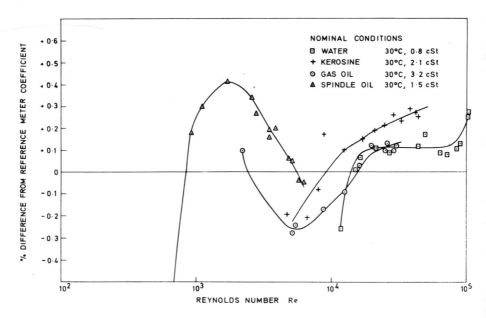

Fig. 10   Meter B: all liquids at 30°C

require formidable numerical techniques to solve the equations developed. Most of these are based upon the "flat plate" fluid dynamic theories, which are not necessarily applicable to the complex conditions inside a turbine meter. The authors propose to apply some of these ideas, and some of their own, to their results in the future.

6   CONCLUSIONS

The results presented form a factual basis from which, in terms of accuracy, turbine meters may be compared with other types of meter.

Some conclusions to be drawn from the results reported are:

(a)   The repeatability of all meters tested was better than ±0.05% (95% confidence limits).

(b)   Over the range of conditions covered, there was a very wide variation in the performance of the five different meters, as shown by their sensitivity to test liquid, temperature and flow rate.

(c)   The effect of temperature on meter performance follows more from change in liquid viscosity than from thermal expansion of the meter rotor and body, though the latter effect does explain some of the variation found.

(d)   Preliminary analysis of the experimental results offers an indication that the performance data do follow a universal curve, for each meter, in terms of Reynolds number. However, this is far from conclusive, and the authors propose to analyse the data more fully in the future.

7   ACKNOWLEDGEMENTS

The authors wish to thank the meter manufacturers, whose co-operation was vital to this work.

Permission to publish this paper has been given by The British Petroleum Company Limited.

8   REFERENCES

1   'Measurement of Liquid Hydrocarbons by Turbine Meter Systems', API Standard 2534. American Petroleum Institute, New York, March 1970

2   THOMPSON, R.E., and GREY, J.: 'Turbine Flowmeter Performance Model', Trans ASME J. of Basic Eng., Dec. 1970, pp. 712-723

3   LEE, W.F.Z., and KARLBY, H.: 'A Study of the Viscosity Effect and its Compensation on Turbine Type Flowmeters', Trans. ASME, J. of Basic Eng, 82, 1960, pp. 717-728

4   RUBIN, M., MILLER, R.W., and FOX, W.G.: 'Driving Torques in a Theoretical Model of a Turbine Meter', Trans ASME, J. of Basic Eng., 87, June 1965, p. 413

5   OWER, E.: 'The Theory of the Vane Anemometer', Phil Mag. (7), 2, 1926, p. 881

DISCUSSION

L.A. Salami: I would like to point out one or two things about the rig which was used.

First, at the maximum flowrate, the tank is filled in 30 seconds and I feel this is too short a time to open and shut two valves and to time with any precision (the Authors quote a precision of 0.05%).

Secondly, I think that the flow straightener should be present in the test length even though one may be incorporated in the meter. Fig. A shows the effect of not having a flow straightener on the effect of developing profile on the meter characteristics. The thick graphs are the real profile effects. (The meter is in 'uniform' flow at the start of the 'x' axis and this section is about 100 diameters down-steam of a bend.)

Thirdly, the meter with ball bearing should have been indicated as the type of bearing used affects the meter constant.

Fourthly, an idea of the geometry of the meters tested could have been given as these have a great bearing on the results obtained.

We have also tested turbine flowmeters at the University of Cardiff and some of the aspects covered are:

(a)   effects of developing pipe profile on the meter constant

(b)   the effect of swirl and the effectiveness of different flow straightener designs to remove swirl

(c)   temperature effect using water only up to 60°C

(d)   viscosity effect up to 400 centistokes using different concentrations of 'methocel' in water.

Authors' reply: Experimental tests and theoretical considerations showed that the error introduced by the stationary start and stop principle used in the test rig was only 0.02% for the short run time of 30s.

No flow straighteners were used in the test rig because the pipe-work was designed to minimise swirl in the flow. This was checked experimentally. Indeed, the introduction of a straightening element could introduce more swirl or velocity profile effects than it was intended to remove. Further experimental work is required into the detailed performance, advantages and disadvantages of flow straighteners.

P.A.K. Tan: Although the study of viscosity effects on turbine-meters is not in my research program, I am nevertheless very interested in this aspect and therefore found that part of this paper concerning the viscosity effects very stimulating. It is interesting to note that four out of the five meters had journal bearings, but I am unable to see the advantage of using journal bearings for research into viscosity effects on turbine-meters. The frictional torque of such bearings, although I know that they are widely used in industry, is greatly affected by the lubricating properties of the calibrating liquid used. Any change in bearing friction is very important especially, as Professor Hutton and I have demonstrated in our paper, at the lower range of flow. However to illustrate this point further, I shall consider Fig. 5 of the paper. You will notice that when spindle-oil (which I understand is a lubricating oil) is used, the calibration curve has a very steep rising

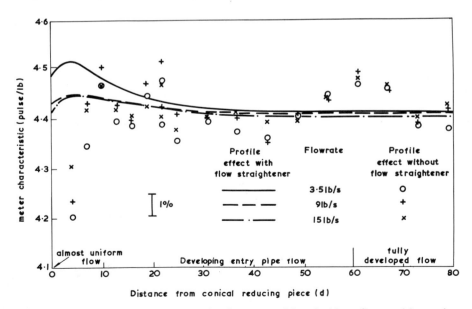

Fig. A   Velocity profile effect on turbine flowmeter with and without flow straightener in the pipe line

318   Withers et al.

section with decreasing flow. From our theory, this is not surprising since the spindle-oil greatly reduces the friction of journal bearings. Particulars concerning the five meters were not given but from the behaviour of the calibration curves shown in Fig. 5, my guess about Meter D is that it has one or more of the following characteristics: –

(a)   It has a large tip clearance
(b)   It has comparatively many blades
(c)   It has a large axial width
(d)   It has a comparatively large hub diameter

From the experimental results, one can see that increasing viscosity increases the meter constant even at high flows where frictional torque is not important. This tends to suggest that the meters have large clearances. To my mind, viscosity can affect calibration curves in at least seven ways but I shall only touch on two of them that can explain such an increase. Prof. Hutton and I have shown in our paper the importance of flow through the tip clearance. Increasing viscosity decreases the Reynolds' number based on the pipe diameter and thus the value of m in the power-law velocity equation. Decreasing m changes the velocity profile and this change is most significant near the wall. On this basis it can be shown by calculation that increasing viscosity increases the meter constant. This type of increase is larger the greater the tip clearance and the larger the hub diameter. However if the velocity profile is ideally flat such an increase will not occur.

The second effect of increasing viscosity is in the increase in resistance to flow through the tip clearance. Such an increase in resistance to flow will decrease the net flow through the tip clearance and thus increase the meter constant.

The effect of temperature on the meter dimensions was mentioned in the paper and I quote, "Thermal expansion causes an increase in both the area of the meter bore and the radius of the meter rotor." It is true that increasing the net area of flow through the meter decreases the meter constant. However, change in the flow area due to temperature is not only the result of change of the meter bore. If temperature has any significant effect on the dimensions of the blade thickness and hub diameter, they too have to be considered in computing the change in the net area of flow. Analysis on the effect of changes of the hub diameter and blade thickness is given in a note by Tan*.

I am not quite sure what is meant by the radius of the meter rotor. Presumably this means the tip radius, but if so the above argument is complicated by the existance of flow between the blade tip and the casing. If one is to consider the effect of changing the blade tip radius of the meter only, one finds that increasing the tip radius decreases the meter constant if the meter is placed, for example, in a fully developed flow region. However, if the velocity profile is flat, increasing the tip radius increases the meter constant instead. Theoretical calculation using the dimensions of our meter shows that increasing the radial clearance from 0.010 in. to 0.050 in. increases the meter constant by 2% for a fully developed velocity profile but decreases the meter constant by 0.2% for a flat velocity profile.

Finally, I should just like to say that it would be of great help, especially for the purpose of analysing the results, if the meter constant were given as an absolute rather than as a percentage value. Also, if particulars concerning the meter, such as the number of blades, bore and rotor dimensions (actually measured and not drawing dimensions) etc. were given as well: it would be much appreciated by those wishing to do quantitative analysis of meter constants.

Authors' reply: The data we present pertain to freely available commercial turbine flow meters and are primarily intended to show the variability in performance which exists in such meters. They all have nominally identical maximum flowrates. We do not claim that these meters are ideally suited for studying the various factors which influence meter behaviour. Thus, we do not claim or infer any advantage or disadvantage in using journal bearings for research into viscosity effects on turbine meters.

Most of the Mr. Tan's contribution deals with his personal interpretation of our work and must in our view, remain conjectural at this stage in turbine meter research.

*TAN, P.A.K.: 'Experimental and Analytical Studies of Turbine-type Flowmeter – Progress Report 2', Dept. of Mech. Eng., University of Southampton, July 1971

We agree that our paper would be more valuable if complete meter details were given. This, of course, would enable the commercial meters to be identified and the conditions of meter loan and publication of results do not permit the identification of the meter.

K.J. Zanker: Could the authors comment on the problems of turbine meter bearings:

(1)   What type of bearings were used?
(2)   What is the order of the bearing friction torque?
(3)   Have the Authors any quantitative methods of assessing bearing performance?

Authors' reply: One meter examined had a ball bearing whilst the others had various designs of journal bearings. We are not able to place the meters in order of bearing friction torques because this is just one of the many factors influencing the overall behaviour of the meters. We do not see how the single effect of bearing performance can be assessed without a full understanding of all the factors involved.

D.L. Smith: Tests we have made on turbine meters with varying amounts of "on line" duty show that they either repeat the original manufacturers' calibration very closely, or they are significantly different due, presumably, to wear. In this context, the larger sizes, e.g. 3 inch, show better linearity and repeatability than the small sizes. Approach configeration is important and flow straighteners are used.

What size of meter was used by Mr. Tan in his work?

Authors' reply: 1½-inch.

See also the Discussion to the paper by Tan and Hutton − Ed.

EXPERIMENTAL, ANALYTICAL AND TIP CLEARANCE LOSS STUDIES IN TURBINE-TYPE FLOWMETERS

P.A.K. Tan and S.P. Hutton

Abstract: This paper shows the importance of tip clearance leakage in turbine-type flowmeters. A theoretical model of the leakage through the tip clearance is presented to describe the behaviour of characteristic curves of turbine flowmeters. When constants in the theoretical equation are determined empirically the plotted equation agrees well with experimental results. Analysis using this leakage model shows that reducing the frictional torque on a turbine meter to a minimum can increase the variation of the meter constant over the operating range rather than its intended purpose of decreasing it.

1   INTRODUCTION

The turbine-type flowmeter is a volumetric instrument which utilises the kinetic and pressure energy of the fluid flow to produce a driving torque on the blades. The rotor, supported on two low-friction bearings, revolves at a rate proportional to the speed of flow.

The practical operating range at low flows depends on the ability of the meter to maintain a constant calibration curve over the range of flow. Rising or falling sections of the calibration curve with decreasing flow rate will introduce non-linearities in the flow measurement.

Lee and Karlby[1] derived equations to account for the rising section of the calibration curve. In their analysis they equated the driving torque to the change of angular momentum of the fluid. Then by using airfoil theory they related the resisting torque to skin friction drag on the rotor blades. Equating the driving and the resisting torque they showed that the rising section of the calibration curve is due to the skin friction drag on the blades.

However, Menkyna[2] suggested that the rising section of the calibration curve is due to wakes formed by the flow straighteners. Because of these wakes, the blades sweep through regions of varying velocities. This gives rise to unsteady flow and thus overestimation.

In the authors' experiments, performed using a turned down rotor (large tip clearance) and a full diameter rotor (small clearance), a rising section was observed in the calibration curve in the case of the former, a falling section in the case of the latter. The two rotors were the same except for differences in the tip clearance. The two models[1,2] mentioned above could not be used to explain the falling section of the full diameter rotor. This suggested that tip clearance may have an effect on the behaviour of the calibration curve. Using airfoil theory a model is presented showing that leakage through the tip clearance can affect the meter constant considerably. This theoretical model can be used to explain not only the rising section for the turned down rotor but also the falling section for the full rotor.

2   THEORETICAL CONSIDERATIONS

The following symbols are used in the analysis:

c	$= r_o - r_t$ = clearance	D_T	= tip diameter of blade = $2r_t$
C	= chord of blade	K	= helical pitch of blades
C_D	= drag coefficient	N	= number of blades
C_1	= lift constant	n	= rotational speed r.p.s.

Q = total volume flow through meter/second

Q_r = volume through rotor blades only/second

Q_c = volume flow through tip clearance/second

r_h = hub radius of rotor

r_t = tip radius of rotor blades

r_o = radius of casing of meter

t = blade thickness

T = retarding torque (excluding drag)

T_d = total driving torque

W = axial width of rotor

α = angle of attack (incident)

β = blade angle with respect to axial direction

θ = angle of relative velocity V_r

γ = $\dfrac{c}{r_o}$ = tip clearance ratio

ϕ = $\dfrac{r_h}{r_o}$ = hub ratio

δ = boundary layer thickness

λ = $\dfrac{\delta}{r_o}$

η = $(1 - \dfrac{r}{r_o})$

U_{mf} = free stream velocity or maximum velocity of power law profile

2.1 Airfoil approach for derivation of rotational speed n

The diagram in Fig. 1 shows the forces acting on an elemental blade of height dr at a radius r. The elemental drag dD acts in the direction of the relative velocity V_r while the elemental lift dL acts perpendicular to V_r.

$$dL = \frac{\tfrac{1}{2} \cdot \rho \cdot U^2 \cdot C_L \cdot C \cdot dr}{\cos^2 \theta}$$

$$dD = \frac{\tfrac{1}{2} \cdot \rho \cdot U^2 \cdot C_D \cdot C \cdot dr}{\cos^2 \theta}$$

Total elemental driving torque dT_d for N blades is

$$dT_d = N \cdot r \cdot (dL \cdot \cos\theta - dD \cdot \sin\theta)$$

$$= \tfrac{1}{2} \cdot \rho \cdot U^2 \cdot N.C. \; r. \; dr. \; (C_L \cdot \cos\theta - C_D \cdot \sin\theta)$$

$$\cos\beta = \frac{W}{C}$$

For small values of angle of incidence

$$C_L = C_1 \cdot \sin\alpha$$

Therefore $dT_d = \tfrac{1}{2} \cdot \rho \cdot N. W. U^2 \cdot r. dr. \left(\dfrac{C_1 \cdot \sin\alpha}{\cos\beta \cdot \cos\theta} - \dfrac{C_D \cdot \tan\theta}{\cos\beta \cdot \cos\theta} \right)$ --- (1)

As shown in Ref. (3)

$$\frac{\sin \alpha}{\cos\beta . \cos\theta} = r. \left(\frac{2\pi}{K} - \frac{2\pi . n}{U}\right) \quad ...(2)$$

For small angles of incidence α, the angle of the relative velocity can be taken to be the blade angle i.e. $\cos \beta = \cos \theta$

Therefore $\qquad \dfrac{\tan \theta}{\cos\beta . \cos\theta} = \dfrac{\tan \theta}{\cos^2\theta}$

$$= \frac{2\pi . r. U. \{U^2 + (2\pi . r. n)^2\}}{U^3} \quad ...(3)$$

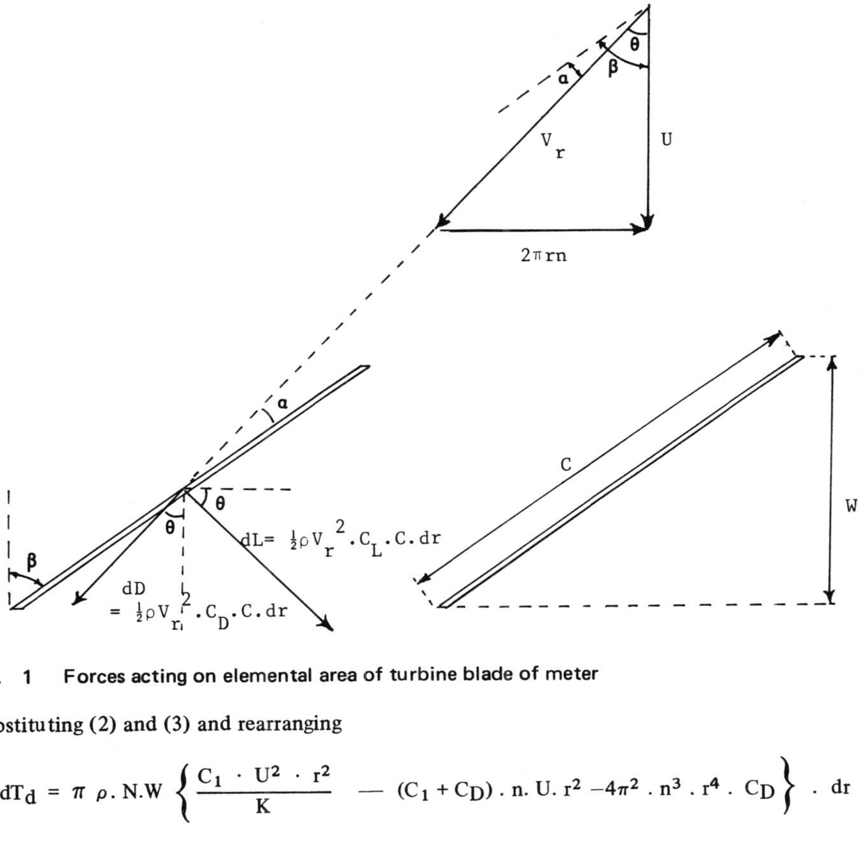

Fig. 1 Forces acting on elemental area of turbine blade of meter

Substituting (2) and (3) and rearranging

$$dT_d = \pi \, \rho. N.W \left\{\frac{C_1 \cdot U^2 \cdot r^2}{K} - (C_1 + C_D). n. U. r^2 - 4\pi^2 . n^3 . r^4 . C_D\right\} . dr$$

Integrating from hub radius to tip radius (i.e. h to t)

$$T_d = \pi . \rho . N . W \left\{\frac{C_1}{K}\int_h^t U^2 r^2 . dr - n\left(C_1 + C_D\right)\int_h^t U r^2 . dr - 4\pi^2 . n^3 . C_D.\int_h^t \frac{r^4}{U} . dr\right\}$$

The total driving torque T_d is equal to the total resisting torque T (this excludes the drag on the blades as it has already been accounted for)

Therefore

$$n (C_1 + C_D) \int_h^t U.r^2 . dr = \frac{C_1.}{K} \int_h^t U^2 .r^2 . dr - 4\pi^2 . n^3 . C_D. \int_h^t \frac{r^4}{U} dr - \frac{T}{\pi\rho.NW}$$

Therefore

$$n = \frac{\dfrac{C_1}{K} \displaystyle\int_h^t U^2. r^2 \, dr - 4\pi^2 n^3 \cdot C_D \int_h^t \dfrac{r^4}{U} \cdot dr - \dfrac{T}{\pi \cdot \rho \cdot NW}}{(C_1 + C_D) \cdot \displaystyle\int_h^t U \cdot r^2 \cdot dr} \qquad \text{... (4)}$$

2.2 Derivation of meter constant equation

Q is the total volume flow rate through the meter while Q_r is the volume flow rate through the rotor blades only. The difference between the two flows is the tip clearance leakage flow Q_c

$$Q = \int_h^t U . (2\pi r - N.t) . dr + \int_t^{r_0} U . 2\pi r . dr \qquad \text{... (5)}$$

$$Q_r = \int_h^t U . (2\pi r - N.t) \, dr \qquad \text{... (6)}$$

t is the blade thickness

n/Q is the meter constant measured experimentally while n/Q_r is the meter constant that the rotor itself is registering.

From eqns. (4) and (6),

$$\frac{n}{Q_r} = \frac{\dfrac{C_1}{K} \displaystyle\int_h^t U^2 r^2 . dr - 4\pi^2 . n^3 . C_D . \int_h^t \dfrac{r^4}{U} . dr - \dfrac{T}{\pi . \rho . NW}}{(C_1 + C_D) \displaystyle\int_h^t U . r^2 . dr . \int_h^t U. (2\pi r - N . t) . dr} \qquad \text{... (7)}$$

In the calibration rig, there is a conical contraction from a 10 in. diam. pipe to a 2 in. diam. pipe. Due to viscosity effects, the velocity profile changes from a flat profile at the conical entrace of the 2 in. diam. pipe to the fully developed flow in the downstream position. Therefore the boundary layer thickness δ grows from zero at the entrance to r_0 at the fully developed region. The boundary layer is assumed turbulent and the velocity profile is assumed to obey the power law equation

i.e.
$$\frac{U}{U_{mf}} = \left(\frac{r_o - r}{\delta}\right)^{\frac{1}{m}}$$
...(8)

Thus in the evaluation of the integral of the registered meter constant n/Q_r, three conditions of boundary layer thickness exist:

(a) When the boundary layer thickness δ is less than the tip clearance, i.e. $0 < \delta < (r_o - r_t)$ where $r_o - r_t = c$

(b) When the boundary layer thickness is greater than the tip clearance c but less than $(r_o - r_h)$ where r_h is the hub radius. This is the case when the boundary layer thickness is within the rotor blade i.e. $(r_o - r_t) < \delta < (r_o - r_h)$

(c) When the boundary layer thickness is greater than $(r_o - r_h)$ and for fully developed flow i.e. $\delta > r_o - r_h$ and $\delta = r_o$.

Evaluations of the integrals of equation 7 for different conditions are given in Ref. (3). The equation becomes

$$\frac{n}{Q_r} = \frac{\dfrac{C_1}{C_1 + C_D} \cdot \dfrac{DT}{K} \cdot A(n)}{2\pi \cdot r_t \cdot r_o^2 \cdot B(n) \cdot C(n)} - \frac{4\pi^4 \cdot r_o^6 \cdot \dfrac{C_D}{C_1 + C_D} \cdot \left(\dfrac{n}{Q}\right)^3 \left\{C(n) + D(n)\right\}^3 \cdot Z(n)}{B(n) \cdot C(n)}$$

$$- \frac{\dfrac{T}{C_1 + C_D} \cdot \left[C(n) + D(n)\right]^2}{2\rho \cdot \left(\dfrac{NW}{DT}\right) \cdot r_o \cdot r_t \cdot B(n) \cdot C(n) \cdot Q^2}$$
... (9)

where n is 1, 2 or 3 depending whether it is condition (1), (2) or (3) as defined above. Their values for all conditions are given in Appendix 6.1.

Instead of using the power law equation in the evaluation, a more accurate assumption would be a modified version of this equation.

i.e.
$$\frac{U}{U_{mf}} = \left(\frac{r_o - r}{\delta} \cdot \frac{r_o}{r_o - r_h}\right)^{\frac{1}{m}}$$
... (10)

The effect of this is to shift proportionally the power law velocity profile in the pipe to that within the annulus between the rotor and the meter casing. Evaluation of the integrals is still the same except that a constant $\left\{ r_o/(r_o - r_h) \right\}^{1/m}$ needs to be included as and when required.

Assuming the velocity in the tip clearance follows the power law equation, then the registered volume is

$$\frac{Q_r}{Q} = \frac{C(n)}{C(n) + D(n)}$$
... (11)

and the leakage ratio

$$\frac{Q_c}{Q} = \frac{D(n)}{C(n) + D(n)}$$
... (12)

In the theoretical case where there is no tip clearance leakage, the registered meter constant is

$$\left(\frac{n}{Q_r}\right)_{n\ell} = \frac{\dfrac{C_1}{C_1 + C_D} \cdot \dfrac{D_T}{K} \cdot A(n)}{2\pi \cdot r_t \cdot r_o{}^2 \cdot B(n) \cdot C(n)} - \frac{4\pi^4 \ r_o{}^6 \cdot \left(\dfrac{n}{Q}\right)^3 \cdot \dfrac{C_D}{C_1 + C_D} \cdot C(n)^2 \cdot Z(n)}{B(n)}$$

$$ - \frac{\dfrac{T}{C_1 + C_D} \cdot C(n)}{2\rho \cdot \left(\dfrac{NW}{D_T}\right) \cdot r_o \cdot r_t \cdot B(n) \cdot Q^2} \qquad\qquad \dots (13)$$

2.3 Analysis of derived equation

Equation 9 has three terms and each term is discussed briefly below

2.3.1 Term 1

$$\frac{\dfrac{C_1}{C_1 + C_D} \cdot \dfrac{D_T}{K} \cdot A(n)}{2 \ r_t \cdot r_o{}^2 \cdot B(n) \cdot C(n)}$$

For a particular meter placed in a position with constant power law exponent $(1/m)$, this term has a constant value.

$$\frac{C_1}{C_1 + C_D} = \frac{1}{1 + (C_D/C_1)}$$

C_D/C_1 is very small and can be neglected. Therefore $C_1/(C_1 + C_D) = 1$.

Hence term 1 can be evaluated. As the flow rate increases, the registered meter constant n/Q_r tends toward the value of term 1.

2.3.2 Term 2

$$\frac{4\pi^4 \cdot r_o{}^6 \cdot \left(\dfrac{C_D}{C_1 + C_D}\right)\left(\dfrac{n}{Q}\right)^3 \cdot \left\{C(n) + D(n)\right\}^3 \cdot Z(n)}{B(n) \cdot C(n)}$$

For low viscosity like in water, term 2 is very small when compared with term 1 and can be neglected. Using $C_D = 0.005$, $C_1 = 2\pi$ and $n/Q = 3278.1$ rev./M^3 for full rotor at upstream position, the value of term 2 is 1.57 rev./M^3 as compared with 3330.9 rev./M^3 for term 1. Term 2 is neglected in all calculations in this paper.

This assumption is true only for low viscosity. In the case of high viscosity fluids the value of term 2 is large enough to predominate over the term of the leakage model causing the rising section of meter calibration curve. For evaluation of this term at high viscosity when it is no longer possible to neglect it, C_D can be calculated as for a flat plate and its value increases with decreasing flow rate. Such an analysis done on a commercial meter of rated flow of 72M^3/hr, shows that the value of term 2 expressed as a percentage of term 1 increases from 0.13% at flow rate of 30M^3/hr to 0.32% at 5M^3/hr for viscosity of 1 c Stoke. For 60 c Stoke, its value increases from 0.99% at 30 M^3/hr to 2.43% at 5 M^3/hr. Thus a meter having a rising section when calibrated in water may have a falling section when calibrated in a liquid of high viscosity.

2.3.3 Term 3

$$\frac{\left(\dfrac{T}{C_1 + C_D}\right) \cdot \left\{C(n) + D(n)\right\}^2}{2\rho \cdot \left(\dfrac{NW}{D_T}\right) \cdot r_o \cdot r_t \cdot B(n) \cdot C(n) \cdot Q^2}$$

The magnitude of term 3 varies inversely with the square of the volume flow rate Q, other factors being constant. Therefore its value is large at small flow rates but tends toward zero as the flow rate increases. This is the term responsible for pulling the meter constant down in the calibration curve i.e. it is responsible for the falling section of calibration curve at low flow rates.

Before this term can be calculated the value $T(C_1 + C_D)$ must first be known. The value $T(C_1 + C_D)$ is calculated for the case of full diameter rotor placed in a flat velocity profile flow (upstream position).

From experimental measurement the meter constant for a flow rate of $2.642 \, M^3/hr$ was $3204.7 \, rev/M^3$. Using these values $T(C_1 + C_D)$ was approximated to be $0.123 \, gm \, cm$. This value was used for all calculations in this paper. Taking C_1 as 2π and C_D as 0.005, the value of T is $0.787 \, gm \, cm$.

The resisting torque, besides the all important bearing torque, also consists of the magnetic pickup torque if any, rotor drag, blade tip drag, etc. This means that item 3 can further be split up to include a term that varies inversely with the volume flow rate. Such an analysis would provide a good mathematical exercise but it only complicates matters further. For this paper it is good enough to assume that T is constant throughout the range of flow.

2.4 Leakage model

Equation 9 gives the registered meter constant n/Q_r and it is based on the volume flow rate Q_r through the rotor blades. The meter constant measured experimentally is n/Q and is based on the total volume flow rate Q. The two meter constants differ in value because of leakage through the tip clearance.

Taking Q_c as the tip clearance leakage flow, then

$$Q = Q_r + Q_c$$

$$\frac{n}{Q} = \frac{n}{Q_r} \cdot \frac{Q_r}{Q}$$

$$= \frac{n}{Q_r} \cdot (1 - \frac{Q_c}{Q}) \qquad \qquad \text{... (14)}$$

Equation 14 shows the relationship between the measured meter constant n/Q and the registered meter constant n/Q_r where Q_c/Q is the leakage ratio. At any flow rate it is evident from the equation that the measured meter constant will depend on the percentage leakage through the tip clearance. The higher the value of the leakage ratio, the smaller will be the measured meter constant. If the velocity in the tip clearance flow is the same as in the main stream, then Q_c/Q is a constant value throughout the flow range and has the value given in equation 12. In such a case the shape of the calibration curve follows that of equation 9. A plot of leakage versus flow rate is shown in Fig. 3. It is a straight line passing through the origin. The gradient of the line is g and equals the leakage ratio as given in equation 12. The value of intercept k is zero. Because k is zero, the calibration curve does not have a rising section with decreasing flow rate. The meter constant decreases with decreasing flow rate as predicted in equation 9. Curve 2 in Fig. 2 shows the calibration curve in such a case where the leakage velocity is flat as in the main stream. For the turned down rotor the leakage ratio using equation 12 is 0.091.

However if the leakage ratio Q_c/Q varies with flow rate, the measured meter constant also varies accordingly. It will be shown empirically that the leakage ratio is of the form

$$\frac{Q_c}{Q} = g - \frac{k}{Q} \qquad \qquad \text{... (15)}$$

where g and k are constants. A plot of leakage against flow rate is also a straight line–the difference being that the straight line does not pass through the origin but has a negative intercept k. Fig. 3 shows the case for the turned down rotor.

Combining equations 14 and 15

$$\frac{n}{Q} = \frac{n}{Q_r}\left(1 - g + \frac{k}{Q}\right) \qquad \ldots (16)$$

Combining with equation 9, equation 16 can be written in the form

$$\frac{n}{Q} = \left(P - \frac{R}{Q^2}\right) \cdot \left(1 - g + \frac{k}{Q}\right) \qquad \ldots (17)$$

where

$$P = \frac{\left(\dfrac{C_1}{C_1 + C_D}\right) \cdot \left(\dfrac{D_T}{K}\right) \cdot A(n)}{2\pi . r_t . r_o^2 . B(n) . C(n)}$$

and

$$R = \frac{\left(\dfrac{T}{C_1 + C_D}\right) \cdot \left\{C(n) + D(n)\right\}^2}{2\rho . \left(\dfrac{NW}{D_T}\right) . r_o . r_t . B(n) . C(n)}$$

From equation 17 it is evident that two interacting terms come into play, namely

$$(1) \quad P . \frac{k}{Q}$$

and

$$(2) \quad \frac{R . (1 - g)}{Q^2}$$

The values of these two terms are small at very high flow rates Q and the measured meter constant tends to the value P. $(1 - g)$. As the flow rate decreases $(P . k)/Q$ tries to pull the calibration curve up while R. $(1 - g)$ tries to pull the calibration curve down. These will depend on the values of k and T respectively. At medium flow rates, $(P . k)/Q$ predominates thus giving the rising section of the calibration curve. However below a certain flow rate, $R(1 - g)/Q^2$ predominates and its value increases rapidly with decreasing flow rate Q. As a result the calibration curve falls very rapidly at very low flow rates.

The rising section of the calibration curve will only occur if the value of k is sufficiently big in comparison with the resisting torque T. This is so in the case of big clearance (e.g. the turned down rotor). In the case of the full diameter rotor whiere the value of k is small, no rising section of the calibration curve is obtained (see Fig. 7).

2.5 Determination of g and k

Before equation 16 can be evaluated theoretically for plotting, the values of g and k need to be known. These values are constants relating the leakage ratio to the flow rate and they can be evaluated in three ways.

2.5.1 Method 1 (semi-experimental deduction)

In this method, the theoretical calibration curves for no tip clearance leakage and for constant

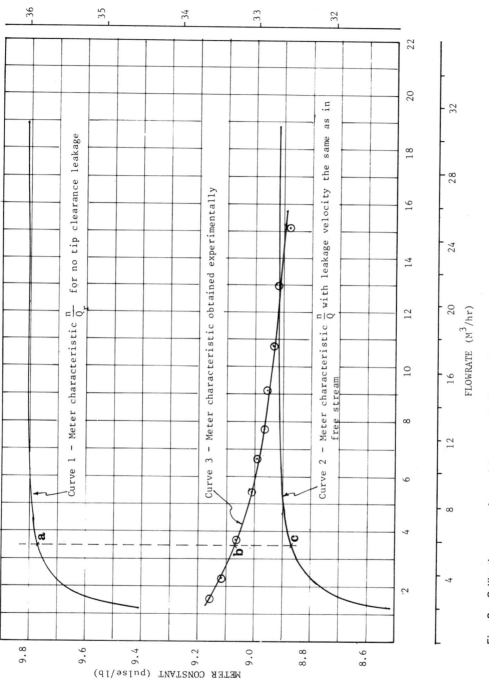

Fig. 2 Calibration curves of turned down rotor in flat velocity profile for deducing leakage

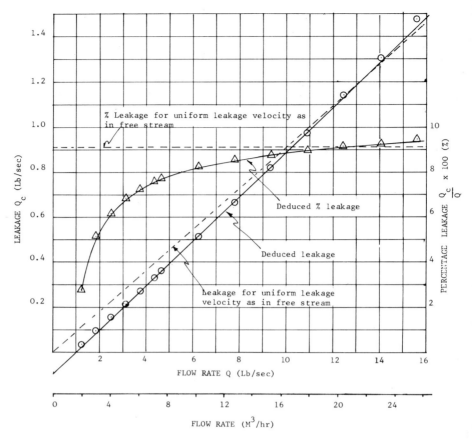

Fig. 3 Variation of tip clearance leakage and percentage leakage with flow rate for turned down rotor in flat velocity profile

leakage ratio Q_c/Q are plotted using equations 13 and 14 respectively. Then by using proportionality the leakage ratio Q_c/Q and the tip clearance leakage, Q_c at any flow rate can be deduced. Q_c, so obtained, is plotted against flow rate Q. The gradient of the plot gives the value of g and the intercept the value of k. An example is given in the following paragraph.

Fig. 2 shows the case for the turned down rotor at the upstream position where the velocity profile is flat. Curve 1 has no tip clearance leakage while curve 2 has a constant leakage ratio of $D(1)/\{C(1) + D(1)\} = 0.091$. Curve 3 is the experimental calibration curve.

At any flow rate, the proportion $\{(a - b) / (a - c)\}$ x 0.091 gives the leakage ratio at that ,w flow rate. The leakage would then be $\{(a - b) / (a - c)\}$ x 0.091 x Q. This is done for other flow rates. Graphs of percentage leakage against flow rate and tip clearance leakage against flow rate can be plotted as shown in Fig.3.

The gradient of the graph of Q_c against Q gives the value g and the intercept gives the value k. Using these values of g and k equation 17 can be plotted. Fig. 4 shows the plot by this method.

2.5.2 Method 2 (Lee and Karlby method)

This method was used by Lee and Karlby[1] to evaluate the constants in each of the three equations they derived to describe the different sections of the calibration curve. In all they took six experimental points since each equation has two unknown constants.

However, using the leakage model, only one equation (17) is required to describe the calibration throughout the whole range of flow. Hence only two experimental points need be used to determine g and k. For the case of the turned down rotor at the upstream position, the two points taken are:

	Flow Rate	Experimental meter constant	Theoretical n/Q_r
1	7.136 M³/hr	3317.3 M³/hr	3588.4 M³/hr
2	15.291 M³/hr	3283.1 M³/hr	3598.7 M³/hr

With these two points the values of g and k evaluated with equation 17 are 0.0983 and 0.0994 respectively. Fig. 5 shows the plot by this method.

2.5.3 Method 3 (Graph fitting by method of least squares)

To fit an empirical equation to a set of experimental data, a good fit can be obtained if the form of the equation used is correct. If the numerical equation is correct, it is possible to extrapolate so as to predict the behaviour of experimental points outside the range actually performed experimentally.

Equation 17 is of the form

$$\frac{n}{Q} = a_1 + \frac{a_2}{Q} + \frac{a_3}{Q^2} + \frac{a_3}{Q^3}$$

The term a_4/Q^3 is the product of $-R/Q^2$ and k/Q. Since these two terms are small in comparison with a_1, their product can be neglected and the equation becomes

$$\frac{n}{Q} = a_1 + \frac{a_2}{Q} + \frac{a_3}{Q^2} \qquad \text{... (18)}$$

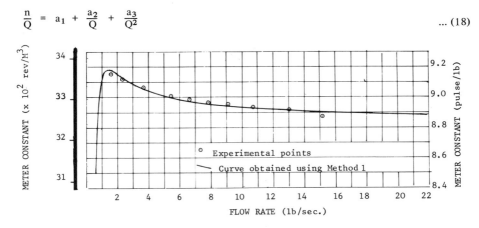

Fig. 4   Calibration curve of turned down rotor in flat velocity profile using Method 1

By the method of Least Squares three equations can be obtained from the experimental data involving a_1, a_2 and a_3. Solving the three equations the values a_1, a_2 and a_3 are obtained. By comparing these values with equation 17, the values of g and k are 0.098 and 0.104 respectively. Fig. 6 shows the plot by this method.

Method 2 is the least accurate of the three methods because the accuracy depends on how well the two points were chosen. By Method 1, care needs to be taken to draw the mean straight line through the deduced points. This is because the value of the intercept k is comparatively small and any slight shift of the line alters the value k considerably.

2.6 Effect of variation of resisting torque

It has sometimes been thought that by decreasing the resisting torque, it is possible to reduce the variation in the meter constant over the operating range. However analysis, on the assumption that g and k remain constant, shows that this may not necessarily by the case. Analysis for the full diameter and the turned down rotor are shown in Fig. 10 and 11 respectively. For the full diameter rotor where there is no rising section of the calibration curve, the decrease in resisting torque does reduce the variation. On the other hand for the turned down rotor, increasing the resisting torque can reduce the variation of meter instead of increasing it. There is then an optimum resisting torque where this variation is minimal. Any further increase in resisting torque will increase the variation. For both the rotors, decreasing the resisting torque does decrease the minimum flow that the meter can measure but does not necessarily improve the uniformity of the calibration curve.

3   EXPERIMENTAL RESULTS

Experiments were carried out using a closed circuit rig with water as the metered liquid. The description of this rig is given by Salami[4]. Calibration for both full diameter and turned down rotors were performed at the conical entrance of the 2'' diameter pipe where the velocity profile is virtually flat and at 84 diameters downstream where the flow is fully developed. Fig. 7 shows

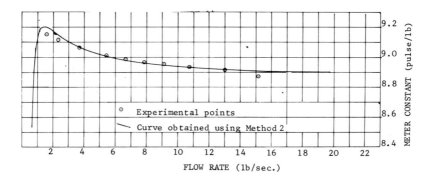

Fig. 5   Calibration curve of turned down rotor in flat velocity profile using Method 2

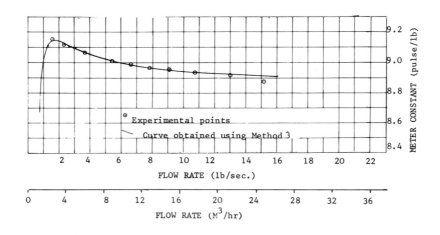

Fig. 6   Calibration curve of turned down rotor in flat velocity profile using Method 3

the experimental results and their comparison with theoretical values obtained using Method 1. Experiments were also carried out on a commercial meter. Figs. 8 and 9 show the comparison between the experimental results and the theoretical values by using Method 3 and Method 1 respectively.

It is seen that the experimental calibration curves for all three meters agree well with the theoretical model developed in this paper.

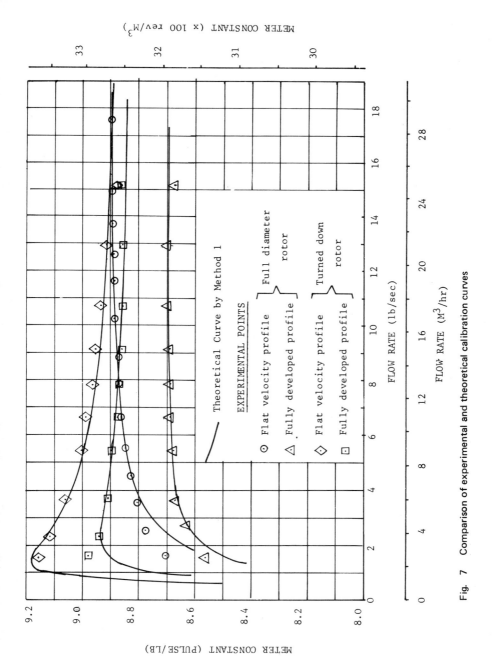

Fig. 7 Comparison of experimental and theoretical calibration curves

4   ACKNOWLEDGEMENTS

The authors express their gratitude to:

The Science Research Council for financing this research

The Esso Company for the research studentship grant to the first author

Elliots Ltd. for the loan of a turbine flowmeter

Mr. L.A. Salami for the use of an experimental calibration curve of the Elliots meter

Mrs. D.F. Griffiths for typing this paper.

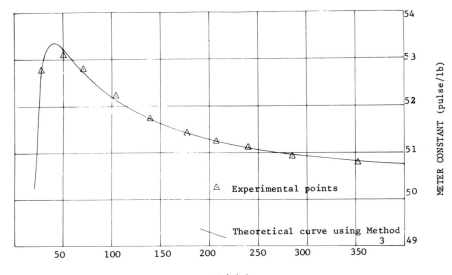

FLOW RATE (lb/min)

Fig. 8   Calibration curve of Elliots meter using leakage model: Method 3

5   REFERENCES

1   LEE, W.F., and KARLBY, H.: 'Study of Viscosity Effect and its Compensation on Turbine Type Flowmeters', J. Bas. Engg., Trans. Am. Soc. Mech. Engrs, 1960, 82, pp. 717-728

2   MENKYNA, V.L.: 'Correction of the Error Curve of Axial Turbine Liquid Flow Meters', Proc. 3rd Conf. on Fluid Mech. and Fluid Machy, 1969, Akademai Kiado, Budapest, pp. 373-379

3   TAN, P.A.K.: 'Experimental and Analytical Studies of Turbine-Type Flowmeter — Progress Report I', Dept, Mech. Engg., University of Southampton, Feb. 1971

4   SALAMI, L.A.: 'Turbine-type Flowmeter Calibration Rig', Dept. Mech. Eng., University of Southampton, April 1971

6   APPENDIX

6.1 The values of A(n), B(n), C(n), D(n) and Z(n) are as follows:

$$A(1) \quad = \quad \frac{1}{3} \left[(1 - \lambda\eta)^3 \right]_{\eta = \frac{1}{\lambda}(1 - \phi)}^{\eta = \frac{\gamma}{\lambda}}$$

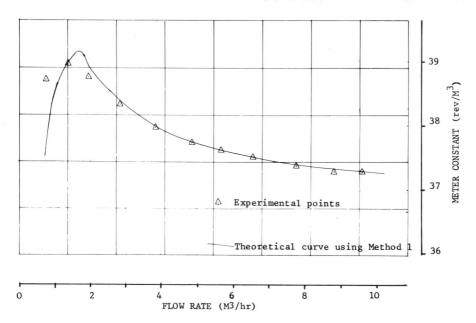

Fig. 9 Calibration curve of Elliot meter using leakage model: Method 1 (experimental points obtained by L.A. Salami)

$$B(1) = \frac{1}{3} \left[(1 - \lambda\eta)^3 \right]_{\frac{1}{\lambda}(1-\phi)}^{\frac{\gamma}{\lambda}}$$

$$C(1) = \left[(1 - \lambda\eta)^2 + \frac{N.t.\lambda.\eta.}{\pi.r_0} \right]_{\frac{1}{\lambda}(1-\phi)}^{\frac{\gamma}{\lambda}}$$

$$D(1) = \left[(1 - \lambda\eta)^2 \right]_{\frac{\gamma}{\lambda}}^{1} + \left[2 \cdot \sum_{n=1}^{n=2} \frac{\eta^{(1/m)+n}(1-\lambda\eta)^{2-n}\left(\frac{1}{m}\right)! \lambda^n}{\left(\frac{1}{m}+n\right)!} \right]_{0}^{1}$$

$$Z(1) = \frac{1}{5} \left[(1 - \lambda\eta)^5 \right]_{\frac{1}{\lambda}(1-\phi)}^{\frac{\gamma}{\lambda}}$$

$$A(2) = \frac{1}{3} \left[(1 - \lambda\eta)^3 \right]_{\frac{1}{\lambda}(1-\phi)}^{1} + \left[\sum_{n=1}^{n=3} \frac{\eta^{(2/m)+n}.(1-\lambda\eta)^{3-n}. 2! (2/m)! \lambda^n}{(3-n)! \left\{(2/m)+n\right\}!} \right]_{\frac{\gamma}{\lambda}}^{1}$$

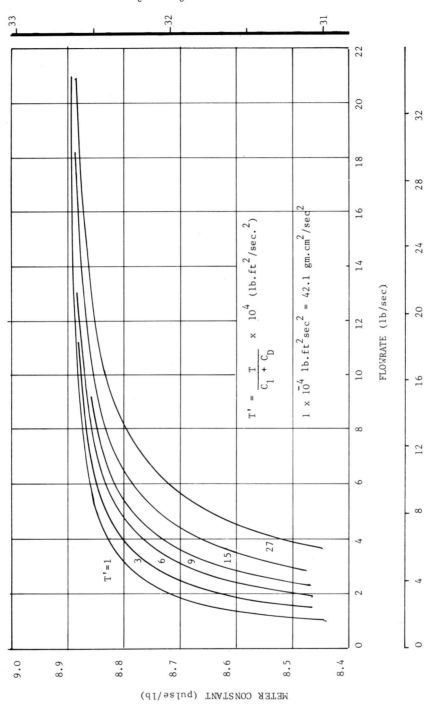

Fig. 10 Variation of meter calibration curve with retarding torque for full diameter rotor at flat velocity profile

$$T' = \frac{T}{C_1 + C_D} \times 10^4 \quad (\text{lb.ft}^2/\text{sec.}^2)$$

$1 \times 10^{-4} \text{ lb.ft}^2 \text{sec}^2 = 42.1 \text{ gm.cm}^2/\text{sec}^2$

$$B(2) = \frac{1}{3}\left[(1-\lambda\eta)^3\right]^1_{\frac{1}{\lambda}(1-\phi)} + \left[\sum_{n=1}^{n=3} \frac{\eta^{(1/m)+n} \cdot (1-\lambda\eta)^{3-n} \cdot 2! \left(\frac{1}{m}\right)! \lambda^n}{(3-n)! \ \{(1/m)+n\}!}\right]^1_{\frac{\gamma}{\lambda}}$$

$$C(2) = \left[(1-\lambda\eta)^2 + \frac{N.t.\lambda.\eta.}{\pi.r_0}\right]^1_{\frac{1}{\lambda}(1-\phi)} + \left[2\sum_{n=1}^{n=2} \frac{\eta^{(1/m)+n} \cdot (1-\lambda\eta)^{2-n} \cdot \left(\frac{1}{m}\right)! \lambda^n}{\{(1/m)+n\}!}\right.$$

$$\left. - \frac{N.t.\lambda.\eta^{(1/m)+1}}{\pi.r_0 \cdot \{(1/m)+1\}}\right]^1_{\frac{\gamma}{\lambda}}$$

$$D(2) = \left[2 \cdot \sum_{n=1}^{n=2} \frac{\eta^{(1/m)+n} \cdot (1-\lambda\eta)^{2-n} \cdot (1/m)! \lambda^n}{\{(1/m)+n\}!}\right]^{\frac{\gamma}{\lambda}}_0$$

$$Z(2) = \frac{1}{5}\left[(1-\lambda\eta)^5\right]^1_{\frac{1}{\lambda}(1-\phi)} + \left[\sum_{n=1}^{n=5} \frac{\eta^{n-(1/m)} \cdot (1-\lambda\eta)^{5-n} \cdot (-1/m)! 4! \lambda^n}{\{n-(1/m)\}! (5-n)!}\right]^1_{\frac{\gamma}{\lambda}}$$

$$A(3) = \left[\sum_{n=1}^{n=3} \frac{\eta^{(2/m)+n} \cdot (1-\lambda\eta)^{3-n} \cdot (2/m)! 2! \lambda^n}{\{(2/m)+n\}! (3-n)!}\right]^{\frac{1}{\lambda}(1-\phi)}_{\frac{\gamma}{\lambda}}$$

$$B(3) = \left[\sum_{n=1}^{n=3} \frac{\eta^{(1/m)+n} \cdot (1-\lambda\eta)^{3-n} \cdot (1/m)! 2! \lambda^n}{\{(1/m)+n\}! (3-n)!}\right]^{\frac{1}{\lambda}(1-\phi)}_{\frac{\gamma}{\lambda}}$$

$$C(3) = \left[2 \cdot \sum_{n=1}^{n=2} \frac{\eta^{(1/m+n)} \cdot (1-\lambda\eta)^{2-n} \cdot (1/m)! \lambda^n}{\{(1/m)+n\}!} - \frac{N.t.\lambda.\eta^{(1/m)+1}}{\pi.r_0. \{(1/m)+1\}}\right]^{\frac{1}{\lambda}(1-\phi)}_{\frac{\gamma}{\lambda}}$$

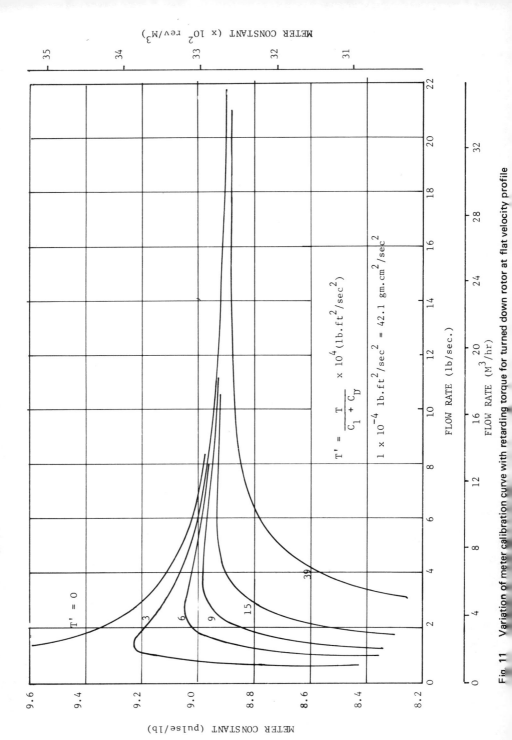

Fig. 11   Variation of meter calibration curve with retarding torque for turned down rotor at flat velocity profile

$$D(3) = \left[2 \cdot \sum_{n=1}^{n=2} \frac{\eta^{(1/m)+n} \cdot (1-\lambda\eta)^{2-n} \cdot (1/m)! \, \lambda^n}{\{(1/m)+n\}!} \right]_0^{\frac{\gamma}{\lambda}}$$

$$Z(3) = \left[\sum_{n=1}^{n=5} \frac{\eta^{n-(1/m)} \cdot (1-\lambda\eta)^{5-n} \cdot (-1/m)! \, 4! \, \lambda^n}{\{n-(1/m)\}! \, (5-n)!} \right]_{\frac{\gamma}{\lambda}}^{\frac{1}{\lambda}(1-\phi)}$$

6.2 Physical dimensions of turbine flowmeter

Full diameter rotor:

Number of blades	N	$= 3$
Thickness of blades	t	$= 0.05$ in.
Axial width of rotor	W	$= 0.437$ in.
Helical pitch of blades	K	$= 6.284$ in.
Hub diameter	$2r_h$	$= 0.25$ in.
Casing diameter	$2r_o$	$= 2$ in.
Blade tip diameter	$2r_t$	$= 1.985$ in.

Turned down rotor:

All dimensions are the same except
Blade tip diameter $2r_t$ = 1.912 in.

DISCUSSION

V.R. Withers: The Authors appear to dismiss the theory of Lee and Karlby. Is this because the idea of changes from laminar to turbulent flow across the rotor blade is wrong or because it is not applicable to the Authors' particular tests with water? Lee and Karlby's theory is intended to explain the effects of viscous liquids on turbine meter performance, and it assumes that the viscous drag on the blade surfaces is the predominant source of total resisting torque. Do the Authors disagree with this assumption, firstly for low viscosity fluids (e.g. water), and, secondly, for higher viscosity fluids?

Authors' reply: Why Lee and Karlby's theory cannot be used to explain the behaviour of all calibration curves has been explained in the introduction to this paper. However, for the benefit of the questioner, we shall elaborate in greater detail. We do not dispute the idea of changes from laminar to turbulent flow across the rotor blade, neither do we dispute the effects of viscosity upon turbine meter performance. In fact the coefficient of drag C_D is present in the equation derived in our theory and if desired, its value at any flow rate can be calculated as indicated in the paper. However, we maintain that the behaviour of the calibration curve is not solely due to the variation of viscous drag on the rotor blade. As stated in this paper, a rotor with a big tip clearance has a calibration curve that rises initially with decreasing flow rate but finally falls again. According to Lee and Karlby, this behaviour is due to the variation of drag on the rotor blades depending whether the flow over them is fully turbulent, laminar or transitional. If this is correct, then a similar rotor but with a small tip clearance should behave in a similar manner since the rotor blades experience the same variation of drag. As shown in this paper, this was not so as the calibration curve for small clearance falls all the way with decreasing flow rate. These variations with tip clearance can be explained very satisfactorily with our theory but not by that of Lee and Karlby.

Our full diameter rotor was tested by Dr. L.A. Salami, not just in water but also in liquid solutions with viscosities ranging higher than those used by the questioner (to about 400 cStoke).

In this paper, we have already stated that it is necessary to consider viscous drag on the blades for high viscosity liquids. For low viscosity liquids we maintain that the bearing torque is the main source of total resisting torque T. As shown in our theory, the term containing the resisting torque is important at low flow rate. To illustrate this experimentally, we reproduce here Fig. 10 of Ref. (1) as Fig. A. The figure shows a great change in the calibration curve at low flow rates when the rotor has been run for 75 hours. This change is due to an increase in bearing torque of the ball bearings used. Thus, even for ball bearings, the bearing torque is so important; it would be even more so for the questioner's meters where journal bearings were used in four out of five of his meters.

D.L. Smith: Tests we have made on turbine-meters with varying amounts of 'on-line' duty show that they either repeat the original manufacturer's calibration very closely, or they are significantly different due, presumably, to wear. In this context, the larger sizes, e.g. 3 in. show better linearity and repeatability than the smaller sizes. Approach configuration is important and flow straighteners are used. What size of meter was used in this work?

Authors' reply: The turbine-meter can be a very accurate instrument if it is used intelligently. It is important that when installing a turbine-meter, it must be done correctly; otherwise its calibration may be altered. Misalignment of the pipe with respect to the meter or having the gasket joint protruding into the flow can give rise to errors in the meter reading. This can best be demonstrated by some experiments performed by the authors. A concentric orifice disc was placed directly in front of the meter rotor such that 0.050 in. of the disc protruded into the flow. It was found that such obstruction of flow near the blade tip can cause a considerable change in the meter constant (as much as 2% in certain cases). Results of these experiments are given in Ref.(1). It is not difficult to visualise that errors resulting from such incorrect installation would be greater for the smaller meters.

It is now unwise to have implicit faith in the calibration curve supplied by the manufacturer without first having some idea of the calibration rig used and the method employed in the calibration. For instance, it is not uncommon that dynamic weighing is used by some manufacturers in calibrating their meters. In the rig used by the authors, static weighing is used for all research work but the rig can also be used for dynamic weighing. A series of tests were performed to cross-check the error of dynamic weighing with respect to static weighing. An average error of about 2 lb was recorded between the two methods of weighing but this error can go as high as 9 lb. Though the percentage error is not substantial if a large volume is collected, it can become very significant if a small volume is collected as would be the case in calibrating smaller size meters. It is true that dynamic weighing is much faster and simpler but it is good practice to know how far one can stretch the limits of accuracy.

Linearity of the calibration curve is a question of design and one can get the best possible linearity by manipulating the necessary variables. An elementary attempt to this type of design based on the theory presented in the paper is given in Ref. (1). However the authors would not argue if one were to say that it is easier to get a better linearity for a bigger size. Much remains to be discovered about how the value of the leakage factor k varies, especially with regard to the size of meters.

If one intends to compare the performance of turbine-meters of different sizes or for that matter, meters of the same size but of different blade angles or helical pitch, different hub diameter, etc., obviously one requires some basis for comparison. It is no good just comparing the calibration curves. Going through the available literature, no general method has yet been put forward for such a comparison. One method which the authors suggest is to divide the flow rate by the effective area of the meter and thus give the flowrate in terms of the average axial velocity of fluid through the meter rotor. For the case of turbine-meters with helical blades, the meter constant n/Q is multiplied by the helical pitch K and the effective area of the meter, i.e.

$$\frac{n}{Q} \cdot K.\pi \left\{ r_o^2 - r_h^2 - N.t. (r_t - r_o) \right\}$$

For turbine-meters with constant angle blade, the meter constant is multiplied as follows:

$$\frac{n}{Q} \cdot \frac{\frac{4}{3}\pi \cdot \left\{ r_o{}^2 - r_h{}^2 - N.t.\,(r_t - r_o) \right\}}{\tan \beta} \cdot \frac{(r_t{}^3 - r_h{}^3)}{(r_t{}^2 - r_h{}^2)}$$

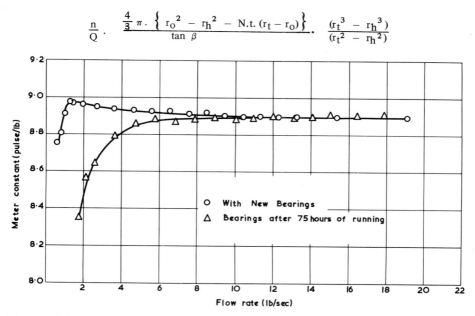

Fig. A   Effect of running time on calibration curve of turned down rotor at downstream position (Fig. 10 from Ref. 1)

Alternatively, to eliminate any effects of the presence of flow straighteners, the meter constant can be expressed as a percentage of that at the highest flow rate. On the basis of the first method proposed, the theoretical performance of different sizes of gas turbine-meters are given in Fig. B. To permit this theoretical comparison, certain variables, as given in the figure, are kept constant. Gas turbine-meters are used in this exercise because the magnitude of the torque term is relatively much bigger than the leakage factor term because of the much smaller density of gas. Thus the leakage factor k is taken as zero. The value of g is as calculated from Equation 12. It is obvious from Fig. B, the linearity is worse the smaller the meter size. In fact at any particular flow velocity, the drop in the modified meter constant is proportional to the square of the meter size. The authors do not wish to present a similar comparison for liquid turbine-meters because little is yet known of the variation of leakage factor k with meter size. Because leakage factor is so important for liquid meters, a comparison by keeping k constant would be most misleading.

When discussing 'approach configuration', one must distinguish between the three types of 'approach configuration' possible, namely:

(a)   presence of swirl in the flow

(b)   the shape of inlet velocity profile

(c)   asymmetry in the inlet velocity profile

The use of flow straighteners can eliminate swirl but it is of extreme importance that if a flow straightener is to be used as an integral part of the meter, the meter must be calibrated together with it. One of the reasons is because it is very difficult to align the elements of the straightener completely axially. Also, one cannot change the straightener with another of the same make and still expect the same accuracy without further calibration of the meter.

In reply to the question raised, the authors used 1½ in. and 2 in. meters in their research.

K.J. Zanker: Could the authors comment on the problems of turbine-meter bearings: what type of bearings were used? What order is the bearing friction torque? Have they any quantitative methods of assessing bearing performance?

Authors' reply: The authors are very glad that the writer brings out the question of turbine-meter bearings. This indeed is one of the major problems of turbine-meters.

For the case of our own turbine-meters, the bearings used are miniature ball-bearings. We have done a series of tests to look at the repeatability of our meters using both shielded and open bearings. The tests were done over a period of more than 70 hours of running time. Results of these tests are given in Ref. [3] of the paper and Ref. (A) at the end of this comment. The order of the resisting torque of our meter is estimated in the paper.

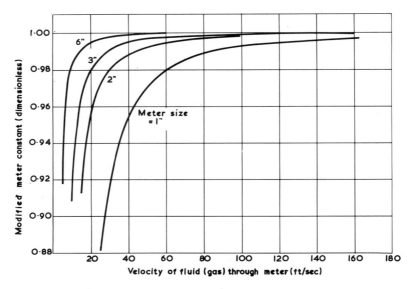

Fig. B   Comparison of performance of gas turbine meter for different meter sizes

Resisting torque T	$= 1.5 \times 10^{-3}$ ft^2 lb/sec^2
Density of fluid (air) ℓ	$= 0.0758$ lb/ft^3
Blade angle β	$= 15°$
No. of blades N	$= 6$
Axial width of blade W	$= 1$ in
Blade tip ratio $\frac{r_t}{r_o}$	$= 0.990$
Hub ratio $\frac{r_h}{r_o}$	$= 0.500$

Assessment of bearing performance is difficult; more so if journal bearings are used. For gas turbine-meters 'spin-test' was proposed by Lee and Evans in Ref. (B). However this method would be difficult for liquid turbine-meters. If ball-bearings are used, perhaps one can dry the meter first and then do the spin test in air. However it would be difficult for journal bearings not only because of their higher magnitude of frictional torque but also because the bearings would then not be lubricated by the metered liquid.

In our case, we can assess the performance of our turbine-meter bearings by recalibrating the meter and comparing it with its original calibration curve. It is possible from this comparison to estimate the increase in the frictional torque. Such an exercise is given in Ref. (A). However this method would not be suitable for the general users of turbine-meters as it would require a precision calibration rig.

The problem of frictional torque becomes more and more acute as one tries to extend the meter range to lower and lower flows. It would not be out of place to include here a brief discussion of turbine-meters using journal bearings. In journal bearings when conditions are such that the value $\mu n/p$ is relatively high, hydrodynamic lubrication will prevail and the theoretical relationship between frictional torque, viscosity and speed of rotation is simple. In this context

μ is the absolute viscosity of the lubricant, n the rotational speed and p the average bearing pressure. Thus, the coefficient of friction will depend upon the viscosity of the lubricant and the speed of rotation. However, when the value $\mu n/p$ is relatively low, boundary lubrication prevails and the value of the coefficient of friction is now very much higher. Its value in this region depends upon lubricating properties of the lubricant other than its viscosity. Hence the coefficient of friction is no longer dependent upon the viscosity and a change to a different type of lubricant (in turbine-meters a different metered liquid) having the same viscosity may result in a different coefficient of friction. In between the two regions of boundary lubrication and hydrodynamic lubrication, there also exists a region of mixed lubrication.

In turbine-meters, the effect of frictional torque increases with the square of decreasing flowrate. However, hydrodynamic lubrication will occur if the speed of rotation (hence the flow-rate) is sufficiently high, otherwise boundary lubrication will prevail. Thus one can see the existence of a vicious circle; low coefficient of friction when frictional torque is not important and high coefficient of friction with frictional torque is very important.

If boundary lubrication prevails when friction torque is important, one can see the disadvantage of the use of journal bearings. Calibration of turbine-meters is normally done with water although the meters are nearly always used to meter another liquid whose lubricating properties may be very different from the calibrating liquid. We know that journal bearings are widely used commercially and we appreciate the advantages of journal bearings over ball bearings. However in research where one is interested to look at the fundamentals of turbine-meters, one would not be ill-advised to use ball bearings.

REFERENCES

(A)   TAN, P.A.K.: 'Experimental and Analytical Studies of Turbine-type Flowmeters – Progress Report 2'. Dept. of Mech. Eng'g., University of Southampton, July 1971

(B)   LEE, W.F.Z., and EVANS, H.J.: 'A Field Method of Determining Gas Turbine Meter Performance'. ASME Publication, Paper No. 69–WA/FM-1

L.A. Salami: The experimental results for the full diameter and the turned-down rotor meter on which the theory is based were first obtained by me whilst I was at Cardiff University. This lead me to traverse the velocity profile at various sections of the test pipe at three different flowrates to see whether the results were due to perhaps changes in the upstream velocity profile with flowrate.

A theory based on aerofoil theory and on the velocity profile just upstream of the rotor explains my results satisfactorily, in Ref. (C). The velocity traverse carried out leads to the assertion that irrespective of whether the pipe flow is developing or fully developed the velocity becomes "peakier" with decrease in flowrate or Reynolds' number or increase in viscosity.

Thus, the meter constant for a meter with large tip clearance increases with decrease in Reynolds' number because the meter responds to the relative increase in the flowrate near the pipe centre. For a meter with a small tip clearance, the meter drops because there is a negative angle of attack at the blade tip which increases with decrease in flowrate with consequent increased drag losses in this region.

It is well known that it is not possible to investigate different flow parameters in isolation and a computer programme has been developed which allows other changes in the flow condition to be taken into account while the effect of one parameter is being investigated.

Figure C shows the typically good agreement between experimental and theoretical prediction of the effect of developing profile on a meter with large clearance.

Good predictions of the calibration curves and of the swirl effects on a meter with small tip clearance is also obtained with this theory.

Figure D shows the fair agreement of meter constant that can be obtained when the data on temperature effects are plotted on the basis of Reynolds number, except when the temperature also considerably affects the profile upstream of the meter.

Figure E shows the temperature effect data in Fig. 3 in the paper by Withers, et al. (see p. 305) plotted on the basis of Reynolds' number. The meter constant at the largest flowrate has been ignored as I feel the rise in this constant could be due to cavitation in parts of the test rig.

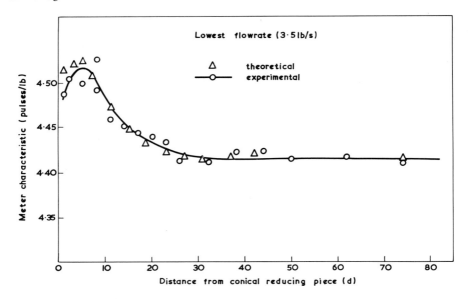

Fig. C   Effect of developing pipe velocity profile on meter with large tip clearance at lowest selected flowrate

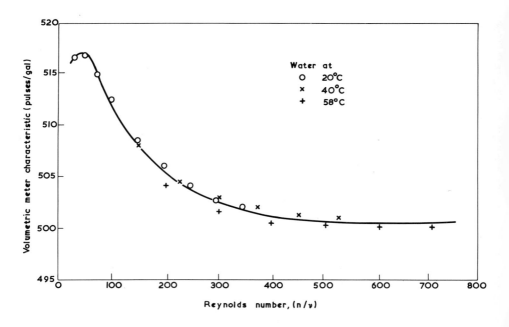

Fig. D   Temperature effect on a commercial meter

Figure F shows the viscosity effect on the meter with small clearance at a point in the test pipe where the turbulent velocity profile is fully developed in water. Note the effect of the pipe flow transition from turbulent to laminar on the meter constant thus confirming the profound influence of the upstream velocity distribution.

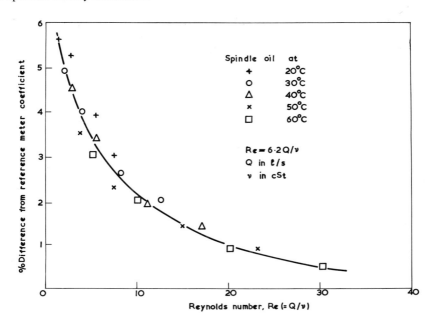

Fig. E   Correlation of temperature effect on meter D in the paper by Withers et al. (see p. 305)

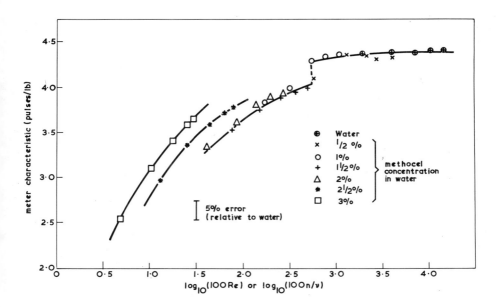

Fig. F   Viscosity effect on a turbine flowmeter showing the influence of the upstream velocity profile

REFERENCE

(C)   SALAMI, L.A.: 'Effect of Velocity Profile just upstream of a turbine flowmeter in its characteristic', Southampton University, Report TM3, July 1971 — as yet unpublished (part of a 1971 Ph.D. Thesis)

MODERN DEVELOPMENTS AND NEW APPLICATIONS OF MAGNETIC FLOW-METERS

R. Theenhaus

Abstract: The effects of variations in the shape of the velocity profile on the performance of the electromagnetic flowmeter are discussed theoretically and it is shown that an ideal shape of magnetic field cannot be produced which gives independence of flow profile if the highest accuracy (< 1% approximately) is required. Methods are discussed by which the influence of the velocity profile can be minimised, but, for highest accuracy, in-situ calibration is required. Comments are also made on the effect of deposits on the electrodes and on the design of the associated electronic equipment.

1   INTRODUCTION

The inductive flow measurement, which was discovered by Faraday approximately 150 years ago and applied by him to measure the flow of the river Thames, has, in the last 10 years achieved increasing importance in industrial measurement. This is caused by the requirements of modern process-automation as well as by the use of modern electronics which are essentially necessary for the inductive flow measurement.

The main advantages of the inductive flow measurement compared to other methods of flow measurement are the following:−

Independence of pressure, temperature, viscosity, conductivity, flowprofile (with certain exceptions), very aggressive fluids and fluids with solid particles can be measured without difficulty; exact linearity between flow rate and output signal; no moving parts within the measuring system; no pressure loss, measurement in both directions; high accuracy and repeatability; high reliability; small servicing requirements.

In this article firstly the influence of flowprofiles will be discussed. Theoretically it will be shown that it is not possible to have an ideal primary with cylindrical cross-section− if one considers an ideal primary being totally independent from flowprofile changes. Methods are discussed with which influences of flowprofile changes are minimized.

Deposits on electrodes and their influences are discussed. The electronics and the construction of modern instruments are shown. In addition special developments for special problems are discussed.

2   THE 2-DIMENSIONAL PROBLEM

The 2-dimensional case, that means

$$\mathbf{u} = v(x,y) \, \hat{\mathbf{u}}_z$$

shall be considered, where $\hat{\mathbf{u}}_z$ is the unit vector in the z direction and $\underline{\mathbf{u}}$ is the fluid velocity.

Starting from Maxwell's equations for moving material, we can write

$$\text{curl } \mathbf{E} = 0 \qquad \text{div } \mathbf{B} = 0$$
$$\text{curl } \mathbf{H} = i \qquad \text{div } \mathbf{D} = \rho \qquad \qquad ...(1)$$

The Editor is indebted to Dr. P. Hutchinson, A.E.R.E. Harwell, for his help in changing the original Gothic symbols used in many of the equations in this paper to conventional English notations.

$$D = \epsilon E + (u \times B)K \qquad\qquad K = \epsilon - \epsilon_o$$

$$B = \mu_0 \, (H - Ku \times E) \qquad\qquad\qquad\qquad\qquad ...(2)$$

$$i = \rho u + \sigma \, (E + u \times B)$$

and the equation of continuity

$$\text{div } u = 0 \qquad\qquad\qquad\qquad ...(3)$$

From (1, 2, 3) one obtains the differential equation

$$\triangle V = \text{div } (u \times B) \qquad\qquad\qquad\qquad ...(4)$$

where $\triangle$ is the Laplacian operator

To obtain this differential equation it has to be assumed, that $v\epsilon/\sigma$ is very much smaller than every other length $\triangle L$, in which a considerable change of the magnetic field will occur. From this assumption, it follows that,

$$\text{div } (E + v \times B) = 0 \qquad\qquad\qquad\qquad ...(5)$$

Taking into consideration the boundary conditions for an isolating wall

$$\frac{\partial V}{\partial r} = 0 \qquad V_i = V_a \qquad\qquad\qquad\qquad ...(6)$$

the result of equation (4) is equation (7), the detailed derivation of which is omitted.

$$V(Q) = \int (\text{Grad } G \, (P,Q) \times B) \, v \, dx \, dy \, dz \qquad\qquad ...(7)$$

[G(P,Q) means the Green's function.]

For the difference of the potentials $V(Q1) - V(Q2) = U$ one obtains the result

$$U = (W \times B)v \, dx \, dy \, dz \qquad\qquad\qquad\qquad ...(8)$$

with

$$W = \text{Grad}_p \left\{ G(P,Q_1) - G(P,Q_2) \right\} \qquad\qquad ...(9)$$

From equation (8) some simple special cases immediately can be considered:

2.1 We assume that the velocity only has one component in the direction of the axis of the tube and the magnetic field is 3-dimensionally inhomogeneous:

$$u = u \left\{ 0,0, v(x,y) \right\} \qquad B = B(x, y, z)$$

Then we obtain from (8):

$$U = \iint dx \, dy \, v \int dz \, (W_x \, B_y - W_y \, B_x)$$

2.2 We assume that the velocity has only one component in the direction of the axis of the tube and the magnetic field is 2-dimensionally inhomogeneous:

$$u = u \left\{ 0,0, v(x,y) \right\} \qquad B = B \left\{ B_x(x,y), B_y(x,y), 0 \right\}$$

We then obtain

$$U = \iint (W_x \, B_y - W_y \, B_x) \, v dx \, dy$$

I want to emphasize already at this point that for a case of a homogeneous magnetic field ($B_y = 0$), a circular cross-section of the tube and 2 electrodes perpendicular to the field, one component of the weight vector is identical with the wellknown Shercliff's weight function.

3   GREEN'S FUNCTION FOR THE 2-DIMENSIONAL PROBLEM

We assume the velocity only having one component in the direction of the axis of the tube and the magnetic field having the shape

$$\mathbf{B} = B_x (x,y)\, \hat{u}_x + B_y (x,y)\, \hat{u}_y$$

Then equation (10) is valid

$$U = \int (W_x\, B_y - W_y\, B_x)\, v\, dx\, dy \qquad \qquad ...(10)$$

The Green's function then can be written [1], [2] as

$$G = \frac{1}{4\pi}\, w\, \left[\{r^2 + \rho^2 - 2r\rho\cos(\phi - \psi)\}\{r^2\rho^2 + a^4 - 2a^2\, r\rho\cos(\phi - \psi)\} \right] \qquad ...(11)$$

$$\text{(see Fig. 1)}$$

If point Q lies on the peripheral circle with radius a, equation (11) reduces to

$$G = \frac{1}{2\pi}\, w\, \{r^2 + a^2 - 2ra\,\cos(\phi - \psi)\} \qquad \qquad ...(12)$$

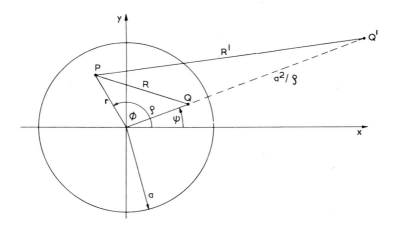

Fig. 1   System of the axis

Assuming 2 point-electrodes perpendicular to the field at $\psi = \pi/2$ (usual construction) we obtain for the components of the weight vector

$$W_x\, (\pi/2) = \frac{2}{a\pi}\, \frac{a^2\, 2\, x\, y}{a^4 + 2a^2\, (x^2 - y^2) + (x^2 + y^2)^2} \qquad ...(13)$$

$$W_y\, (\pi/2) = \frac{2}{a\pi}\, \frac{a^4 + a^2\, (x^2 - y^2)}{a^4 + 2a^2\, (x^2 - y^2) + (x^2 + y^2)^2} \qquad ...(14)$$

This weight vector has to be set into equation (10).

The components of the weight vector are shown in Figs. 2 and 3.

Here W_y is equal to Shercliff's weight function [3] except for a constant factor. (The electrode connection is in the direction of the Y-axis.)

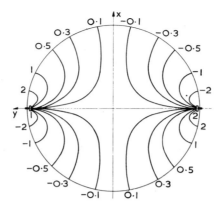

Fig. 2   Weight function Wx

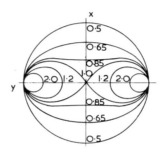

Fig. 3   Weight function Wy

3.1 Homogeneous magnetic field with axial-symmetrical flow profiles

Taking a homogeneous magnetic field in the direction of the Y-axis (perpendicular to the electrode connection) and having a flow profile which is symmetrical to the Z-axis

$$\mathbf{u} = v(r)\,\widehat{\mathbf{u}}_z$$

We obtain from equation (10) equation (15)

$$U = \frac{2}{a\pi}\quad \mathbf{B} \cdot v(r)\cdot \frac{a^4 + a^2 r^2 \cos 2}{a^4 + 2a^2 r^2 \cos 2\phi + r^4}\quad r\,dr\,d\phi \qquad \qquad ...(15)$$

$$= \frac{4}{a}\int_0^a \mathbf{B} \cdot v(r)\,r\,dr$$

$$= 2aB \cdot \frac{2\,v(r)dr}{a^2}$$

$$U = 2aB \cdot v_m$$

(The integration of the weight function on ϕ is equal to 2π, independent of the radius.)

From equation (15), which by other means has already been found in 1941 by Thürlemann [4],

350   Theenhaus

it follows immediately, that inductive flow heads with a circular cross-section, point-electrodes and a homogeneous magnetic field corresponding to the above assumptions will in any case measure the mean velocity and therefore the volume flow rate if the flow profiles are symmetrical to the Z-axis, independent of turbulent or laminar flow. For these classes of flow profiles these primaries are "ideal primary heads".

Primaries with a homogeneous field are still manufactured by some companies. Applying these primaries one has to take into account a straight line before the primaries which is 5 times the diameter of the tube in order to be sure that within the primary near the electrodes a flow profile will exist which is symmetrical to the axis.

3.2 Homogeneous magnetic field with an arbitrary flow profile

From equation (10) one obtains equation (16) (see also [3]).

$$U = \frac{2}{a\pi} \int B \cdot \frac{a^4 + a^2 (x^2 - y^2) \cdot v(x,y)}{a^4 + 2a^2 (x^2 - y^2) + (x^2 + y^2)^2} \, dxdy \qquad ...(16)$$

From equation (16) one immediately can see, that in the general case (arbitrary flow profile) one does not measure in all cases the mean velocity. That means that depending on flow profile changes, which are no longer symmetrical to the axis, these primaries with a homogeneous field will give deviations and therefore errors.

I want to point out, that besides flows with axial-symmetrical flow profiles also other flow profiles can be measured exactly by primaries with homogeneous fields, if those flow profiles fulfill certain conditions concerning the symmetry to the X- and Y-axis. In these cases one obtains an exact result in the same way as for axial-symmetrical flow profiles. (See [5]).

4   GREEN'S FUNCTION FOR THE THREE-DIMENSIONAL CASE

For a three-dimensional velocity

$$u = v_r(r,\phi,z) \, \hat{u}_r + v_\phi(r,\phi,z) \, \hat{u}_\phi + v_z(r,\phi,z) \, \hat{u}_z$$

and for a magnetic field

$$B = B_x (r,\phi,z) \, \hat{u}_x + B_y(r,\phi,z) \, \hat{u}_y$$

one obtains for the potential

$$V = \int \left\{ v_z \frac{\partial}{\partial y} (\epsilon B_x) + v_z \frac{\partial}{\partial x} (\epsilon B_y) + \epsilon B_x \frac{\partial v_y}{\partial z} + \epsilon B_y \frac{\partial v_x}{\partial z} \right\} r dr d\phi dz \qquad ...(17)$$

(see also [6]).

The voltage between the electrodes (potential-difference) can be obtained by taking the potential-difference between the two corresponding points on the circle with radius a.

I do not want at this point to develop in detail the Green's function for this problem, because it would take too much time. In [2], [6], Green's function is written as a sum of Bessel functions:

$$G = \sum_{m=0}^{\infty} \sum_{s=1}^{\infty} f \left\{ I_{m(\beta)} \right\} \cdot \left\{ e^{-\beta|z_0 + z|} + e^{-\beta|z_0 - z|} \right\} \qquad ...(18)$$

with $\beta = \dfrac{\pi \cdot \alpha_{m,s}}{a}$ a = radius.

In this equation (18) $\pi \cdot \alpha_{m,s}$ means the S root of equation

$$I'_m (K_a) = 0 \qquad ...(19)$$

A very important conclusion following from equation (18) is the dependence of the Green's function from Z, furthermore that the function decreases as well in the negative Z-direction as in the positive Z-direction.

Naturally, the Green's function changes if the boundary conditions change. That means that the length of the flow head has a certain lower limit if one wants to avoid the influence of the flanges of the tubes on the Green's function and therefore on the sensitivity of the flow head.

Following investigations of A.C. Haacke [6] the length of an inductive primary head shall be at least equal to the diameter of the tube to be sure that the conditions of installation do not influence the calibration of the flow head.

5   CAN AN IDEAL PRIMARY BE VERIFIED?

We have seen that an inductive flow head with a cylindrical cross-section, 2 diametric point-electrodes, a homogeneous field, measures exactly the mean velocity if the flow profile is symmetrical to the axis of the tube; that means, if the velocity only depends on the radius R.

From equation (16) can be derived, that with an arbitrary velocity profile v_z (x,y), a homogeneous field and the above mentioned conditions normally this construction of inductive flow head does not allow the mean velocity to be measured exactly.

Furthermore, it can be seen from equation (16), that the contribution of a certain flow volume to the voltage between the electrodes has to be multiplied with a certain "weight" which is given by the weight function. By this, more or less, a deviation of the signal is caused compared to the mean velocity or the volume flow.

One also has to consider, that because of div $\mathbf{u}$ = 0 (because of the incompressibility of fluids) one has flow components perpendicular to the axis of the tube if the flow profile deviates from symmetry to the Z-axis. Because of this one has to take into account equation (17), instead of equation (16), if the flow profile is an arbitrary one and not symmetrical to the tube-axis.

From equation (17) it can be seen directly, that an adaption of the magnetic field to the Green's function with the aim of being independent of flow profile generally is not possible (or better: useful!), because there are two further disturbing terms in which are as a factor the derivatives of the perpendicular flow-components. (The adaption of the magnetic field to the Green's function means in the 1-dimensional case (equation (16)) the constancy of the product field x weight function and in the 2-dimensional case [equation (10)] the constancy of the difference $W_x B_y - W_y B_x$)

One can therefore conclude immediately from equation (17) that it is not possible by means of a specially shaped magnetic field to measure exactly the mean velocity and therefore the volume flow rate with a primary having a cylindrical cross-section, 2 point-electrodes for all arbitrary flow profiles.

For that reason the different companies also manufacture inductive flow heads with homogeneous magnetic field as well as with inhomogeneous or modified magnetic fields. The homogeneous fields have the advantage of being able to measure exactly the mean velocity and the volume flow rate in the case of symmetrical profiles (symmetrical to the axis of the tube), they have the disadvantage of needing a relatively large straight length before the flow head and they cause relatively great errors in the case of deviations from the mentioned symmetry.

On the other hand, the inhomogeneous and modified magnetic fields are solutions which are compromises. The theory for these solutions says that one will have errors if a flow profile changes from one symmetrical profile to another because of changes of viscosity. These errors can be kept far below 1%, so that this does not mean a disadvantage for application in process-measurements.

Again, these modified or inhomogeneous fields have the advantage of decreasing errors for arbitrary flow profiles, which might occur behind bends, valves or if the fluid contains a high percentage of solid particles (hydraulic transport). In these cases a tolerance can still be obtained of better than 1%.

(These inhomogeneous and modified fields are special shaped 3-dimensional fields).

Flow profiles with a spin, in which there are strong velocity components perpendicular to the tube axis can cause errors even in the case of inhomogeneous and modified magnetic fields.

One other possibility to prevent, to a certain extent, the influence of flow profile-changes is given by shaping the electrodes, in other words by increasing the point-electrodes to those with a certain area. These electrodes with an area of approximately some $100\ cm^2$ will cause an integration (mean value) to a certain extent. They have the big disadvantage of being very sensitive to deposits, even if these deposits are only partial. By deposits on these electrodes the weight

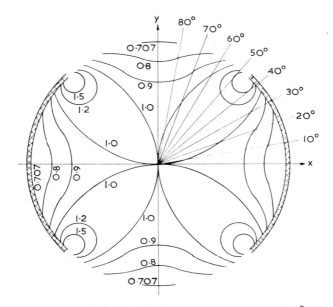

Fig. 4   Weight function for real electrodes having an angular aperture of 90°

function changes — which causes a change of the sensitivity of the primary and therefore of the calibration. That means errors will occur.

In Figs. 4 and 5 one-dimensional weight functions are shown for an electrode angle of $\alpha = 90°$ and $\alpha = 45°$ [7].

This disadvantage does not happen with point electrodes!

I want to mention furthermore, that Engl [2] for the 2-dimensional problem has found the very interesting result that the mean value of the product of the potential multiplied by $\sin \psi$, taken on the periphery of the tube, is exactly proportional to the flow rate in the case of a homogeneous magnetic field and arbitrary velocity profile as well as in case of an arbitrary inhomogeneous magnetic field and a velocity profile with axial symmetry. This allows the conclusion that flow heads with cylindrical cross-section and with more than two electrodes on the periphery of the tube can be another possibility to minimize the influence of flow profile changes. For practical applications on the other hand there has to be considered the disadvantage of higher costs, especially if one has to use platinum or some other expensive material as the electrode.

6   THE TECHNIQUE OF ALTERNATING SIGNALS AND ITS CONSEQUENCES

Because of the application of direct fields (d.c. fields) many different effects between electrode and fluid cause different disturbance potentials, which are summarized in [8]. To avoid these disturbances alternating fields with 50 or 60 cycles are normally applied. The corresponding amplifier-technique is therefore a low frequency technique.

Applying alternating fields means that beside the measuring voltage also quadrature voltages and capacitive voltages will be measured between the electrodes. These disturbing voltages can have the same order of size (mV) as the measuring voltages. The disturbing voltages depend to a very important extent on the grounding conditions and on the impedance of the primary, therefore also on the resistance between electrode and fluid (which can change if deposits are on the electrodes). As the disturbing voltages can be measured at the electrodes with the fluid velocity V = 0 they can be compensated in the amplifier by an equal negative voltage. The user of an inductive flow-instrument can therefore, by means of a potentiometer on the panel of the amplifier, set the "Zero" before he opens the valves. If after some time deposits have built up on the electrodes which will change the internal resistance of the primary, deviations of the Zero and therefore errors will occur. Because of this fact the user has to control from time to time the Zero-set and eventually has to correct the Zero. Years of experience with this technique have shown that after some time a saturation effect will be reached so that eventually deviations of the Zero caused by deposits will come to a steady value, so that the Zero will become constant.

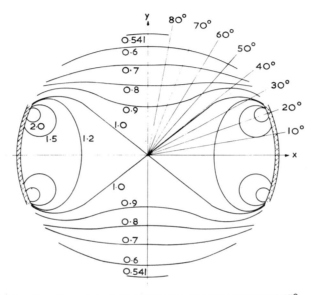

Fig. 5   Weight function for real electrodes having an angular aperture of 45°

Different methods of electrode cleaning have been discussed and investigated during the last 3 - 5 years. In many cases mechanical cleaning methods have been applied. In one construction, for instance, a sort of knife cleans the surface of the electrode by turning round a crank outside the housing of the primary. This can be done without interrupting the measurements. For the bigger sizes of primaries interchangeable electrodes can be applied. These electrodes can be pulled out and cleaned without taking off the primary.

In the past two years ultrasonic cleaning devices and electro-chemical cleaning devices have been discussed intensively.

Experience so far with these two methods are too small to be sure in which cases the methods are practicable. It can be said generally that deposits on point electrodes are non-existent in more than 95% of applications. Even in cases of measuring slurries, electrode deposits only have been observed to a very small extent.

A further consequence following from the alternating signal technique is the limit of the conductivity. The primary (flow head) can be considered as a generator with an internal impedance which is reciprocal proportional to the conductivity of the fluid. With standard point electrodes this internal impedance for instance is appr. 1 kOhm with a conductivity of 1 m Mhos/cm.

This internal impedance is loaded by the stray-capacities and the capacity of the cable. The stray-capacities cause a limitation of the conductivity at about 10 μMhos/cm. With some effort (preamplification, shielding and so on) this limit can be decreased to 1 μMhos/cm.

The physical limit caused by dielectric currents in the fluid is some orders smaller than the limit caused by the above mentioned capacities.

7   MODERN INSTRUMENTS

In Fig. 6, a flow head system ALTOFLUX is shown, constructed as a short type primary head with a modified magnetic field; Fig. 7 shows the corresponding amplifier, system ALTOFLUX, type TIV 30.

The main requirements for the electronic amplifier are:

High differential input impedance (higher than 100MΩ), exact linearity between input and output, compensation of changes of magnetic field caused by changes of the supply voltage or other reasons, independence of the load within certain limits, long life, high stability, applicable under strong conditions, high reliability.

Fig. 8 shows a block diagram of a modern amplifier (ALTOFLUX, TIV 30):

After the preamplifier (2) and the range set (5) the signal comes into the input amplifier (6), then into the phase-sesnsitive demodulator, the integrator and the current converter. The phase-·sensitive rectifier separates the measurement voltage (which is proportional to the mean velocity) from the quadrature voltages, which have a phase shift of about 90°. The multiplier (13) provides the feed-back of the product of measurement voltage and reference signal, which is an image of the magnetic-field.

By this means the independence of the output-signal from changes of field is attained.

In this system shown in Fig. 8 the reference signal is produced by a loop being mounted

Fig. 6   Primary flow head: ALTOFLUX system

directly into the magnetic field so that all changes of magnetic field independent from their reasons are compensated. Besides this, not only is the measurement voltage fed back but the whole mixture of signals containing the measuring signal, harmonics and all quadrature voltages. This is done by starting with the feed back direct behind the Zero-amplifier (6), where the whole mixture of signals exists. The result of this is an amplifier, which is extremely independent of quadrature voltages and harmonics.

Normally present-day electronics is constructed in the form of printed circuits, applying integrated operational amplifiers. For special applications several solutions have been developed which can be added to the main amplifier using printed circuits. For instance the "range switch", which chooses the optimal range, automatically between two ranges, the "frequency output" to feed a pulse counter, "Zero-drop-out" to drop out the Zero, "empty-pipe-indicator" to avoid errors in case of an empty-primary head, and so on.

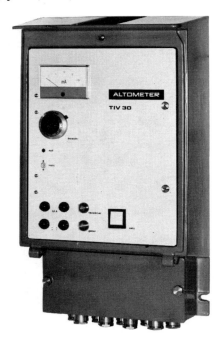

Fig. 7   Amplifier: ALTOFLUX system, TIV 30

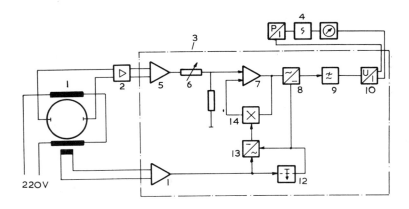

Fig. 8   Block diagram of modern amplifier

All inductive flow measuring devices from 2 mm up to the greatest dimensions of 2 m should be calibrated experimentally in a special calibration device, because so far it has not been possible to achieve a tolerance less than ±0.5% by calibrating the primary heads theoretically by measuring the magnetic field.

8   CONCLUSION

It was the aim of this paper to give a short survey of the technique of inductive flow measurement and of the newest theoretic investigations. It was pointed out that it can be seen from the latest theoretical calculations that it is impossible to produce a primary head which allows a measurement of flow rate independent of flow profiles. In practice, good results have been achieved by applying inhomogeneous and modified magnetic fields which are approximations. A tolerance better than 1% can be realized in practice for nearly all flow profiles.

Amplifiers has been developed intensively so that inductive flow measurement has become a very exact, reliable and well constructed industrial measurement method.

Further developments are attempting to reduce costs by optimizing the construction; on the other hand attempts are being made to increase the precision and decrease the limit of conductivity. Materials should be found for high temperature and high pressure applications (higher than 200°C and higher than 100 bars). As the primaries applied to electrolytic fluids must have an isolating liner the problem of operation at high temperature and pressure requires the use of plastic materials.

LITERATURE

[1]   Frank-Mises, Die Differential- und Integral-gleichungen der Mechanik und Physik, Band 1, Kapitel VI

[2]   W.L. ENGL: Der induktive Durchflußmesser mit inhomogenem Magnetfeld, Archiv fur Elektrotechnik 53, 1970, S. 344 - 359

[3]   J.A. SHERCLIFF: The Theory of Electromagnetic Flowmeasurement, Cambridge University Press, 1962

[4]   B. THÜRLEMANN: Methode zur elektrischen Geschwindigkeitsmessung von Flüssigkeiten, Helv. Phys. Acta, 14, 1941, S. 383 - 419

[5]   P. POLLY: Duisburg, Private Communication

[6]   A.C. HAACKE: Sensitivity of the Electromagnetic Flowmeter to Fluid Velocity Profile, Symposium on Flow Measurement, Pittsburg, 1971

[7]   C. SMYTH: J. of Physics E. Scientific Instruments, Band 4, Nr. 1, 1971

[8]   K.W. BONFIG: Zur Theorie, Problematik und Verwirklichung der induktiven Durch-flußmessung mit getastetem Gleichfeld, Dissertation, TH - München, 1971

DISCUSSION

A.M. Crossley: At zero flow, the fluid in the pipeline forms a transformer of one turn in an alternating magnetic field. This effect is compensated during the initial calibration of the equipment, but is it possible for this effect to alter and cause drift from zero when the flow velocity is zero?

Author's reply: The disturbance voltage caused by the alternating magnetic field has a phase shift of approximately 90° to the measurement voltage which is proportional to the mean velocity of the fluid. Modern amplifiers are constructed so that all quadrature voltages with a phase shift of 90° are suppressed. This means that even changes of these interfering voltages do not influence the measurement.

On the other hand, the danger of deposits building up on the electrode during zero flow is increased. If these deposits are isolating, the capacitive voltages might change and cause a certain zero drift in special cases (polymerisation or crystallisation effects).

A.M. Crossley: What errors are caused by the pipe flowing "half-full" and are there any recommended methods of overcoming this problem? What errors are caused by the fluid giving off gas?

Author's reply: If the pipe is only partially filled but the level is higher than the electrode level (both electrodes still in contact with the fluid) the error corresponds approximately to the ratio of the non-filled volume to the total volume of the pipe. To some extent this difficulty can be overcome by simultaneously measuring the level of the fluid within the pipe.

If the level of the fluid is below the electrode level, the interfering voltages, and therefore the output of the amplifier, is indefinite: it might be more than 20 mA, for example.

The Krohne Company, especially, has developed an electronic device which short circuits the output of the amplifier and simultaneously gives an alarm signal. So all following instruments (recorders, controllers, etc.) are set to zero.

Gas bubbles with a homogeneous distribution within the fluid do not disturb the measurement if the electrodes have contact with the fluid. The total volume flow rate, including the gas volume, is measured.

A.M. Crossley: Does the variation in conductivity of the fluid cause variations in the capacitance of the path between the electrodes and what effect does this have on quadrature?

Author's reply: The variation in conductivity does not influence either the path between the electrodes or the quadrature voltages and therefore the output of the amplifier is not affected so long as the conductivity is homogeneous.

Non-uniformity as well as anisotropy of conductivity influence the sensitivity of an electromagnetic flowmeter and its quadrature voltages. The first theoretical investigations were published by M.K. Bevir in J. Phys. D: Appl. Phys., 1971 Vol. 4.

M.K. Bevir: In view of the better theoretical performance that can be obtained with large or strip electrodes and the additional complications of dirt and deposit on them which will upset their performance, does your organisation intend to stick to small (point) electrodes or to use the large ones?

Author's reply: Most manufacturers of "magmeters" (our company included) use small point electrodes, because large or strip electrodes are very sensitive to deposits on the electrode surface, even if these deposits are only on parts of the electrode surface.

ELECTROMAGNETIC FLOWMETERS FOR LIQUID METALS

G. Thatcher

Abstract: This paper describes two different types of flowmeter suitable for use in large-scale liquid metal systems, such as liquid-metal, fast breeder reactors. Theoretical and experimental evidence is presented to show how transverse field electromagnetic flowmeters become non-linear when applied to large liquid metal systems. The saddle coil flowmeter, designed to overcome this problem and to obviate the need for calibration, is fully described, together with exhaustive tests on a prototype. The second flowmeter to be described is the flux distortion type. A numerical analysis of this type is presented, followed by the results of extensive laboratory tests on production models.

1   INTRODUCTION

The two forms of flowmeters described in this paper were designed for use in a prototype fast-neutron breeder reactor (PFR), using molten sodium as the heat transfer medium. There are two sodium circuits, primary and secondary. The primary circuit sodium is pumped from a pool. In the intermediate heat exchangers heat is transferred to the secondary sodium circuit which is pumped via 14 in. diam. pipes to the steam generators.

For safety purposes the primary and secondary sodium flowrates have to be continuously monitored, and for subsequent design check calculations the secondary flow has to be measured as accurately as possible. The first half of this paper gives an account of the secondary circuit flowmeter, and the second half describes two flowmeters used in the primary circuit.

2   SECONDARY CIRCUIT FLOWMETER (SADDLE COIL)

During reactor operation the secondary sodium flows into the intermediate heat exchangers in 14-in. diam. pipes at a rate of 7,500 gallons per minute and a temperature of ∼400°C. An accurate measurement of this flow is required.

Because the fluid is a liquid metal it seems obvious that a transverse-field electromagnetic flowmeter would be the most suitable device. However, a conventional form of such a flowmeter, using a permanent magnet, was used on the Enrico Fermi reactor in the U.S.A., and this was found to have a very non-linear output/flow characteristic[1]. Errors greater than 50% occurred. This accelerated an investigation of the performance of large scale transverse-field e.m. flowmeters, covering their linearity and the effects of non-uniform applied magnetic fields.

2.1 Linearity of large-scale e.m. flowmeters

A schematic diagram of a transverse e.m. flowmeter is shown in Fig. 1. Its operation depends upon the interaction between the magnetic field and the moving liquid conductor, producing an electrical potential distribution throughout the fluid. Because the magnetic field has finite extent in the flow direction, two effects occur. Firstly the liquid metal outside the magnetic field short-circuits the electrodes, reducing the sensitivity. Secondly, potential gradients in the end regions of the flowmeter cause circulating currents to flow, producing additional fields which in turn affect the sensitivity.

These currents, the field distortion they produce and their effect on sensitivity were investigated for the field profiles shown in Fig. 2, namely:

(a)   step
(b)   rectangular
(c)   trapezoidal.

The step profile has been treated elsewhere[2], but as it is the basis for work on the other profiles it is important to restate the assumptions made in the analysis.

Referring to Fig. 3:

(a)   The conducting fluid flows with uniform velocity in the z-direction.

(b)   For z less than 0 the applied field is uniform over all x and only has a y-component.

(c)   For z greater than 0 the applied field is zero over all x.

(d)   The channel is rectangular in section with insulating walls. Its dimensions in the x and y directions are 2a and 2b respectively.

(e)   Electric currents only flow in the x-z planes.

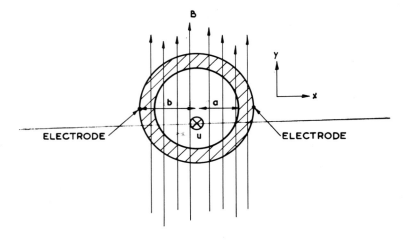

Fig. 1   Principle of electromagnetic flowmeter

From Maxwell's equations and Ohm's law the following differential equation for the potential V can be derived

$$\frac{\partial^2 V}{\partial x^2} + \frac{\partial^2 V}{\partial z^2} = \mu_0 \sigma u \frac{\partial V}{\partial z}$$

where μ_0 = $4\pi \times 10^{-7}$ (H/m)

          σ =   electric conductivity of fluid (S/m)

          u   =   fluid velocity (m/s)

From the assumptions and model used, the following boundary conditions on V apply:

$$\frac{\partial V}{\partial x} = \frac{\partial V}{\partial z} = 0 \qquad \text{at } z = +\infty$$

$$\frac{\partial V}{\partial x} = -uB_0, \frac{\partial V}{\partial z} = 0 \quad \text{at } z = -\infty$$

$$\frac{\partial V}{\partial x} = -uB_0 \quad \text{at } x = \pm a \text{ for } z < 0$$

$$\frac{\partial V}{\partial x} = 0 \qquad \text{at } x = \pm a \text{ for } z > 0$$

B_0 is the intensity of the applied field.

Boucher and Ames[2] obtained a solution to the above equation in the form:

$$V = \frac{2uB_0}{a} \left\{ \sum_{n=1}^{\infty} \frac{(-1)^n \, \delta_n \, e^{\gamma_n z} \sin (a_n x)}{a_n^2 \, (\gamma_n - \delta_n)} - \frac{ax}{2} \right\} \qquad \text{for } z < 0$$

$$V = \frac{2uB_0}{a} \sum_{n=1}^{\infty} \frac{(-1)^n \, \gamma_n \, e^{\delta_n z} \sin (a_n x)}{a_n^2 \, (\gamma_n - \delta_n)} \qquad \text{for } z > 0$$

where
$$\gamma_n a = [R_m + \{R_m^2 + 4 (a_n a)^2\}^{1/2}]/2$$

$$\delta_n a = [R_m - \{R_m^2 + 4 (a_n a)^2\}^{1/2}]/2$$

and
$$a_n a = (2n-1)\pi/2$$

The dimensionless quantity R_m is the magnetic Reynolds' number (= $\mu_0 \sigma u a$), and is a measure of the electromagnetic scale of the system.

These results for a step field profile are not applicable to short magnetic systems. The simplest approximation to a real field is the rectangular profile. This can be formed by superposing two semi-infinite fields shown in Fig. 4. If sa is the half-length of the field and $z' = z/a$, $x' = x/a$ the potential distribution is given by:

$$V = 2B_0 ua \sum_{n=1}^{\infty} \frac{(-1)^n \, \delta_n a \sin (a_n a x') \{e^{\gamma_n a (z' + s)} - e^{\gamma_n a (z' - s)}\}}{(a_n a)^2 \, (\gamma_n a - \delta_n a)} \qquad \text{for } z' < -s$$

$$V = 2B_0 ua \left\{ \sum_{n=1}^{\infty} \frac{(-1)^n \, \delta_n a \sin (a_n a x') \, e^{\gamma_n a (z' + s)}}{(a_n a)^2 \, (\gamma_n a - \delta_n a)} - \frac{x'}{2} \right.$$

$$\left. - \sum_{n=1}^{\infty} \frac{(-1)^n \gamma_n a \sin (a_n a x') \, e^{\delta_n a (z' - s)}}{(a_n a)^2 \, (\gamma_n a - \delta_n a)} \right\} \qquad \text{for } -s < z' < s$$

$$V = 2B_0 ua \sum_{n=1}^{\infty} \frac{(-1)^n \, \gamma_n a \sin (a_n a x') \, e^{\delta_n a z'} (e^{\delta_n a s} - e^{-\delta_n a s})}{(a_n a)^2 \, (\gamma_n a - \delta_n a)}$$

For flowmetering it is customary to position the potential taps on the axis of symmetry of the magnetic field, i.e. $z' = 0$. The potential difference V_0 between the taps is given by:

$$\frac{V_0}{2uB_0 a} = 2 \left\{ \sum_{n=1}^{\infty} \frac{\delta_n a e^{\gamma_n a s} - \gamma_n a e^{-\delta_n a s}}{(a_n a)^2 \, (\gamma_n a - \delta_n a)} + \frac{1}{2} \right\}$$

This relationship is shown in Fig. 5 for various values of s, the aspect ratio. As R_m approaches zero this expression reduces to

$$\frac{V_0}{2uB_0a} = 1 - 2 \sum_{n=1}^{\infty} \frac{e^{a_n as}}{(a_n a)^2}$$

$$= 1 - \frac{8}{\pi^2} \sum \frac{\exp\left\{(2n-1)\pi/2\right\}}{(2n-1)^2}$$

which agrees with an expression derived by Shercliff[3]

The infinite field gradients at the ends of the rectangular field profile result in an overestimate of the field distortion effect. To gain a more realistic estimate without involving very tedious mathematics, a trapezoidal field profile was considered, again using the superposition method. The main result is that the potential difference, V_0, between the electrodes (at z = 0) can be expressed as:

$$V_0 = 2S_1 S_2 uB_0 a \qquad\qquad \text{... (1)}$$

where S_1 is a correction factor for wall shorting and S_2 is the end-shorting/field distortion factor given by:

$$S_2 = 1 - \frac{32}{\pi^4 t} \sum_{n=1}^{\infty} \frac{\alpha_n^2 e^{\beta_n s} - \alpha_n^2 e^{\beta_n(s+t)} + \beta_n^2 e^{-\alpha_n s} - \beta_n^2 e^{-\alpha_n(s+t)}}{(2n-1)^4 (\alpha_n - \beta_n)}$$

where $\alpha_n = \gamma_n a$

$\beta_n = \delta_n a$

s = aspect ratio (ratio of half-length of uniform section of field to half-channel width*)

For the large pipes used in PFR the wall shorting is negligible, i.e. $S_1 = 1$.

Re-arranging equation (1) gives

$$\frac{V_0 \mu_0 \sigma}{2B_0} = S_2 R_m$$

S_2 itself being a function of R_m. This relationship, which represents the output/flow characteristic is shown in Fig. 6. Figs. 7 and 8 show further results of the analysis. Fig. 7 shows the field distortion along x = 0 and Fig. 8 shows the potential difference across the channel at points along the channel. Tables of S_2 versus R_m for a wide range of aspect and taper ratios have been produced.

2.2 Practical applications

The assumptions made in the above analysis are not sufficiently valid to allow derived values of S_2 to be used to correct for gross field distortion, where the departure from linearity is greater than a few percent; but these values of S_2 do give a good indication of the length of field required to reduce the non-linearity to an acceptable level. This allows a magnet system to be designed, and the saddle coil flowmeter will serve as a good example of the design procedure.

* For circular pipes, [3] radius r, replace a by $\pi r/4$

2.3 Saddle coil flowmeter

The electromagnetic flowmeters to be used on the secondary circuits of PFR must have output/ flow characteristics which do not depart from linearity by more than 3% at maximum flow. This means that S_2 must be greater than .97 at maximum R_m. This also ensures that S_2 can be calculated with reasonable confidence. The maximum R_m for the 14 in. pipes is 5.2 and from the

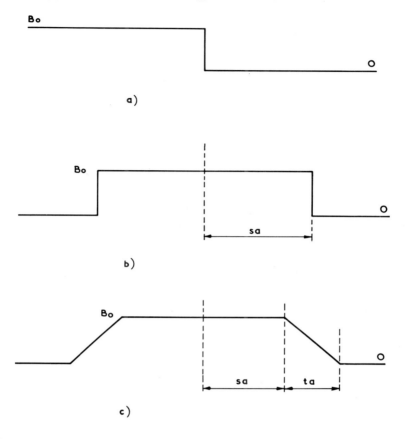

Fig. 2   Field profiles
   a   Step
   b   Rectangular
   c   Trapezoidal

previously produced tables of S_2, the aspect ratio of the equivalent trapezoidal field must be greater than 5.5 for a typical taper ratio of 4.5. This means that the length of the uniform region of the field must be 1.0 times the mean semi-depth of the pipe, i.e. 58 in. (1.47 m).

A permanent magnet system employed to produce a uniform field over such a large volume would be unwieldy. A much more satisfactory alternative is a DC-energised air-cored coil.

The requirements of such a coil system are:

(a)   The field strength should not fall by more than 1% between the pipe axis and the electrodes*

* When these requirements were first laid down, (a) was a cautious estimate, but recent work on the effects of non-uniform fields on flowmeter output, reported below indicates that condition (a) is unnecessarily stringent and that a fall of 15% at the pipe walls affects the output by, e.g. only 3% for the laminar flow.case.

(b)   The aspect ratio should be greater than 5.5.

(c)   The field strength at the centre of the coil system should be 5mT, to give an easily measurable output in the nominal range 0 to 10mV for a flow range 0-7000 gallons per minute ($.53m^3/sec$).

For efficacy and convenience of installation the coils take the form of two saddles as shown in Fig. 9. This enables the coils to be assembled round the pipe very simply after the pipe has been installed.

2.3.1 Coil design

It was necessary to arrive at an optimum winding configuration which would satisfy the above conditions with the smallest coil.

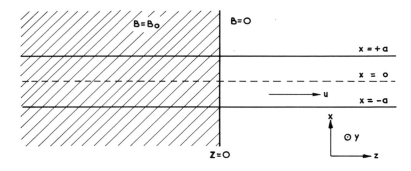

Fig. 3   Model for Boucher and Ames analysis

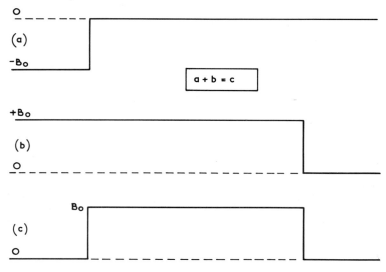

Fig. 4   Formation of rectangular field profile

The field produced by the coils is uniform only over the central portion of the region enclosed by the coils. The extent of the uniform region in the x-direction depends on the angle (ϕ) subtended by the longitudinal windings at the pipe axis. For an optimum ϕ the uniform region reaches its greatest extent of about 40% coil diameter. This means that the minimum coil diameter should be 2.5 times the pipe I.D. to fulfil condition (a).

364   Thatcher

This large coil has the considerable advantage that it can be mounted outside the pipe's secondary containment (leak jacket) and lagging, thus eliminating any problems due to heating of the windings.

The required aspect ratio for a given pipe diameter determines the length of the coil. The number of turns is determined by the required field strength, the absolute size of the coil, and the current-carrying capacity of the windings.

To indicate an optimum value of ϕ and gain an estimate of the current required to give 5mT at the pipe axis, preliminary calculations were made for an infinitely long 'coil' with the windings represented by continuous current sheets. The optimum ϕ fixes the product of wire diameter, number of turns and reciprocal of the coil diameter. For a given size of coil and field strength, the power loss in the windings is proportional to the number of turns. To keep this to a minimum and yet retain ease of winding, a cable diameter of 0.5 in. was chosen for the prototype (and subsequent production models).

The field distribution for a finite-length coil was then calculated by summing contributions from individual turns starting with parameters obtained from the infinite case. The length was varied to obtain the required aspect ratio. The results indicated that 110 amps would be required to produce 5mT. The prototype has 26 turns per component coil, and they overlap at the ends, primarily to save space, but in fact resulting in a better field profile. The overall length of the coil is 7 ft. 8 in (2.34m) and the total length of cable is 1200 ft(366m). When the coils are assembled round the pipe, a link turn has to be inserted between the two halves to complete the electric circuit. All the materials used in the construction are non-magnetic; mainly aluminium and copper alloys.

2.3.2 Tests of prototype

The tests on the prototype consisted of two parts; field distribution measurements and calibration on a 6,000 gallons per minute sodium loop.

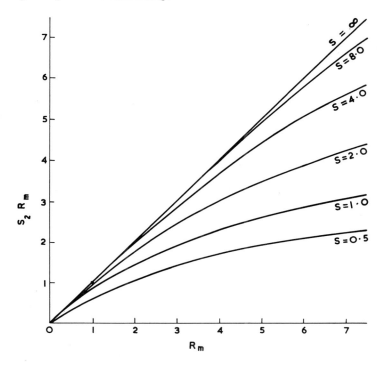

Fig. 5   Output versus flow for rectangular profile with aspect ratio s

The field measurements were carried out to check the calculations and to investigate the effects on the field distribution of structural ironwork external to the flowmeter. An example of a field plot is shown in Fig. 10 together with computed values, between which there is good agreement.

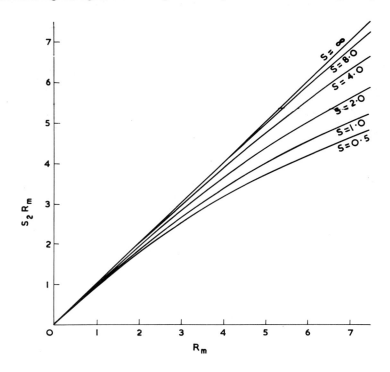

Fig. 6   Output versus flow for trapezoidal profile with taper ratio of 4.5

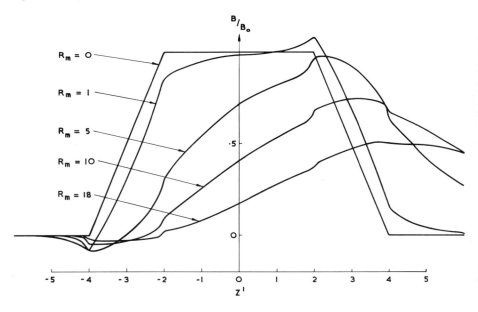

Fig. 7   Distortion of trapezoidal field for s=t=2

An impression of the effects of external ironwork was gained by placing steel beams at various positions around the coil and observing the effect on the field distribution. For the worst configuration that could occur in practice, perturbation of 1-2% were observed. These effects are very difficult, if not impossible to estimate and on PFR they are avoided by positioning the flowmeters so that their electrodes are as far away as possible from substantial ironwork.

2.3.3 Calibration

A number of tests were made on the saddle coil flowmeter when it was installed on the 6000 gallons per minute liquid sodium test loop at REML. These tests were:

(a)   to endorse the theoretical calibration over the range 2,000 to 7,000 gallons per minute;

(b)   to observe any effect on electrical output due to the proximity of a loop bend;

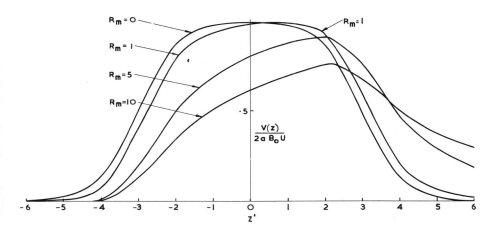

Fig. 8   Potential distribution distortion for s=t=2

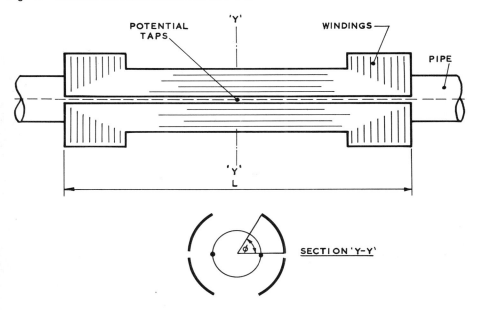

Fig. 9   Schematic representation of the saddle coil flowmeter

(c)   to measure the electric potential distribution along the pipe in the region of the flow-meter to check the theory of end-shorting presented earlier.

The main test loop consisted of a 12 in. stainless steel pipe having ¼ in. wall thickness surrounded by lagging and a secondary containment. The coil was mounted outside this and during loop operation at 400°C, the windings were cool enough to be touched by hand.

Seven pairs of mineral-insulated stainless steel cables were welded to the pipe at 12 in. intervals to act as electrodes. These were led out to a compact terminal block to minimise any temperature differences, thus minimising e.m.f.'s.

The coil was energised from a rectified three-phase supply, the current being monitored by measuring the potential difference across a 0.001Ω shunt connected in series with the coil and power supply.

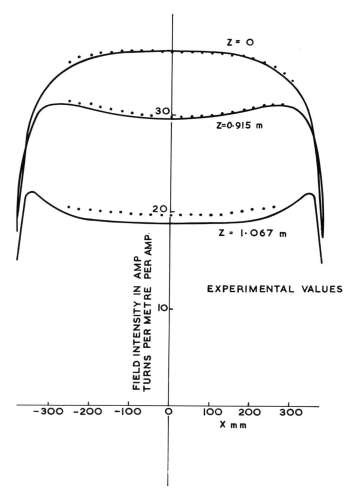

Fig. 10   Field distribution in electrode plane

For the calibration the coil was mounted with its centre over the central pair of electrodes, and the potential difference between these was measured, with the same potentiometer that was used to measure the current-shunt potential difference, for a range of flows. A venturi flowmeter, which had previously been calibrated in water over the whole flow range to an uncertainty of ±½% by the 'weigh-tank' method, was used as reference. Its calibration had to be corrected to account for density and viscosity differences when used for sodium, introducing a further uncertainty estimated at ±½%. The pressure drop across the Venturi was measured absolutely with a sodium-filled manometer to an accuracy of ±½%. The overall error in flow measurement amounted to ±1% r.m.s.

Values of S_2 versus R_m for the trapezoidal field profile which most closely approaches that of the saddle coil on the 12 in. pipe are given in Table 1. The highest flow at 400°C corresponded to an R_m of 5.0. The uncorrected and corrected flows as calculated from equation (1) and the flowmeter output are compared with the Venturi results in Table 2. From these results it is apparent that the output/flow characteristic is becoming non-linear by about 2%. The corrected flow, however agrees with the Venturi-measured flow to ±1%.

TABLE 1 Values of S_2 v. R_m for saddle coil during calibration

RM	0.	0.1	0.2	0.3	0.4	0.5	0.6	0.7	0.8	0.9
0	1.0000	1.0000	1.0000	1.0000	1.0000	1.0000	1.0000	1.0000	1.0000	1.0000
1	0.9999	0.9999	0.9999	0.9999	0.9999	0.9998	0.9998	0.9997	0.9996	0.9996
2	0.9995	0.9994	0.9992	0.9991	0.9989	0.9988	0.9986	0.9983	0.9981	0.9978
3	0.9975	0.9972	0.9968	0.9964	0.9960	0.9955	0.9950	0.9945	0.9939	0.9933
4	0.9927	0.9920	0.9913	0.9906	0.9890	0.9890	0.9881	0.9872	0.9863	0.9854
5	0.9844	0.9833	0.9823	0.9812	0.9800	0.9789	0.9777	0.9764	0.9752	0.9739
6	0.9725	0.9712	0.9698	0.9684	0.9670	0.9655	0.9640	0.9625	0.9610	0.9594
7	0.9579	0.9563	0.9546	0.9530	0.9513	0.9497	0.9480	0.9463	0.9445	0.9428
8	0.9411	0.9393	0.9375	0.9357	0.9339	0.9321	0.9303	0.9284	0.9266	0.9247
9	0.9229	0.9210	0.9191	0.9172	0.9153	0.9134	0.9115	0.9096	0.9077	0.9058
10	0.9039	0.9019	0.9000	0.8981	0.8961	0.8942	0.8923	0.8903	0.8884	0.8865

TABLE 2 Calibration results

Flow by Venturi gall/min	Saddle Coil Uncorrected Flow gall/min	Saddle Coil Corrected Flow gall/min
3236±30	3239±60	3243±60
4595±40	4525±80	4546±80
5290±50	5180±100	5222±100
6434±60	6304±120	6403±120
7120±70	6900±140	7050±140

To determine the effect of the pipe-bend on the flowmeter output, the coil was moved so that its centre lay over the line joining the pair of electrodes 1½ m from the bend. The output in this position differed from that in the normal position by about ¼% which was comparable to the standard deviation of the V/I measurements. This means that a pipe bend in the magnetic field plane, which is greater than four pipe diameters downstream from the electrodes, does not affect the output. There was also no change when the coil was moved so that the measuring electrodes were six pipe diameters downstream from the Venturi orifice.

A change in sodium temperature is accompanied by changes in the pipe diameter and sodium electric conductivity. As the coefficient of expansion of stainless steel is 2×10^{-5} per degC, changes in pipe diameter are negligible. Significant changes in sodium conductivity affect R_m which in turn affects S_2. Unless the magnetic field is short and the velocity high, the output is reasonably temperature independent. This was verified on the test loop when the pump was run at full flow for temperatures of 400°C and 258°C, producing less than a 1% change in output.

The potential differences across the seven pairs of electrodes were measured for each position of the saddle coil to observe the effects of field distortion on the potential distribution. The results of this are shown in Fig. 11, where they are compared with the theoretical results. Considering the simplicity of the model, reasonable agreement is obtained.

3   PERFORMANCE OF E.M. FLOWMETERS HAVING NON-UNIFORM APPLIED MAGNETIC FIELDS

Simple e.m. transverse field flowmeter theory assumes that the fluid velocity distribution in a circular pipe is axisymmetric, and that the applied magnetic field is uniform over the pipe cross-section and infinitely long (in the flow direction). An earlier part of this paper studies the effects of a finite length field which, however, remains uniform over the pipe cross-section. This section accounts for the case where the field is infinitely long in the flow direction but is non-uniform over a plane perpendicular to this direction. It is applied in particular to the saddle-coil flowmeter.

The questions arising are: (i) how does the fall in field strength near the pipe walls affect the flowmeter output? and, (ii) what is the smallest saddle coil that would give a specified accuracy?

Assume that the velocity distribution is axisymmetric and is unchanged by the magnetic field. Let the velocity profile be given by the expression

$$u = u_0 (1 - \rho^n)$$

where $\rho = r/a$, a is the pipe radius and n is a variable parameter which determines the shape of the profile, e.g. for laminar flow n = 2.

Referring to the co-ordinate system shown in Fig. 12, if u has only a z component and the applied field B has no z component, the following differential equation can be derived:

$$\rho \frac{\partial^2 V}{\partial \rho^2} + \rho u' \frac{\partial B'_\theta}{\partial \rho} + \frac{\partial V}{\partial \rho} - u'B'_\theta + \frac{1}{\rho} \frac{\partial^2 V}{\partial \theta^2} - u' \frac{\partial B'_r}{\partial \theta} + \rho B'_\theta \frac{\partial u'}{\partial \rho} = 0$$

where
$$u' = u/u_m$$
$$B'_r = B_r/B_0$$
$$B'_\theta = B_\theta/B_0$$
$$V = \frac{\text{actual potential}}{u_m B_0 a}$$

where u_m is the mean fluid velocity and B_0 is the field strength at the pipe centre.

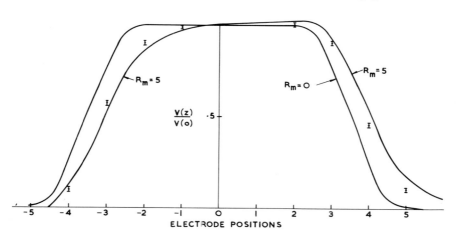

I EXPERIMENTAL POINTS FOR $R_m = 5$
— THEORETICAL RESULTS FOR
$S = 6 \cdot 5$
$t = 5 \cdot 0$

Fig. 11   Effects of field distortion on electric potential distribution

This equation was solved numerically by replacing the derivatives by finite differences between points on a mesh covering one quadrant of the pipe. Before the equation can be solved it is necessary to calculate u for each mesh point and to have values of B_r and B_o for each mesh point. As there can be up to 10,000 mesh points, measurement of the B s for an arbitrary field is not feasible. Fortunately, for the saddle coil these can be calculated, and the analysis which had previously been used for the infinitely long coil was used here.

Values of u'. B'_r and B_θ were stored in the computer before an iterative procedure for solving the difference equations was started.

The programme was checked by running it with a coil having a diameter 9 times that of the pipe, to ensure a uniform field across the pipe, and noting the deviations from unity of the u' for various flow profiles. These results are shown in Table 3, indicating that errors up to 1% can occur. This is for a 20 x 20 mesh. Increasing the number of nodes would improve the accuracy but this has not been tried to date.

Fig. 13 and Table 4 show results for a saddle coil strapped closely to the pipe, the worst case. Although there is a reduction in sensitivity the important thing to note is the small difference between sensitivities for different velocity profiles.

4   CONCLUSIONS

The saddle coil flowmeter presents a simple method of measuring the flow rates of conducting fluids in large ducts, to an absolute accuracy of ±2% and requiring no calibration. The theoretical analysis of end-shorting enables the length of the flowmeter, required to give a specified linearity, to be chosen. A computer programme has been written to predict, with a good degree of certainty, the field distribution produced by coils of any size, having the same geometry as the prototype. Work on the effects of non-uniform fields on flowmeter output indicates that saddle coils of smaller diameter than the prototype could be used without serious loss of accuracy, if required.

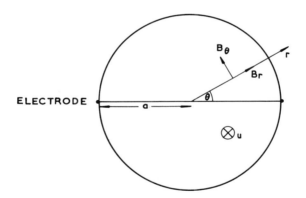

Fig. 12   Co-ordinate system for non-uniform field analysis

TABLE 3 Flowmeter sensitivity — uniform field

n	$\dfrac{v}{2B_ou_ma}$
2	.9977
3	1.004
4	1.007
5	1.010

TABLE 4 Flowmeter sensitivity — saddle coil strapped to pipe

n	$\dfrac{v}{2B_ou_ma}$
2	.967
3	.965
4	.960
5	.955

5   PRIMARY CIRCUIT FLOW MONITORS (FLUX DISTORTION FLOWMETERS)

There are two flow-monitoring requirements on the primary circuit of the PFR; one to monitor the flow from the pumps and the other to monitor flow through the core and to trip the reactor should the flow fall below a set level or at a rate greater than a given limit. The environmental conditions existing in the primary circuit exclude conventional forms of flowmeters, such as permanent magnet e.m. flowmeters. The type adopted is based on the Lehde and Lang[4] flux distortion flowmeter. Its principle of operation is illustrated in Fig. 14.

A.C. passing through a central primary coil produces a magnetic field which penetrates the sodium near the end regions of the coil. The radial component of the field interacts with the flowing sodium to produce circulating currents which, in turn produce their own magnetic fields, and these appear to distort the applied field. If two secondary coils, connected in opposition are placed symmetrically about the primary, the distorted field will induce an e.m.f. in them. This can be monitored, giving a measure of the sodium flow rate.

6   THEORETICAL STUDIES

Theoretical studies[5-7] of certain idealized forms of this flowmeter have been made, but, in general, their results have not been applicable to the practical case. Recently the author has been attempting a numerical analysis of a system approaching the real one with a certain measure of success so far. This will now be described.

6.1 Voltage induced by non-uniform field

Firstly, because the external magnetic field of the primary is very non-uniform, although axisymmetric, it is necessary to determine the voltage induced in a coil by a non-uniform field. The component of field which causes induced voltages is the z-component, B_z, and this varies with r and z. The total flux passing through any surface s is $\phi = \iint_s B_z(r,z).ds$. But $\mathbf{B} = \text{curl } \mathbf{A}$ where $\mathbf{A}$ is the magnetic vector potential. In the axisymmetric case under consideration, $\mathbf{A}$ only has an azimuthal component A_θ. Applying Stokes' theorem to the above integral gives:

$$\phi = \int_\ell A_\theta \ (a,z).dl$$

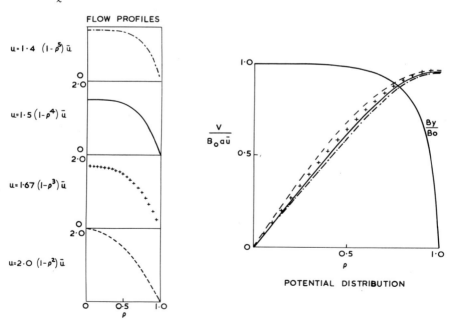

Fig. 13   Potential distribution for saddle coil strapped on pipe

where l is along the coil winding and a is the coil radius. If α is the pitch angle of the coil then

$$dl = \frac{dz}{\sin\alpha}$$

And $\mathbf{A}.\mathbf{dl} = A \dfrac{dz}{\sin\alpha}.\cos\alpha$

Therefore $\phi = \cot\alpha \displaystyle\int_L A(a,z)dz$

where L is the coil length.

Therefore total induced voltage V is given by:

$$V = \cot\alpha \int_L \frac{\partial A(a,z)}{\partial t}.dz$$

In practice A will vary sinusoidally with time and change phase along the coil. It can be represented by

$$A(a,z) = \left\{A_0(a,z) + i A_1(a,z)\right\} e^{i\omega t}$$

$$V = \omega\ e^{i\omega t}\ \cot\alpha \left\{ - \int_L A_1(a,z)dz + i \int_L A_0(a,z)dz\right\}$$

If the coil has n turns per unit length then

$$\cot\alpha = 2\pi na$$

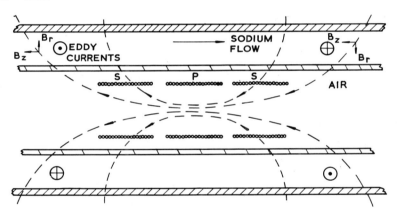

Fig. 14   Principle of flux distortion flowmeter

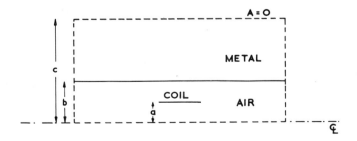

Fig. 15   Model for numerical analysis

Therefore $V = 2\pi na\, e^{i\omega t}\left\{-\int_L A_1\,(a,z)dz - i\int_L A_0\,(a,z)dz\right\}$

This shows that the magnetic potential distribution has to be obtained at least for $r = a$.

6.2 Potential distribution

The theoretical model is shown in Fig. 15. The primary coil is replaced by a current sheet of density Js amps/m, and the secondary coils are not required for this analysis. The governing equations are:

curl H	$= j$		where	$u =$	fluid velocity (assumed in these initial studies) m/s
divB	$= 0$				
B	$= \mu\mu_0 H$			$\sigma =$	fluid conductivity s/m
divj	$= 0$			$\mu =$	relative magnetic permeability
j	$= \sigma(E + unB)$			$\mu_0 =$	permeability of free space
curlE	$= -\dfrac{\partial B}{\partial t}$			$j =$	current density in fluid amp/m^2

Expressing these equations in terms of the magnetic potential gives

$$\text{curl}\left(\frac{1}{\mu\mu_0}\,\text{curlA}\right) = \sigma\,(E + u.B)$$

But $E = -\dfrac{\partial A}{\partial t}$

$$\frac{1}{\mu\mu_0}\,\text{curl curlA} = \sigma\left(-\frac{\partial A}{\partial t} + u.\text{curlA}\right)$$

In cylindrical co-ordinates where there is no azimuthal variation

$$\text{curlA} = \left(-\frac{\partial A}{\partial z}\right)k + \left(\frac{A}{r} + \frac{\partial A}{\partial r}\right)1$$

where k and l are unit vectors in the r and z directions.

Substituting in the above equation gives

$$\frac{1}{r}\frac{\partial A}{\partial r} - \frac{A}{r^2} + \frac{\partial^2 A}{\partial r^2} + \frac{\partial^2 A}{\partial z^2} = \mu\mu_0\sigma\frac{\partial A}{\partial t} + \mu\mu_0\sigma u\frac{\partial A}{\partial z} \qquad \ldots (3)$$

Because the exciting current varies sinusoidally with time A can be represented $A'e^{i\omega t}$ where A' is a complex function, of (v,z) and the above differential equation becomes purely spatial.

$$\frac{1}{r}\frac{\partial A'}{\partial r} - \frac{A'}{r^2} + \frac{\partial^2 A'}{\partial r^2} + \frac{\partial^2 A'}{\partial z^2} = i\mu\mu_0\sigma\omega A' + \mu\mu_0\sigma u\frac{\partial A'}{\partial z}$$

For the region $0 < r < b$, this equation becomes

$$\frac{1}{r}\frac{\partial A'}{\partial r} - \frac{A'}{r^2} + \frac{\partial^2 A'}{\partial r^2} + \frac{\partial^2 A'}{\partial z^2} = 0$$

The magnetic field is discontinuous across the current sheet and in fact the differences in field across the sheet is equal to the current density

$$(H_2)_{r=a} - (H_1)_{r=a} = Js$$

But $\mu\mu_0 H = \text{curl} A$; therefore

$$\left(\frac{A}{r} + \frac{\partial A}{\partial r}\right)_2 - \left(\frac{A}{r} + \frac{\partial A}{\partial r}\right)_1 = \mu\mu_0 Js$$

As J_s is A.C. $\quad J_s = J_s' e^{i\omega t}$ and

$$\left(\frac{A'}{r} + \frac{\partial A'}{\partial r}\right)_2 - \left(\frac{A'}{r} + \frac{\partial A'}{\partial r}\right)_1 = \mu\mu_0 J_s'$$

where J_s' is a real constant.

At the boundary between (2) and (3) A' and H are continuous.

If r and z are expressed in terms of the sodium skin depth $\dfrac{1}{\sqrt{(\mu\mu.\sigma\omega)}}$

so that $r' = \sqrt{(\mu\mu_0\sigma\omega)}\, r$

and $\quad z' = \sqrt{(\mu\mu_0\sigma\omega)}\, z$

equation (3) becomes

$$\frac{1}{r'}\frac{\partial A'}{\partial r'} - \frac{A'}{(r')^2} + \frac{\partial^2 A}{\partial (r')^2} + \frac{\partial^2 A}{\partial (z')^2} = iA' + R_m \frac{\partial A'}{a'\partial z}$$

where $a' = \dfrac{\text{coil radius}}{\text{skin depth}}$ and R_m is the magnetic Reynolds' number based on the primary coil radius.

It was thought that this equation could not be solved analytically and a numerical method was adopted. This amounted to covering the three regions with finite-difference meshes, replacing the derivatives by finite differences. The resulting algebraic equations were solved by the extrapolated Liebman method.

Because a numerical method can only cover a finite number of points an artificial boundary has to be used. This is shown as a dotted line in Fig. 15. Because of the axial symmetry only the upper half of the field need be considered and the magnetic potential along the axis is constant (Br = 0). For convenience this is taken to be zero. If the boundary is taken remote enough from the coils and the interface between regions (2) and (3) then the field strength here can be assumed zero with the result that A = 0 around the whole boundary.

Only preliminary results have so far been obtained. Fig. 16 shows the sum of the magnetic potentials* (magnitude only) at r = a over the region between the ends of the primary coil and the boundary, for the case where

   a = 1.0 skin depths
   b = 1.5 and 2.0 skin depths
   c = 5.0 skin depths

Length of primary coil = 2.0 skin depths over a range of R_m. For this particular configuration the output appears to be linear with flow for low R_m but falling off at higher R_m, consistent with the work of Cowley[5].

6.3 Axial magnetic field

A simple method of ascertaining the validity of the numerical method is to compare the field distributions along the primary coil axis, derived both from an exact analysis and the numerical

* taking account of the fact that any secondary coils will be differentially connected.

results. This is shown in Fig. 17. Considering that the numerical results are for a coil surrounded by a distant annulus of sodium, and the analytic results for a coil only, agreement is reasonable.

6.4 Future work

It is hoped to extend this work to cases where the fluid flow profile is non-uniform and the coils are wound on a former of magnetic material.

7 EXPERIMENTAL EVALUATION

The operation of the real flowmeters is more complicated than the above theoretical model because, firstly, the coils are partially screened from the sodium by a stainless steel pocket which reduces sensitivity, the amount depending on the operating frequency, and secondly the velocity

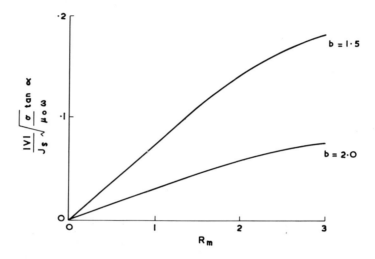

Fig. 16   Output versus R_m

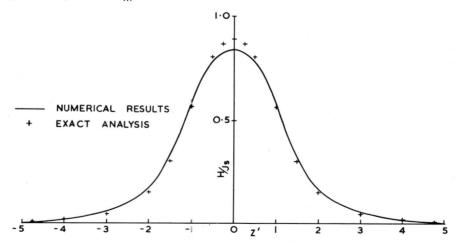

Fig. 17   Axial field profile

       —   numerical results
       +   exact analysis

profile near the sensor is unknown and is dependent on flowrate and the position of the sensor relative to pipe wall etc. Extensive experimental evaluations of production models of the two PFR flowmeter types, have been made under reactor conditions[8].

7.1 Pump flowmeter

The sensor is designed so that it can be pushed down a 40ft long tortuous containment pocket, terminating centrally in a 10 in. diameter discharge pipe from each pump, and operate satisfactorily at 450°C under moderate irradiation. It also has to withstand temperature cycling between 0°C and 450°C without mechanical or electrical failure and without unpredictable changes in its electric and magnetic properties.

The construction is shown in Fig. 18. Anodised aluminium wire, wound on a stainless steel former, is connected to aluminium-cored, mineral-insulated, aluminium-sheathed cable which acts both as electric connection to the measuring station and as a means of inserting and withdrawing the sensor. A stainless steel straining wire is also provided as a precaution against breakage.

The primary coil is driven with a constant current of 400mA at 700Hz. At this frequency it was observed that there is a minimum variation of output with temperature. Empirical calibration on the 6,000 gall/min sodium loop at the Engineering and Materials Laboratory of the U.K.A.E.A., at Risley (REML), showed the output/flow characteristic to be linear.

When these flowmeters are installed in the reactor, they will initially be calibrated using the technique of measuring the transit time of naturally occurring temperature fluctuations[9].

7.2 Core flowmeter

The form of this flowmeter and the way in which it is inserted into a control rod guide tube is shown in Fig. 19. The coils and guide tube have diameters of 1½ in. and 2 in. respectively, leaving ¼ in. annulus for the sodium. This sensor must operate in 700°C sodium and high radiation levels, and be capable of withdrawal. All parts of the sensor are machined from 18.8.1 stainless steel and the windings are stainless steel-sheathed, magnesia-insulated nichrome. The windings are brazed to the sensor body for stability.

Production models have been successfully tested in a special high temperature sodium loop at REML. The output is linear with flow as shown in Fig. 20. As the R_m associated with these sensors never exceeded about ½, this is in agreement with the above theoretical work.

Details of the electronic back-up systems to these sensors has been described in reference 8.

8   CONCLUSIONS

Although the dependence of the flowmeters on their environment precludes their use as accurate flow measuring devices, they present a very reliable, fast response method of reactor coolant flow monitoring and alarm condition detection.

9   ACKNOWLEDGEMENTS

The author wishes to acknowledge the efforts of Mr G McGonigal (REML) for his initial work on the saddle coil flowmeter and of Mr H Pugh (REML) for transforming the theoretical design of coil into physical reality. Useful suggestions and encouragement on the flux distortion flowmeter from Messrs S A Dean and E Harrison (REML) are greatly appreciated. This paper is published by permission of the Managing Director of the Reactor Group, U.K.A.E.A.

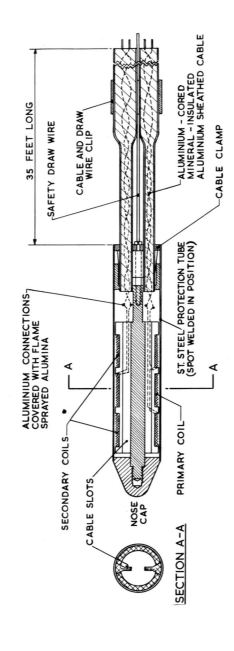

SECTION A-A

SAFETY DRAW WIRE

CABLE AND DRAW WIRE CLIP

35 FEET LONG

ALUMINIUM - CORED MINERAL - INSULATED ALUMINIUM SHEATHED CABLE

CABLE CLAMP

ALUMINIUM CONNECTIONS COVERED WITH FLAME SPRAYED ALUMINA

ST. STEEL PROTECTION TUBE (SPOT WELDED IN POSITION)

SECONDARY COILS

CABLE SLOTS

NOSE CAP

PRIMARY COIL

Fig. 18   Pump failure flowmeter
Scale 1:1

10   REFERENCES

1   MESHII, T, and FORD, J.A.: 'Calibration of Electromagnetic Flowmeters in the Enrico
    Fermi Atomic Power Plant'. Nuclear Applications and Technology, 1969, 7, No.1,
    pp. 76-83

2   BOUCHER, R.A., and AMES, D.B.: 'End Effect Losses in DC M.H.D. Generators',
    J. Appl. Phys. 1961, 32, No.5, pp. 755-759

3   SHERCLIFF, J.A.: 'The Theory of Electromagnetic Flow Measurement', (Cambridge
    University Press, 1962), p.38

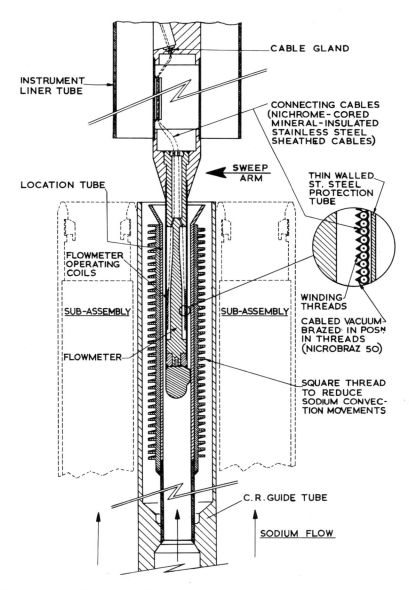

Fig. 19   Above core flowmeter and location tubes
          Scale 1:4

4   LEHDE, H., and LANG, W.T.: U.S. Patent No. 2435043, 1948

5   COWLEY, M.D.: 'On some Kinematic Problems in Magnetohydrodynamics', Quart. J. Mech. Appl. Math, 1961, **14**, Pt.3, pp. 319-333

6   COWLEY, M.D.: 'Flowmetering by a Motion-induced Magnetic Field', J. Sci. Instrum, 1965, **42**, No.6, pp. 406-409

7   BANDELET, C.: 'Two Theoretical Analyses of the Output Signal from an Eddy Current Sensor', Rev. Sci. Inst., 1971, **42**, No.4, pp. 458-461

8   DEAN, S.A., HARRISON, E., and STEAD, A.: 'Sodium Flow Monitoring', Nucl. Eng. Int., 1970, **15**, No.174, pp. 1003-1007

9   THATCHER, G., BENTLEY, P.G., and McGONIGAL, G.: 'Sodium Flow Measurement in PFR', Nucl. Eng. Int., 1970, **15**, No.173, pp. 822-825

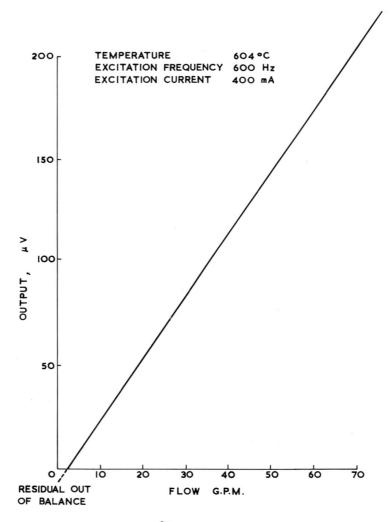

Fig. 20   Output versus flowrate at 604°C

ERRORS IN THE VELOCITY-AREA METHOD OF MEASURING ASYMMETRIC FLOWS IN CIRCULAR PIPES

L.A. Salami

Abstract: The paper compares some of the existing well known velocity-area methods of flow integration by using theoretical asymmetric axial velocity distribution functions which can be integrated exactly over the pipe section. It also shows that at least six traverse points per radius are required to give consistently good results (error not greater than 0.5%) and recommends that traverses be carried out on at least six radii. For a given total number of traverse points of above 24 it is demonstrated that it is better to use say four traverse points per radius on six traverse radii than eight points per radius on three radii.

1   INTRODUCTION

There are three difficulties in measuring flow in circular pipes using pitot tubes. Firstly because the pitot tube will only measure point velocities, theoretically, one must use an infinite number of traverse points to obtain the flow exactly. As this is neither possible nor practicable one has to use a finite number of traverse points and then apply an 'accepted' velocity-area integration method to obtain flowrate. Secondly, there is a series of errors associated with a given pitot-static tube reading. Thirdly all the methods in the national and international codes are based on axisymmetric velocity profiles but in practice they usually are asymmetric.

It is only the problem of using a finite number of traversing points that will be tackled here and four now widely used velocity-area methods of integration will be considered. To simplify the discussion, therefore, the velocity obtained from the reading of the pitot tube will be assumed to be error-free.

The basic assumption in deriving most of the velocity-area methods for velocity profile integration is that the curve assumed approaches the real velocity profile in the line of traverse. This is reasonable for axisymmetric flow profiles. However, in asymmetric flow the velocity profile varies from one radius to the other so that, even if the velocity profiles on a few selected traversing lines are correctly represented there is still the possibility that these lines will not be representative of the whole pipe section.

Thus, the problem of asymmetric flow measurement reduces to making suitable assumptions about both the radial and the circumferential velocity distributions.

2   THE WORK AT SOUTHAMPTON

The ultimate aim of the work now being undertaken at the University of Southampton is to optimise the number of traverse points necessary to give a certain percentage of accuracy for the flow measurement in circular pipes by means of a pitot-static tube (or pitot-wall-static combination) for an asymmetric flow profile. As a prelude to this, the first stage of the work is to assess the existing velocity-area methods most of which have been evolved to deal with axisymmetric and near axisymmetric flows. The number of traversing points required when the errors obtained in applying these methods to asymmetric profiles become either constant or negligible is of importance. Also important is the number of radii required to make the standard deviation small or constant. The product of these two latter numbers may give an indication of the optimum number of traverse points sought although it may not necessarily be that number.

For the comparison of the velocity area methods it is important that the integration be performed along exactly the same paths. These paths in this paper are seventy-two equally spaced radii on the circular pipe section. The criteria to be employed in deciding which is the best method will be the method that gives the lowest average error, this being the difference between the exact flowrate and that from the total number of traverse points on the seventy-two radii.

Unlike in Ref. (1) where the velocity-area methods are compared by using the graphical integration method which is in itself susceptible to errors, theoretical velocity profile functions will be used here which the author feels are quite realistic and which can be integrated exactly over the pipe section. Instances will be quoted from past literature where velocity profiles similar to the ones used can be found.

3   VELOCITY-AREA METHODS

The theoretical assumptions and the application of the various velocity integration methods used will now be briefly considered. For fuller details the appropriate reference shown against the method should be consulted.

3.1 Log-linear[1]

This method is based on the fact that the velocity profile along the radius of a pipe can be represented by

$$U = A + B \log \frac{1}{2} (1 - R) + \frac{1}{2} C (1 - R)$$

where

$$U = \frac{u}{U_0} \qquad u = | \text{ the velocity at any radial point}$$

$$U_0 = | \text{ the velocity at the pipe centre}$$

$$R = \frac{r}{R_0} \qquad R_0 | = | \text{the pipe radius}$$

A, B and C are constant

The integration of this profile becomes independent of the constants A, B and C when the location of the traverse points are as given in the Appendix of the above reference. The area of the pipe is divided into n equal areas i.e. a central circle and (n − 1) annuli. The flow through these areas is assumed to be given by two points in each portion. Thus in the mth annuli the first and the second radial distances R_{m1} and R_{m2} are given by

$$R_{m1} = 1 - \frac{1}{2} \left\{ a_m + \sqrt{\left(a_m^2 - b_m^2 \right)} \right\}$$

$$R_{m2} = 1 - \frac{1}{2} \left\{ a_m - \sqrt{\left(a_m^2 - b_m^2 \right)} \right\}$$

where $a_m = 1 - n \left[(2/3)t^2 \ t \right]_{t^2 \ = \ 1 - m/n}^{1 \ - \ m/n \ + \ 1/n}$

and

$$\log b_m = n \left[(t^2 - 1) \log (1 - t) + \frac{1}{2} \log \epsilon (1 - t) (3 + t) \right]_{t^2 \ = \ 1 \ - \ m/n}^{1 \ - \ m/n \ + \ 1/n}$$

The above procedure will obviously give even numbers of traverse points, 2n per radius. For odd numbers of points per radius, $2n - 1$, the area is divided into a central circle and $n - 2$ annuli of equal area and an outermost annulus of area equal to half of each of the other portions. Only one traverse point is taken from this outer annulus and its radial distance is given by

$$\log (1 - R) = (2n - 1) \left\{ \frac{1}{2n-1} \log \left(1 - \sqrt{\frac{2n-2}{2n-1}} \right) - \frac{1}{2} \log \epsilon \left(1 - \sqrt{\frac{2n-2}{2n-1}} \right) \left(3 + \sqrt{\frac{2n-2}{2n-1}} \right) \right\}$$

The radial positions of the traverse points in the mth annulus is given by a procedure similar to that already outlined above but now

$$\log b_m = \frac{1}{2} (2n - 1) \left[(t^2 - 1) \log (1 - t) + \frac{1}{2} \log \epsilon \quad (1 - t) \right.$$

$$\left. \times \quad (3 + t) \right] \sqrt{\frac{2(n - m + 1)}{2n - 1}}_{\displaystyle \sqrt{\frac{2(n - m)}{2n - 1}}}$$

and $a_m = 1 - \frac{1}{2}(2n - 1) \left[\frac{2t^3}{3} \right] {\Large t = \sqrt{\frac{2(n - m + 1)}{2n - 1}} \atop t = \sqrt{\frac{2(n - m)}{2n - 1}}}$

To apply the method the number of points required per radius, n, automatically fixes the location of the traverse points. The average of the velocities at these points gives the average velocity for the profile.

It should be pointed out that the profile on which this method is based is unrealistic near the wall where U tend to $- \infty$.

3.2 Methods of cubics[2]

This method replaces the actual velocity profile by a series of cubics and then integrates the area under them to give the flow rate. Suppose now that m ordinates are taken on the profile along a radius between the pipe centre and the wall. Let the radial distances be $R_1, R_2 \ldots R_m$, then the cubics replacing the profile between two adjacent points at R_i and R_{i+1} for $1 \leqslant i \leqslant m - 1$ have the property that the tangents at R_i and R_{i+1} are parallel to the lines joining the tops of the ordinates at R_{i-1} and R_{i+1} and R_i and R_{i+2} respectively.

The cubic joining the ordinate at R_1 to the centre is given by

$$u = U_o - \left\{ \frac{u_2 - U_o}{R_2} \quad R_1 + 3(U_o - u_1) \right\} \left(\frac{R}{R_1} \right)^2$$

$$+ \left\{ \frac{u_2 - U_o}{R_2} \cdot R_1 + 2 (U_o - u_1) \right\} \left(\frac{R}{R_1} \right)^3$$

where subscript 'o' denotes the centre. From this equation it is clear that the slope of the tangent cubic at the centre is zero.

The extrapolation curve joining the last traverse point, co-ordinate on the (v, R^2) system R_{m-1}, to the wall is based on von Karman's theory, i.e.

$$u = \frac{u_m (1 - R^2)^{1/n}}{(1 - R_m^2)^{1/n}}$$

Finally, the property of the cubic joining the ordinates at R_{m-1} and R_m is such that the slope at the tangent at R_m is equal to the slope of the tangent to the extrapolation curve above at R_m.

With these assumptions the mean velocity is given exactly by

$$U_M = \frac{1}{4} R_1^2 \left\{ (U_0 - u_2) \frac{\dot{u}_1}{3R_2} - (U_0 - u_1) \right\}$$

$$+ \frac{1}{12} \sum_{i=2}^{i=m} (u_i - u_{i-2}) \left\{ (R_i^2 - R_{i-1}^2) - (R_{i-1}^2 + R_{i-2}^2) \right\}$$

$$+ \frac{1}{2} \sum_{i=1}^{i=m} (u_i + u_{i-1}) (R_i^2 - R_{i-1}^2)$$

$$+ u_m \left\{ \frac{n}{n+1} (1 - R_m^2) + \frac{(R_m^2 - R_{m-1}^2)^2}{12n(1 - R_m^2)} \right\}$$

In the application of the above procedure the reference says it is enough to use between four to six points per radius; four points for small pipes and six points for large ones. It does not give limits to the size of pipe when only four points can be used nor does it indicate what factors decide how many radii should be employed. The location of the traverse points is arbitrarily chosen but it seems from the specimen calculations that R_1^2 is maintained constant at 0.07 for the number of points per radius. The last traverse point at R_m seems to be arbitrarily chosen but the location of intermediary points is given by

$$R_i^2 = \frac{i(R_m^2 - R_1^2)}{m - 1} \qquad i = 2, 3 \ldots \ldots m$$

It also seems the value of n in Von Karman's equation already given and which appears in the expression for U_M is also arbitrarily chosen to be either 7 or 9.

3.3 Tangential method[1]

This is now a well known method in which the area of the pipe is divided into a central circle and n - 1 annuli all of equal areas. The traverse points are given by the intersection of the radius and the concentric circles with divided each of these n portions into two equal areas. Thus

$$R_i = \sqrt{\left(\frac{2i - 1}{2n} \right)} \qquad i = 1, 2 \ldots \ldots n$$

The arithmetic average of the velocities obtained at these points along a radius gives the average velocity for the profile at that radius.

3.4 Simpson's rule[3]

This method with 10(=2n) traverse points per diameter is recommended in Ref. (3), clause 37. It appears the traverse points have been obtained as follows:

Starting with the centre of a diameter as the origin the square of the non-dimensional pipe radius is plotted on a new 'x' axis from -1 to +1. The range -0.9 ($= -(2n-1)/2n$) $\leqslant R \leqslant 0.9$ ($= (2n-1)/2n$) is further sub-divided into 10(=2n) equal parts and these give the traverse points. (The fraction or number in brackets is assumed here to be the general expression as this has not been given in the above reference). The corresponding traverse points in the pipe is deduced by taking the square root of the distance of the traverse points on the new axis i.e.

$$R_i = (i-1)\sqrt{\left(\frac{2n-1}{2n}\right)} \qquad 1 \leqslant i \leqslant n+1$$

Over this region of R the velocity is integrated just as any other function would be using the Simpson's Rule and the mean velocity U_M over the whole of the pipe section is given by

$$U_{M1} = \frac{2n-1}{12n^2} \left\{ U_1 + U_{2n+1} + 4 \sum_{i=1}^{i=n} U_{2n} + 2 \sum_{i=1}^{i=n-1} U_{2n+1} \right\}$$

which gives for n = 5

$$U_{M1} = \frac{9}{300}(U_1 + 4U_2 + 2U_3 + 4U_4 + 2U_5 + 4U_6 + 2U_7 + 4U_8 + 2U_9 + 4U_{10} + U_{11})$$

In the region $R = \pm0.9\left\{= \pm(n-1)/2n\right\}$ to the pipe wall it is assumed that the velocity can be represented by the law $V = Ky^{1/m}$ where y is the distance from the wall, $(1-R)$. Reference (3) recommends that an attempt be made at estimating the value of m near the wall by taking extra traverse points in this region and this has actually been done here; ten points are taken between the last traverse point and the wall. Using the least squares method a *straight line* is fitted to a logarithmic plot and hence m is obtained. The contribution to the average velocity over the pipe section of this region is given by

$$U_{M2} = 2(U_1 + U_{2n+1}) \left\{ \frac{m}{m+1}(1 = R_{n+1}) - \frac{m}{2m+1}(1 - R_{n+1})^2 \right\}$$

In the case of ten points per diameter, n = 5

$$U_{M2} = 2(U_1 + U_{11})(\frac{m}{m+1} 0.0513 - \frac{m}{2m+1} 0.0513^2)$$

Thus the mean velocity U_M, is $U_{M1} + U_{M2}$

4   CIRCUMFERENTIAL VELOCITY DISTRIBUTION

Hitherto, velocity integration methods have been based on velocity distribution on a radius where the point velocities on this radius are assumed to be the average for a certain number (usually 4 or 8) of equally-spaced radii on the pipe. However, in asymmetric flow circumferential velocity distribution is just as important as radial distribution. While in the future attempts will be made to assess whether equal spacing of the radii is statistically the best arrangement in this paper attention will be limited to demonstrating the effect of using 3, 4, 6 and 8 equally spaced radii on the 'Range of Error' and the 'Standard Deviation' of a complete set of errors.

5   PRESENT BRITISH STANDARD CODE

The BS 1042, part 2, 1970[3] recommends that the log-linear method with 10 points per diameter be used and that the traverse be carried out over at least two diameters at right angles. If the difference between the average velocity on each diameter exceeds 2% of the mean then further traverses should be made over intermediate diameters.

For comparison of the various methods 10 points per diameter will be used.

6   VELOCITY PROFILES

In the past various velocity-area methods have been compared on the bases of pitot traverses,[1] and theoretical profiles along a radius[4]. There are limitations in these procedures. In the former the errors associated with each pitot reading may affect the velocity-area methods differently and this coupled with the fact that no absolute method of flow measurement was employed as the basis of comparison, may obscure the method that gives the best result. Also, thorough investigation will of necessity require a colossal number of pitot traverses. On the other hand, the second procedure of using theoretical profiles along a radius goes some way towards establishing a reference which is unaffected by measurement errors, nevertheless suffers from the disadvantage that this reference is ever-changing for different combinations of radii.

The method here assumes a profile across the section of the pipe which varies from radius to radius, just as in real flow, and can be integrated exactly to give a reference mean velocity for the section.

The profiles have been chosen so that they depict a prominent peak, Profile P1; a pronounced deep, P2; fairly flat central portion, P3; a profile with a slanting side, P4; and a profile with two unequal peaks, P5. (Profile P1 resembles a measured Profile Q in Ref. (5), P2 profile B3 in ref. (6), P3 profile 3 in ref. (1), P4 profile N in Ref. (5) and P5 profile M in ref. (5). Profiles P6 — P10 have been inserted to further investigate whether there is a consistent trend of better results (smaller error range and standard deviation) by using more radii.

The equations of the profiles are as follows:

P1 $\qquad U = (1 - R)^{1/9} + mR(1 - R)^{1/k} e^{-a\theta} \sin\theta$

P2 $\qquad U = (1 - R)^{1/9} + mR(1 - R)^{1/k} e^{-a\theta} \sin\theta$

P3 & P4 $\qquad U = 1 + yR \sin\theta \qquad\qquad$ for $R < b$

$$\text{and } o \leqslant b \leqslant 1$$

and

$$U = (1 + yb \sin\theta)\left\{ \left(\frac{1 - R}{1 - b}\right)^{1/9} + \frac{m(R - b)(1 - R)^{1/k}}{(1 - b)^{1/k}} e^{-a\theta} \sin\theta \right\}$$

P5 $\qquad U = (1 - R)^{1/9} + mR(1-R)^{1/k} e^{-a\theta} \sin^2\theta$

P6 $\qquad U = (1 - R)^{1/9} + mR(1 - R)^{1/k} \sin\theta$

P7 $\qquad U = (1 - R)^{1/9} + mR(1 - R)^{1/k}(1 - \cos^2\theta)$

P8 $\qquad U = (1 - R)^{1/9} + mR(1 - R)^{1/k}(\theta^2 - 1)(1 - \cos\theta)^2$

P9 $\qquad U = (1 - R)^{1/9} + mR(1 - R)^{1/k} \, 2 \, (2\pi - \theta)^2$

P10 $\qquad U = (1 - R)^{1/9} + mR(1 - R)^{1/k} \, (2\pi - \theta) \, \sin^2\theta$

where y, b, m and k are constants chosen to obtain the desired profile characteristics and θ is the position of the radius relative to the flow asymmetry. It should be pointed out that P1 and P2 are special cases of the profile P3 and P4 i.e. when b = 0.

7   COMPUTER PROGRAM

The first part of the program gives the location of the traverse points in some velocity-area methods once their number per radius is given. This number ranges from 2 to 20, not necessarily in even steps.

The program then compares the mean velocity on two diameters (four radii) at right angles to each other and evaluates the difference as a percentage of their mean. This procedure is continued for the complete set of the 4-radius arrays possible from the 72 fixed radii. At the same time the program works out the average error and the standard deviation of this set. The latter task is also carried out using arrays of 8, 3, and 6 radii in that order.

For each velocity-area method a similar approach to the above is used.

8   RESULTS AND DISCUSSIONS

Fig. 1 [(i) − (x)] show the contour map of the velocity distribution over the pipe section and the maximum velocity relative to the velocity at the centre of the section is about 1.2. In the following discussion the notation with respect to the diagrams will be that i corresponds to profile P1, ii to P2, iii to P3, iv P4 . . . etc.

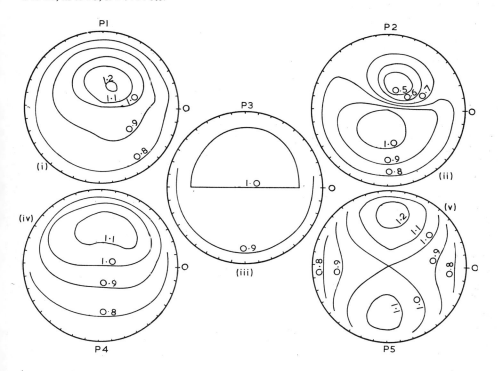

Fig. 1   Velocity profiles

Error as a percentage is defined as

$$100 \times \frac{Q \text{ (from velocity-area method for an array)} - Q \text{ (actual)}}{Q \text{ (actual)}}$$

where Q (actual) is the exact integration of the profiles used.

8.1 Average error

The average errors of a complete set of results given by the four velocity-area methods are plotted in Fig. 2 [(i) − (x)] for each of the profiles against the number of traverse points per radius. Generally all the methods give poor or inconsistent results with few traverse points i.e. below 4 points per radius (except the cubic method which was not investigated in this region), but the results generally rapidly improve with increase in traverse points per radius. However, as the figures illustrate the tangential and the Simpson's Rule methods generally give much poorer results than the other two methods − log-linear and cubic methods − although with very many traverse points, about 24 points per radius say, there is very little difference between them. It should be pointed out though, that the log-linear still gives better results than the Simpson's Rule and this in turn gives better results than the tangential method.

The Simpson's Rule method[3] is cumbersome as extra traverse points are required near the wall, where velocity measurements are most unreliable, to evaluate the value of the exponent, m. Consequently, the value of m should be standardised for instance by using the value of m for a fully developed pipe flow at the same Reynolds number as the flow under consideration. This will inevitably lead to poorer results. This handicap coupled with the fact that both the Simpson's Rule and the tangential methods generally do not converge rapidly enough in the region of 4 to 8 traverse points per radius has demonstrated that they are both inferior to the other two methods from the optimisation point of view.

The striking feature about the log-linear and the method of cubics is that they give low errors for all the profiles used here at relatively low numbers of traverse points per radius. Although for most profiles the method of cubics generally gives poorer results for 4 traverse points per radius they both have almost the same accuracy at 5 traverse points and at 6 points per radius the method

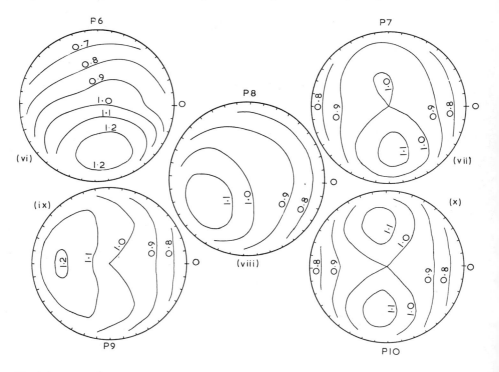

Fig. 1 (continued)

of cubics generally gives slightly better results. Also, at 4 and 5 traverse points per radius the method of cubics gives more consistent results than the log-linear.

For the log-linear method there is only a slight improvement in the result for most of the profiles by using higher numbers of traverse points per radius than 6. At this number of traverse points the log-linear always slightly under-estimates the flow. In contrast, the method of cubics slightly overestimates the flow at this number. It may then be possible in future to investigate the possibility of using both methods simultaneously and taking the mean of the results obtained. This procedure need not increase the number of traverse points as it seems one can use any desired radial spacing of the traverse points for the method of cubics and the log-linear spacing could then be used.

The following points are also noteworthy. First that the average errors for 3 traverse points per radius is still very inconsistent and consequently the writer feels it is unreliable to use so few points. Secondly, that the results at 10 points per diameter for the Simpson's Rule[3] is not

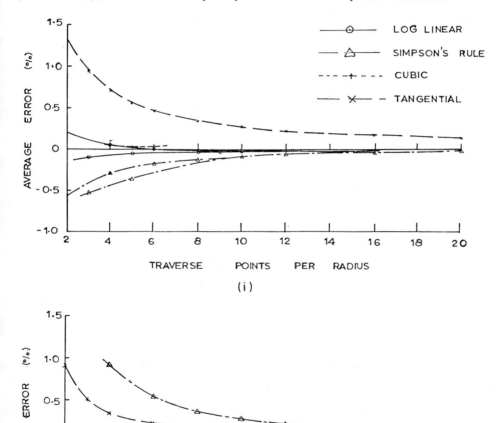

Fig. 2   Comparison of velocity-area methods using average error

consistently good. This is partly due to the effect of 'odd and even' numbered traverse points per radius which is present in both this and the log-linear methods. While in the log-linear method no conclusion can be drawn with regard to this effect for the Simpson's Rule method, even numbers of traverse points per radius seems to give better results. The other reason for doubting the wisdom of this recommendation[3] is that the work here shows that 5 points per radius are still too few for the Simpson's method to give good results. Thirdly, it is likely that the method of cubics will be affected by the exponent m assumed. In the profiles used here the value of m has been around 9 and therefore this number has been used in the computations. In practice the value of m will have to be guessed with possible associated errors.

8.2 Error range

As has already been explained the average error used above is that obtained by using the number of traverse points given by 72 x (the number of traverse points per radius). This partly explains why the average errors are so low in Fig. 2 [(i) − (x)]. In practice however, one decides on using certain arrays of radii and obtains a result for that array.

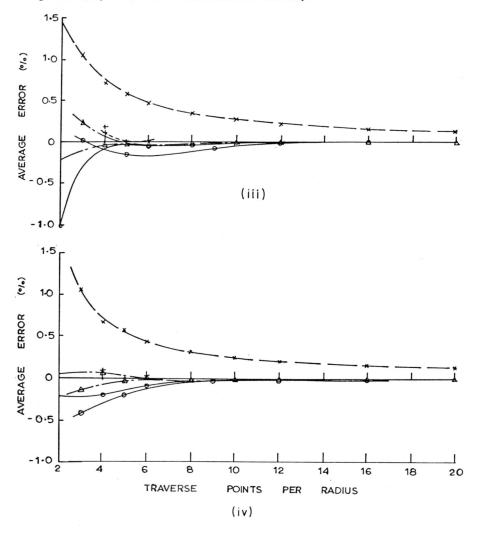

Fig. 2 (continued)

Limiting the traverse paths from which the array is selected to the 72 equally spaced radii the range of error in a complete set of results is plotted in Fig. 3 [(i) − (x)] for the 5 points per radius log-linear method using the ten profiles.

As can be seen the error range diminishes rapidly at first and gradually later on as the number of radii in the array is increased. There is generally only a small improvement in the error range from 6-radii to 8-radii arrays. One may, of course, argue that this is due to the increase in the total traverse points and in order to disprove this Fig. 4[(i) − (x)] has been drawn using the same total traverse points of 48 for each of the ten profiles. This number, 48, has been chosen to ensure that the average error is almost independent of the number of traverse points per radius which occurs at about 6 points per radius.

8.3 Standard deviation

Figure 5 shows the error obtained for different orientations of the array of radii to the asymmetric

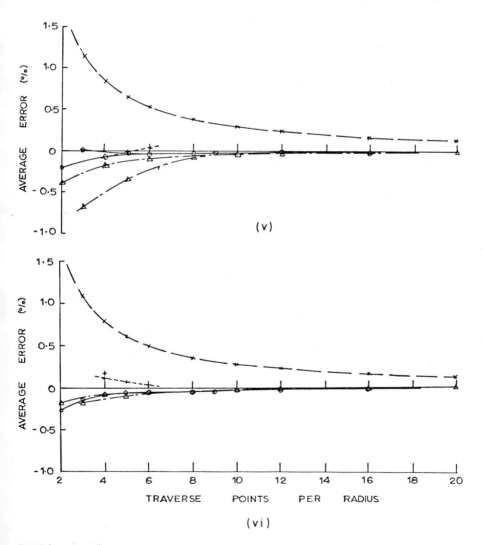

(v)

(vi)

TRAVERSE   POINTS   PER   RADIUS

Fig. 2 (continued)

profile. For the log-linear method as well as for other velocity-area methods used here. As can be easily seen, the distribution of error does not resemble a normal distribution for random variables, it is more like a triangular distribution. Thus, although the statistical parameters associated with normal distributions such as average error and standard deviation will have near enough the same meaning, other derivatives such as confidence interval and tolerance may however be meaningless. The standard deviation here should not be considered in its absolute sense but simply as a means of stating the probability of obtaining the average flow of the set of results by means of a reading corresponding to a particular array. Low values of standard deviation signify a high probability and vice-versa.

Figure 6 (A and B) show the standard deviation of the log-linear method as a function of the traverse points per radius and the array of radii used. The array considerably affects the standard deviation while the influence of the number of points per radius is confined to less than 6 points for Fig. 6A and less than 8 points per radius for Fig. 6B for the 3-radii array. These numbers are smaller for the 6 and 8-radii arrays. Fig. 6A is typical of most profiles used here and Fig. 6B is that obtained for the fairly flat profile, P3.

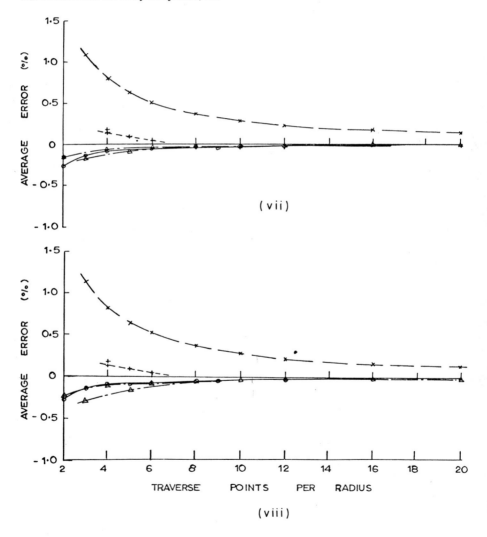

Fig. 2 (continued)

Lines with constant total number of traverse points (broken lines) are also drawn on Fig. 6B which shows the greatest dependence on the number of points per radius. For total traverse points of 24 and above there is a great advantage in using larger numbers of radii in the array. The curve for the total traverse point of 12 show that in this particular case the chances of obtaining the best result is the same whether one uses 6 or 4 radii in the array. However, as already remarked, 2 or 3 points per radius according to the work here are too few to give a reliable flowrate.

Figure 5 also shows that the distributions of error with a given array are very similar in shape for the velocity-area methods. It is not surprising, therefore, that the standard deviations are almost identical for all the methods, e.g. Fig. 7, for profile P4. Consequently, it appears the standard deviation is mainly a function of the velocity profile. This is partly illustrated in Fig. 8 where all the profiles with $e^{-a\theta}$ and one prominent peak, P1 – P4, lie on the same curve. This graph has been plotted non-dimensionally in terms of the standard deviation for the 3-radii array.

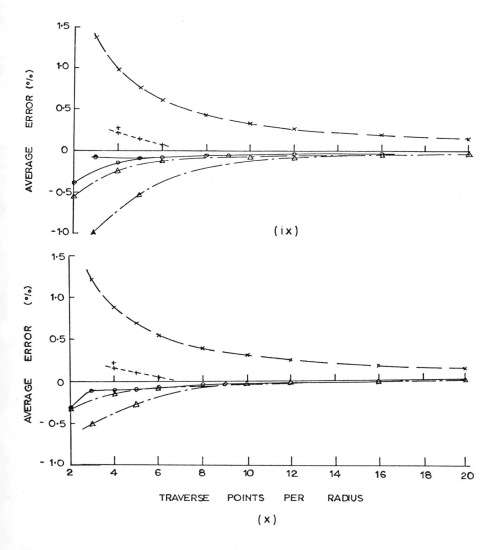

(ix)

(x)

TRAVERSE POINTS PER RADIUS

Fig. 2 (continued)

9   CONCLUSION

Both the tangential and Simpson's Rule methods give poor results even up to 8 traverse points per radius and are not therefore suitable for optimising the number of traverse points. The log-linear and the method of cubics do not suffer from this defect and the average error usually drops to a small value after 6 traverse points per radius.

The log-linear method though slightly superior at around 4 traverse points per radius becomes slightly inferior to the method of cubics at about 6 traverse points per radius. The difference between the results of the two methods is so small that it is a matter of opinion whether the extra calculation involved in using the method of cubics is justified.

By using more radii both the error range and the standard deviation of a complete set of results is considerably reduced. With an array of 6 radii both these curves have begun to flatten out and an array of at least 6 radii is therefore recommended.

For fairly accurate flow measurement (to less than ±0.5%) in a circular pipe by means of a pitot-tube it appears from this theoretical consideration that one requires at least 36 traverse points equally divided amongst six radii.

10   ACKNOWLEDGEMENT

The author is grateful to his supervisor Professor S.P. Hutton of Southampton University, Dr. E A. Spencer and Dr. F.C. Kinghorn both of National Engineering Laboratory (N.E.L.) for their interest in this work. His gratitude also goes to the Department of Trade and Industry who sponsored the work and to the University of Southampton where the programme was developed.

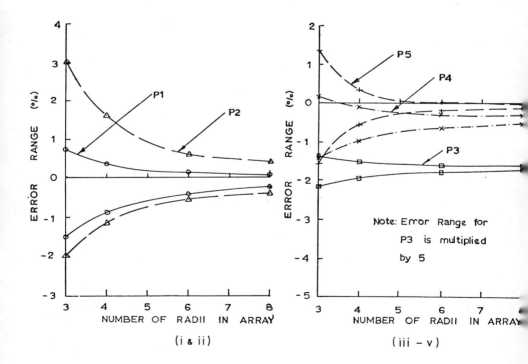

Fig. 3   Error range for log linear method using 5 points per radius

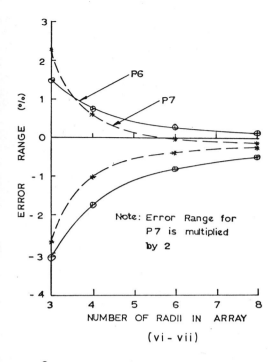

(vi - vii)

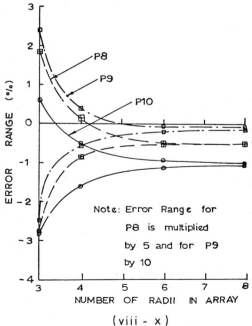

(viii - x)

Fig. 3 (continued)

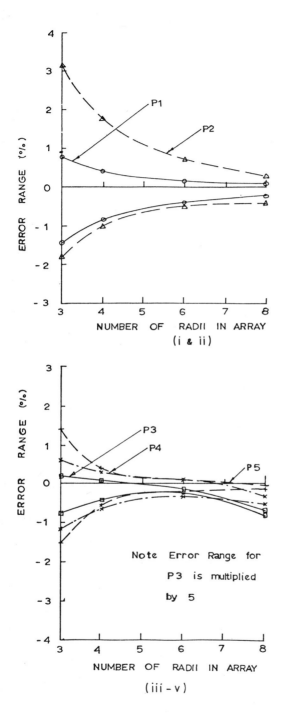

Fig. 4   Error range for log linear method with a total of 48 traverse points

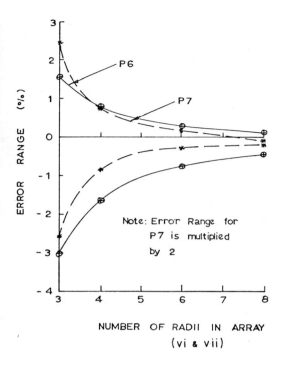

NUMBER OF RADII IN ARRAY

(vi & vii)

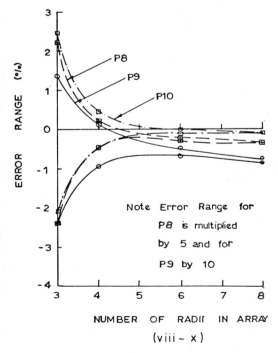

NUMBER OF RADII IN ARRAY

(viii - x)

Fig. 4 (continued)

11 REFERENCES

1 WINTERNITZ, F.A.L., and FISCHL, C.F.: 'A simplified integration technique for pipe flow measurement', Water Power, June 1957, p. 225

2 I.S.O. Technical Committee 'Rapport de la delegation Francaise sur la methode de calcul du debit-volume', ISO/TC 30/GT (France − 3) 15F, Feb. 1969

3 BS1042, pt. 2; 1970: 'Method for the measurement of fluid flow in pipes'

4 SPENCER, E.A.: 'Effects of velocity profile on pump-test measurements', Paper D2-4 Proceedings of Symposium on Pump Design, Testing and Operation (H.M. S.O., Edinburgh, 1966)

5 SHERWOOD, T.K., and SKAPERDAS, G.T.: 'A simplified pitot-tube traverse', Mechanical Engineering, 1939

6 RICHTER: 'Log-Linear Regel − ein einfaches Verfahren zur volumenstrommessung in Rohrleitunger', Heiz-Luft-Haustechn 20 (1969) No. 11 Nove., p. 407-409

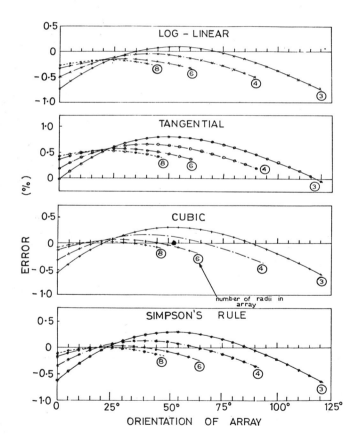

Fig. 5 A complete set of results for profile P4 using 5 points per radius

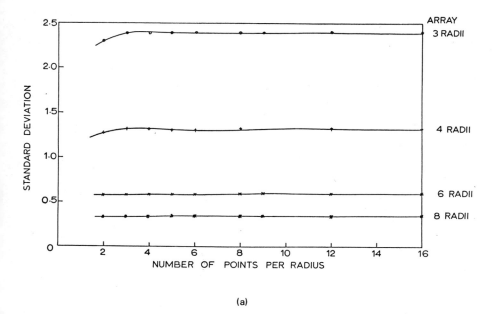

(a)

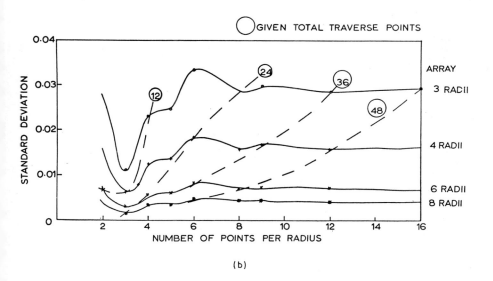

(b)

Fig. 6   a   Standard deviation as a function of the traverse points per radius for profile P2
              (Log—linear)

        b   Standard deviation as a function of the traverse points per radius for profile P3
               (Log—linear)

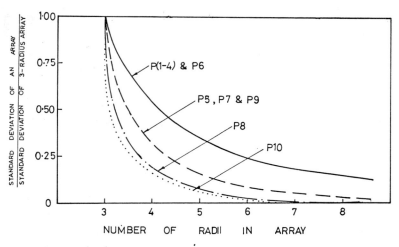

Fig. 7   Standard deviation for log-linear method with 5 points per radius for profiles P1-P10

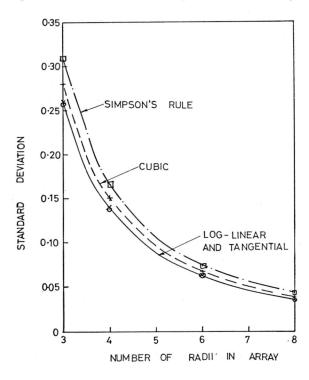

Fig. 8   Comparison of velocity area methods using standard deviation and profile P4

FLOW METERING AND FUTURE AIMS

S.P. Hutton

Abstract: A review is made of the conference papers with the object of assessing the present state of the art, outlining trends and defining aims for the future. The emphasis of the conference has been on modern developments, and the basic question is to decide to what extent new methods will complement or replace existing techniques. Conventional methods have perhaps been developed almost as far as they can go and if higher accuracy is required, it may be necessary to pin our faith on new techniques. Nevertheless, with most new methods, despite their inherent advantages, there are usually new difficulties, and it is not easy to find flowmetering techniques which adequately compromise between low cost, reliability, simplicity and accuracy. Nevertheless, with the aid of integrated circuits and modern electronic techniques, it may be possible to do what was impossible, or at least impractical, a few years ago and to evolve new methods in which complex procedures are carried out accurately and reliably by low-cost minicomputers. Such possibilities certainly extend our horizons.

It is clear from the two review papers which opened the conference that there will be an increasing demand for flow meters of all kinds in the future. Indeed the increase in demand is matched only by the increasing number of papers published on the subject and by the number of new techniques which are invented. Looking backwards in time it is noteworthy how many novel methods have been suggested and yet how many have failed to meet the exacting practical requirements. There is therefore still plenty of scope for new methods but few will make the grade and will be acceptable for wide use and commercial exploitation.

The title of this conference has emphasized new methods, but looking at Clayton's Table (p. 2) showing the dates when various techniques first appeared, it is difficult to decide what is really new. Many good ideas are old ones and they have had to wait until technology has developed sufficiently to exploit them. However very roughly, based on my personal assessment of what is new, about 40% of the papers could be said to be concerned with new methods, 25% provide detailed information and a better understanding of old methods, and 35% are new variations on an old theme.

1   FACTORS GOVERNING ACCEPTABILITY OF A NEW FLOWMETER

It takes time for a new method to be generally accepted and to do this it must satisfy one or more of the following requirements:

repeatability	,	rangeability
reliability	,	accuracy
right price		

"Rangeability" or "turn-down ratio" as it is sometimes called is the flow range over which the meter is accurate within certain limits. All the above terms need to be clearly defined and more will be said about this later.

I have intentionally used "right price" instead of low cost because although low cost is generally important there are some important applications where it is worth paying quite a lot to get a satisfactory meter.

Because of the widely varying demands of different industrial problems the above factors will be ranked in various orders of priority for each situation and there will always therefore be a wide range of meter-types waiting to satisfy the user. For instance, in some process plant applications absolute accuracy may not be so important but repeatability and reliability are essential and may be worth a relatively high capital cost. The operating range may be very narrow, a relatively easy problem, or it may be wide, not so easy. As Spencer pointed out (p. 13) the accuracy claimed for various methods in the past may not be very different from that claimed now, but I suspect that we now have a slightly better idea of the validity of such claims. Users are now taking a more intelligent interest in specifying their own requirements more precisely and in satisfying themselves that the flowmeter manufacturer has met them. It is important to have a sense of proportion about what is required and the need for accuracy, reliability, etc. For those cases where high accuracy is required, and the proportion of these is likely to increase in future for economic reasons, repeatability is the first step towards high accuracy but must not be confused with it.

We must also distinguish between random and systematic errors in the meter being calibrated and in the primary calibration system itself. Engineers in general are just beginning to be tidy, systematic and sensible about such things, but we do need a standardized terminology. For instance I have heard the word precision used several times at this conference but I was not always sure of the intended meaning. Sometimes it meant accuracy and at other times it meant repeatability or perhaps even resolution. This is a confusing situation and the sooner we adopt a standard terminology the better. In this respect I should like to see everyone concerned providing, in a more standard form, all the important performance data required by the intelligent customer. This should include the repeatability, the accuracy at various parts of the operations range, linearity and rangeability. Too often it proves difficult to discover this vital information from sales literature and even research workers in their enthusiasm forget to refer to repeatability, accuracy, etc. Please let us be more specific about such things and quote values. This will help all round in making us more realistic about the practical problems. If a new method is going to succeed it must be fully understood.

2   WEIRS

Ackers paper on weirs (p.21) is a good example of providing new information on an old technique to enable engineers to design a weir with more certainty and to predict its discharge coefficient. However, the data are based mainly on laboratory scale calibration data and one wonders about scale effects and the need to conduct accurate full scale tests to check the coefficients against some other reliable method such as radio-tracer techniques. Weirs like Venturimeters and orifice plates are so sensitive to many external and internal influences that one wishes for some alternative method, but I cannot see any yet which covers such a wide flow range and which is so simple to use in open channels.

3   SOME FLOWMETERS FOR SPECIAL APPLICATIONS

Professor Singer's Nuclear Magnetic Resonance (p.38) is most interesting for very special applications and his velocity function approach is a new idea which may be worth considering in other connections. However, because of its cost and limited accuracy at present, I cannot see the technique being a strong contender for conventional applications. Nevertheless it must regarded as a very special tool which can do things that no other method can do at the moment. The paper by Smith and Said (p.96) about the mini-flow meter to go inside a carburettor is another good example of a special purpose meter. A brave but less successful try at a special problem was the ionization meter of Mustafa, Birchall & Woods (p.223) to measure transient gas flow rates in a pipe in connection with i.c. engine studies. There will always be a need for special developments of this kind and their main justification depends on whether they will solve the particular practical problem.

4   THE LASER DOPPLER METHOD

The Laser Doppler method is most valuable for measuring local fluid velocities and I am sure that it will be widely used for research purposes. It is, however, still some way from being used for accurate flow-metering, except in the laboratory. The method may be further simplified and cheapened but at present it has the limitation of all methods involving point velocity measurement: that traverses along several radii are required to give high accuracy in asymmetric flow fields. For convenience in use the technique must be developed so that the integrated flow can be derived automatically from the traverse readings, and perhaps even the traverses themselves should be made automatically. I was particularly impressed by the compact optical head using back scattering developed by Blake (p.49) of the National Engineering Laboratory and think that this type of attachment has a great future.

The method now seems capable of measuring point velocities to within ±0.25%. This is good, and if the velocity distribution really is axisymmetric then the integrated flow should be obtained to within a similar order of accuracy. However, in most practical cases the flow is asymmetric and it is such asymmetries which give errors in all methods where the flow has to be obtained by spatial integration of velocity. Salami (p.381) has shown that to decrease such errors in circular pipes to a value of ±0.5%, traverses on at least six radii should be used with the velocity measured on at least six points per radius.

The paper by Durst, Melling and Whitelaw (p.78) provided a useful comparison of the various Laser-Doppler methods available and their limitations, but it is likely that the prices of apparatus can in some cases be reduced.

5   ULTRASONIC FLOWMETERS

Similar integration limitations apply to the ultrasonic method although it is one stage better than point velocity measurement in that the ultrasonic pulses give an integrated mean fluid velocity over the path which they traverse. This together with the added power of Gaussian integration methods is perhaps why Fisher and Spink (p. 139) are able to claim such high accuracies for only four chordal sections. Nevertheless it seems that for highest accuracy the velocity distribution must already be known and the plane of the sonic pulses must be perpendicular to the cross-flow to avoid errors due to swirl and secondary flow.

The field experience of Suzuki, Nakabori and Yamamoto (p.115) in Japan is much more extensive, but from their tests they do not claim such high accuracies. In their case their technique has the practical advantage that the transducers can be fixed outside the pipe without needing to drill it, but this may be one of the reasons why their method does not give such high accuracies as those claimed by Fisher and Spink.

Certainly the repeatability of the ultrasonic method is good but I should like to see more explanation of integration errors in asymmetric flow and an attempt to cheapen and simplify the apparatus, if this can be done without sacrificing accuracy. The Alden Laboratory tests using the Westinghouse apparatus downstream of the heat exchanger (p. 147) were an impressive indication of the capabilities of the method in bad flow conditions.

The thermal flowmeter has shown that it can deal with many difficult situations and the paper by Mills and Evans (p.105) gives an example of a straightforward apparatus of this type. There are several other meters now on the market which are claimed to be insensitive to installation. I have heard this claim many times before and would like to see it proven in this case. They may have a chance of integrating and averaging upstream maldistribution and may be quite useful but compared with other techniques their range is not particularly wide and they are likely to be limited to small pipe sizes.

6   PRESSURE DIFFERENCE METHODS

The papers by Cousins (p.160) and Okafor and Turton (p.210) on orifices add something more to the knowledge about orifice meters of small bore at low Reynolds numbers and that by Harris & Magnall (p.180) extends the orifice and the Venturi-tube to non-Newtonian fluids. How far we should stretch such devices is a matter for debate.

The commercial flowmeter described by Ryland (p.200) and Turner (p.191) are most ingenious and as a mechanical engineer I particularly liked the simplicity of Turner's linear and extended scale flowmeter.

This does things mechanically which can be done electronically but I suspect it is more robust and reliable. Nevertheless with the intelligent use of "grey" boxes its working range can be considerably extended. Despite the great improvement and availability of electronic circuitry there is evidently still scope for simple and elegant mechanical solutions.

7   TRACER METHODS

There are many methods which involve tagging the flow and measuring the time of travel of the tagged flow between two points. The cross-correlation technique using turbulence or other naturally existing markers is one of the most powerful and attractive of these. In the paper by Beck, et. al. (p.292) it is claimed to be an absolute method but I would not accept this until it has been shown that the transport velocity of the disturbance is the same as the mean velocity in the pipe. This may possibly be so because there may be an integrating effect across the pipe, but in principle turbulence at the pipe centre-line travels faster than that at the wall and it is therefore not certain just what the mean transport velocity will be. Nevertheless this is one of the methods that I would single out for further study.

I liked Evan's comparison (p.245) between the velocity method and the meter prover. In the case of the meter prover the fluid is tagged in a very definite way by the rapid insertion of a rubber ball in the pipe but even in this case it is not absolutely certain that it travels at the mean fluid velocity.

Tagging by ionization is another variation on the same theme and the pulse method described by Brain (p.236) certainly overcomes many of the problems.

The radiotracer dilution and residence time methods are by now well proven and are both powerful methods as the papers by Clayton (p. 276) and Evans (p. 245) show. Obviously the simpler methods of tagging the flow such as by using turbulence, pressure pulses or other things occurring naturally will save on cost and complexity provided that the measurement can be made with sufficient accuracy. However, the advantage of radioisotopes is in the small amounts of tracer needed and the low concentration levels that can be measured. The dilution technique also avoids the need to measure the volume of pipe along the test length.

8   TURBINE AND ELECTROMAGNETIC FLOWMETERS

Turbine flowmeters are widely used and, provided their bearings are in the same good condition as when they were calibrated, are reliably accurate and have excellent repeatability. However when used with liquids of different viscosities their coefficients vary, especially at low flowrates, and have to be predicted from water tests by empirical methods. The paper by Withers, Inkley and Chesters (p.305) illustrates the relative variability of commercial meters and the paper by Tan & Hutton (p.321) goes some way to explaining the variations in meter coefficients between nominally identical meters. The effects of tip-clearance leakage is important and bearing friction can affect the calibration curve at low flows. Once a better understanding of such factors is available it may be possible to extend the application of such meters still further.

Electromagnetic meters have now been widely used for many years but, as with turbine meters, their characteristics are becoming better understood. It had been claimed by some manufacturers that electromagnetic meters were insensitive to inlet velocity distribution but this is not necessarily so as Theenhaus indicates (p.347). However, with better understanding it may now be possible to design meters which are insensitive to inlet flow conditions. In addition, the electrodes must be maintained in their original condition to avoid drift due to fouling by chemical action. The value of electromagnetic meters in hostile environments such as liquid metals is well illustrated by Thatcher (p.359) and although high accuracy cannot be claimed, the fact that they worked at all and can be used as flow indicators was most valuable in this application.

9   SOME GENERAL COMMENTS ON PRESENT TRENDS AND FUTURE DEVELOPMENTS

As we have already seen an important general problem is the sensitivity of flowmeters to installation which really means to inlet velocity distribution where swirl and asymmetries may be present. As Clayton (p. 1) pointed out, flow is a derived unit and no two flows are the same. Those meters that require axisymmetric flow conditions in which to work reliably will give considerable variations in practice because truly axisymmetric flow is rarely encountered even in long straight pipes. All point velocity techniques require integration of velocity against area to give flow. This may be done in various ways but it must be done accurately. Some methods try to even out asymmetries by using a strongly converging flow as in the case of an orifice or a Venturimeter.

Other methods tend to give an automatically integrated mean value of velocity, for instance the dilution method, if viscous, is 100%, but for others such as the ultrasonic method the integration has to be done in other ways. It is this integrated mean process which is very important if meters are to be insensitive to installation. It seems likely that there will be a demand for meters to work accurately over a wider flow range. To have linearity over a wide range it may be necessary to rely on "black boxes" to correct intrinsic non linearities in the meter itself. This should neither be too difficult nor too expensive with the modern resources of integrated circuitry. The old problems of drift free operation, even long term, we should hope to see disappear. There will probably be more applications requiring high accuracy and there may be more need for on site calibration techniques.

Some techniques today depend for their success on experts to apply them. It is desirable to develop methods which are as afar as possible, foolproof. However having said that, I feel that there will still be a need for a few expert teams to advise and carry out large scale and important flow measurements. One hopes that all concerned will have fewer illusions and a better understanding of matching the practical requirements and the choice of flowmeter.

For the future the situation seems to be similar to what it was ten years ago when I tried to summarise the Flow Measurement Symposium of 1960 held at the National Engineering Laboratory. To what extent will improved versions of existing and often rather old techniques be able to satisfy requirements and to what extent can we hope for new solutions to appear? How much further can we stretch the old techniques by marginally improving them as the result of great effort? Should we call a halt and say no more struggling, let us concentrate all our efforts on new methods? The difficulty with the latter course is that we may often find that new methods not only have similar basic limitations to the old ones but also may have new problems of their own. Old methods we know survive and new methods take time to prove themselves. However, on the whole I think I would put effort into the new methods and would probably concentrate on the cross-correlation technique, the vortex frequency meter, laser-doppler anemometer and the ultrasonic method. These I think are likely to be more rewarding for the same amount of effort compared with concentration on older techniques. One thing is certain for the future, there will be no shortage of problems.